I0309154

Equipment For Bakers

Other Books by the Same Author

Chemistry and Technology of Cereals as Food and Feed

Bakery Technology and Engineering
First and Second Editions

Cereal Technology

Cereal Science

Food Texture

Water in Foods

Ingredients for Bakers

Cookie and Cracker Technology
First and Second Editions

Formulas and Process for Bakers

Snack Food Technology
First and Second Editions; Japanese Edition

* * * *

This book was designed and manufactured to give you maximum value and satisfaction.

The book is printed on a heavy coated paper which contributes to better detail and clarity in the illustrations. A low gloss coating was used to minimize reflection and glare. The typeface is New Century Schoolbook 10/12, an open easy to read style which is esthetically pleasant.

EQUIPMENT FOR BAKERS

by
SAMUEL A. MATZ, PH. D.

President, Pan-Tech International. Formerly, Vice President for Research, Development, and Compliance, Ovaltine Products, Inc. At one time, Vice President for Research and Development, Robert A. Johnston Co.; Technical Director of the Refrigerated Dough Program, Borden Foods Co.; Chief, Cereal and General Products Branch, Quartermaster Food and Container Institute for the Armed Forces; Chief Chemist, Harvest Queen Mill and Elevator Co.; Instructor, Department of Flour and Feed Milling Industry, Kansas State University; Chemist, Iglehart Mills.

PAN-TECH INTERNATIONAL, INC.
P. O. BOX 4548
McALLEN, TEXAS 78502
1988

Library of Congress Cataloging in Publication Data

Matz, Samuel A.
Equipment for bakers / by Samuel A. Matz

p. cm.

Includes bibliographies and index.
ISBN 0-942849-27-2
1. Bakers and bakeries—Equipment and supplies. I. Title
TX763.M334 1988
641.7'1—dc19 88-9747
CIP

PREFACE

The various pieces of equipment used for processing dough products are described in this book. Processing is followed from receipt of ingredients to the finished but unpackaged product. Specifications and functions of the equipment are considered. Variations in models commercially available (e.g., their hourly output) are pointed out. Interaction of ingredients with equipment and effects of changes in machine settings on finished products are discussed.

The emphasis is on equipment made in the U. S. A., as might be expected. Quite a few machines of foreign manufacture are discussed, however.

Only commercial equipment is considered in this volume. Although some space is given to machines suitable for use in the small retail bakery, more extended treatments are given of equipment used for wholesale baking.

I devoted much effort to avoiding duplication of material contained in the first two books of this series, but there were times when some repetition of previously published information on ingredients, formulas, and processes was absolutely necessary in order to make the discussions comprehensible to the reader. I estimate that less than 5% of the material in this book duplicates anything contained in "Ingredients for Bakers" and "Formulas and Processes for Bakers."

Packaging materials and equipment will be covered in the next volume of this series. Sanitation and maintenance procedures and special preservation methods will also be discussed in volume four, "Packaging, Protection, and Product Development for Bakers," which is expected to become available early in 1989.

Very many people have contributed material--text, drawings, photos, etc.—used in this book. The following list of contributors is thought to be complete, or nearly so. If omissions have occurred, please accept my apologies and rest assured that corrections will be made in the next edition.

Descriptions and evaluations of equipment and methods were not in any way biased by financial considerations. No company or person paid compensation of any kind for mention of their equipment and I have never been retained as a consultant for any of the companies mentioned.

Samuel A. Matz
Edinburg, Texas
June 1, 1988

ACKNOWLEDGEMENTS

John Andrew, Director of Mktg.
Vac-U-Max
Belleville, NJ

Bernhard B. Barta
Franz Haas Machinery Co.
Richmond, VA

David E. Betts, Pres.
Pizza Automation, Inc.
Pataskala, OH

Thomas Q. Clock
Clock Associates
Portland, OR

Gil Foulon
Rheon U. S. A.
Irvine, CA

J. Rex Gibson
Business Development Manager
Baker Perkins
Peterborough, England

R. B. Hallman, District Manager
General American Transportation
Chicago, IL

William P. Imhoff
Peters Machinery Co.
Chicago, IL

Walter E. Jorgensen
Spangler Valve Co.
Glendora, CA

James M. Kocher, V. P., Mktg.
AccuRate
Whitewater, WI

Charles Latendorf, Exec. V. P.
Latendorf Conveying Corp.
Kenilworth, NJ

Thomas R. Lugar, Pres.
Thomas L. Green, Inc.
Indianapolis, IN

A. Arnold, Marketing Director
Vicars Group Ltd.
Merseyside, England

Douglas W. Beattie, V. P., Mktg.
Liquid Controls Corp.
North Chicago, IL

A. L. Canham
Baker Perkins
Peterborough, England

Tom Dundas, Sales Mgr.
Oakes Machine Corp.
Islip, NY

Harry Gardner
Union Steel Products, Inc.
Albion, MI

Richard Gray, Pres.
Abbott Scales, Inc.
Houston, TX

Fred Hesley
S. J. Controls, Inc.
Long Beach, CA

John M. Jacus
Bakery Sales and Marketing Manager
APV Crepaco
Chicago, IL

Tom Kice
Kice Industries, Inc.
Wichita, KS

P. Laghetti, Sales Manager
Mondial Forni Benni
Verona, Italy

William F. Lohrer, Advertising Mgr.
Hobart Corp.
Troy, OH

Joseph Lupo, Pres.
Production Line Equipment
Rockaway, NJ

Edward Meise, Executive Vice Pres.
AMF Union Machinery
Richmond, VA

Nancy Minetti
Blodgett Ovens
Burlington, VT

Julian S. Modzeleski, V. P.
Flynn Burner Co.
New Rochelle, NY

Ben W. Muller
Rykaart, Inc.
Hamilton, OH

Thomas Nixon, Pres.
Bakers Pride Oven Co.
New Rochelle, NY

P. Pamart, Sales Manager
Arpin
Gennevilliers, France

Robert L. Pluta
Colborne Manufacturing Co
Glenview, IL

Daniel R. Raio, V. P.
Proprocess Co.
Paramount, CA

Stephan Reimelt
Reimelt Corp
Odessa, FL

Samuel Rodriguez
Western Bakery Imports
San Francisco, CA

Alice Schwarz, Pres.
Advance Food Service Equipment
Westbury, NY

Ken Sharlow
Dawn Equipment Co.
Jackson, MI

Bob Sievert, Manager
APV (Baker Perkins)
Raleigh, NC

Barry Slobodow, Vice President
Vicars Group Limited
Iselin, NJ

Steven J. Smith
Micro Motion, Inc.
Boulder, CO

John Van Laar, Executive Vice Pres.
APV (Werner Lehara)
Grand Rapids, MI

William W. Wade
Breddo, Inc.
Kansas City, MO

Tom Weidenmiller
Weidenmiller Co.
Elk Grove Village, IL

Gordon Wilkinson
Western Bakery Imports
San Francisco, CA

Steven Wright
Hansaloy, Inc.
Davenport, IA

CONTENTS

CHAPTER ONE

BULK HANDLING SYSTEMS

INTRODUCTION

We can define bulk ingredients as those food components which are not packed in containers of a given size, such as 100 lb bags, 30 gal drums, or 50 lb cartons. Water was the first bakery ingredient to be received and handled in bulk. It is still the material received in greatest quantity by bakeries, although much of it is used for non-ingredient purposes, such as steam generation and sanitation. The bulk handling of other ingredients—flour, sugar, syrups, oils, etc.—had to await the development of technologies which made the practice economically justifiable. Implementation of bulk receiving by bakeries involved a collaboration of suppliers and purchasers which was relatively slow to evolve. In many cases, the final step was made when suppliers financed the installation of tanks and other necessary equipment, with payment being included as part of the ingredient cost.

Modern technology makes it possible to store and handle in bulk virtually all of the ingredients normally used in bakeries, but it is usually not considered to be economically advantageous, even in the largest plants, to construct such facilities for more than 7 or 8 of the raw materials. Since bulk transfer is essential for computerized batching, however, a few factories have automated the handling of nearly all the ingredients, including such minor components as ammonium bicarbonate and spices. Some lack of flexibility, in terms of the number of ingredients which can be introduced into the system, is inevitable in these plants.

Receiving, storage, and in-plant conveying of bulk ingredients can often be justified on the basis of economic advantages such as eliminating the cost of disposable containers, reducing waste, and lowering labor costs for material handling. Additional benefits are frequently obtained in the form of improved sanitation, better control over measuring, and reduction in size of storage areas.

The author is a firm believer in the advantages of bulk handling. It must be admitted, however, that hope tends to outrun reality in the designing of some installations. Careful planning, with close attention to all present and future changes in costs, is essential when a proposal is being considered to install a bulk handling line.

GENERAL CONSIDERATIONS

Bulk storage facilities logically fall into two categories based on the physical state of the product—whether it is in liquid or solid (i.e., granular) form. In some cases, the baker has a choice of obtaining the product in either form. Flour is, of course, always handled as a powder and it is more common to transfer sucrose, dextrose, and salt in this form rather than as solutions. Shortenings, corn syrups, and invert sugar are almost always handled as liquids in bulk systems. When milk is received in bulk, the liquid form is often preferred, in spite of such negative factors as increased cost and greater potential for spoilage. Dry milk can be handled successfully, however. Dried eggs are not often handled in bulk systems, but liquid egg products are often treated in this way.

There are general principles applicable to the design of all bulk handling facilities. Sanitation should be a primary consideration in the design. Surfaces which contact ingredients should be nonporous and resistant to corrosion and abrasion; they should be nontoxic and should not transfer odor, taste, color, or particles to the ingredients. They must not accelerate deteriorative changes in the food materials. Physical changes (as in particle size) during transfer or storage should be held to a minimum consistent with necessary design limitations. Tubes carrying powders should be grounded to prevent the buildup of static electricity. Changes in moisture content are to be avoided.

Certain physical characteristics of ingredients control the type of equipment required for transferring and storing these materials. Chief among these physical qualities are the rheological properties (viscosity or apparent viscosity) of liquid or semi-liquid ingredients and the flow properties of powdered and granular ingredients.

Factors Affecting Design of Bins for Powders

Difficulties in handling bulk solids are generally caused by a lack of flowability or a strong tendency to floodability. These two properties are sometimes described as being at two ends of a spectrum with good flowability somewhere between them. A flowable material tends to move smoothly and evenly under the influence of gravity without requiring assisting means. Floodable materials move in a series of separate, irregular avalanches. These properties are manifestations on the macroscopic scale of microscopic factors such as the dimensions of the particle, uniformity of particle size, conformation of the particle (whether spheroidal or cubic or with protuberances, etc.), smoothness or roughness of the particles' surfaces, elasticity of the particles, tendency to entrap air, adhesiveness of the surfaces (whether due to "sticky" layers or to accumulation of static electricity), and actual density.

Established procedures can be used to determine the relative position of an ingredient within the "spectrum" mentioned above, thereby furnishing a basis for predicting handling properties relative to other materials. The following descriptions of tests which can be performed omit many of the procedural details, but give an overall idea of the factors which are important in establishing ingredient flowability. First, the bulk density of the material is estimated by filling a container of known volume, without applying pressure or vibration, and weighing the contents. Second, the "compressibility" of the ingredient is determined by comparing the packed or tapped density of the material in the container used for measuring the bulk density. Compressibility is expressed as the percentage increase in density; if it is high, say above 20%, the material will probably not flow smoothly and will tend to bridge in hoppers. If the compressibility exceeds 40%, there will very likely be problems in discharging the material from the hopper after storage.

Third, the particle size and shape is determined, partially by visual means. Powder and pellets are characterized by particle size, and fibrous or matting or interlocking strands by shape. Powders such as sugar and flour can be roughly characterized by finding the finest mesh sieve that most of the powder will pass through with shaking. Shredded or flaked material such as bran, wheat germ, or coarse rye meal present somewhat more difficult evaluation problems. The fourth test is the measurement of angle of slide. This test consists of pouring a small amount of the ingredient on a polished metal plate and tilting the plate slowly until the powder begins to run downhill under its own weight. The number recorded is the angle formed by the underside of the plate and the table surface. A high angle of slide indicates that the material is probably somewhat sticky and could bridge in bins and hoppers.

The final parameter to be determined in the usual series of preliminary tests is the moisture content, which is measurable by well known methods. Hygroscopicity is also a factor which may have to be considered.

Other tests which can be conducted to give additional information relative to handling properties are the angle of repose, the angle of fall, the angle of difference, the angle of spatula, cohesion, uniformity, and dispersibility. A brief description of these tests is given in the following paragraphs.

The angle of repose is the acute angle formed between the side of a cone-shaped pile of the material and the surface on which it lies. The smaller the angle of repose, the more flowable a material will be and the more floodable a floodable powder will be.

After the angle of repose has been determined, a weight is dropped several times on the surface (plate) holding the material and the angle of the side measured again. The more free-flowing the powder is, the lower its angle of fall will be. A floodable material tends to collapse rather than to assume a uniform angle, that is, the side of the pile is somewhat rounded or irregular. The more floodable it is, the smaller will be its angle of fall.

The angle of difference is obtained by subtracting the degrees of the angle of fall from the degrees of the angle of repose. The greater the angle of difference of a floodable material, the greater is its potential for flooding or fluidizing. It is possible for a floodable material to have a high angle of difference even though it has a low angle of fall.

The angle of spatula is obtained by drawing a spatula of given dimensions up through a mound of material and measuring the resultant angle on each side. It gives an angle of rupture or relative angle of internal friction. A free-flowing material will form a uniform angle based on the internal friction of the particle sizes involved. A material which is not free flowing will form a number of irregular angles on the blade. The greater the angle of spatula, the less flowable the material will be. Except for very free-flowing materials, the angle of spatula is always greater than the angle of repose.

Cohesion is a useful determination when dealing with very fine powders or with materials for which the force binding the particles together can be measured. The cohesion coefficient is a direct measurement of the amount of energy required to pull apart aggregates of cohesive particles in a specified time. A powder is less flowable when it has a high cohesion coefficient.

The uniformity coefficient is value obtained by dividing the width of the sieve opening that will pass 60% of the sample by width of the sieve opening that will pass no more than 10%. It is useful for granular and powdered granular material on which cohesion cannot be measured. The more uniform the size and shape of a mass of particles, the more flowable it probably will be.

Dispersibility can also be measured. As the dispersibility in air increases, dustiness and floodability also increase. Materials with dispersibility ratings of 50% or more, measured by a standard method, are considered to be very floodable and therefore likely to cause handling problems.

Materials which cause problems in bin discharge, either because of their floodability or because they fail to flow, require special bin geometry and/or special appliances such as bin vibrators.

Cost and Convenience Factors

Sources of supply are more restricted for bakers who must rely on bulk delivery. When a bulk system is started, assurance should be obtained that an adequate supply of delivery conveyances will always be available. Shortages of trucks and rail cars have occurred, and are more common in some parts of the country than in others.

Scheduling of deliveries is more difficult with a bulk system because storage space is absolutely limited. It is nearly always possible to find a place to put a few hundred bags of sugar, but, if the syrup tank is too full when a truckload of this material arrives, nothing can be done to remedy the situation. "Running out" is also more serious in bulk systems, since it is often quite difficult to introduce bagged or drummed material into weighing and metering devices designed for bulk systems. Although suppliers will almost always be able to dispatch a truckload of commodity-type material within a day or so after the order is received and get it to the delivery point within a few hours after that, tank cars are subject to the uncertainties of rail traffic and allowing a considerable lead time should be regarded as a normal precaution when they are used.

As is so often the case when changes are advocated for cost saving reasons, the economic advantages of bulk handling, as compared to receiving ingredients in bags or drums, are not as clear-cut as some experts would have us believe. The initial investment required for receiving bulk deliveries and for massive storage facilities is substantial, and seldom conforms to initial estimates. Although such equipment is normally rather long-lived and requires relatively little maintenance, some allowance must still be made for these costs. Offsetting these costs are savings due to eliminating material losses through bag and drum damage, reducing ingredient spoilage through improved inventory control, and minimizing forklift handling of bags and drums. Of course, there are also substantial real savings due to the complete elimination of unit containers such as bags, drums, and cartons. Reimelt (1987) says that these savings make it desirable to receive sugar in bulk even at usage rates as low as 2,000 lbs (of sugar) per day. Companies with smaller consumption rates should stay with bags, small totes, or other containers. In-plant logistics in smaller companies can sometimes be improved by utilizing large bags (capacity of up to

2,000 lbs) which might be incorporated into an automatic conveying and weighing line.

RECEIVING BULK INGREDIENTS

Sampling and Testing

Sampling and testing of bulk loads present some problems. The receiving department is generally anxious to unload the shipment quickly in order to avoid demurrage and other costs and complications, while the quality control department will not want to risk contamination of equipment or old stocks of ingredients with material which may be unsatisfactory, and so will insist on completing an evaluation before unloading is started. There is usually no provision for retrieving defective material which has entered the system—an additional reason for completing quality control tests before unloading.

The often disparate interests of the quality control department and the receiving department should be recognized at the outset, and a clear-cut policy promulgated by top management to establish responsibility and authority for approving unloading of bulk shipments. This policy should be structured so as to promote close coordination of effort and encourage maximum speed in sampling and testing incoming shipments.

Sampling of liquid loads for physical and chemical tests can be done using a simple bottle-on-a-stick type of device inserted through the loading hatch, but it is probably better to collect fluid from an appropriate port in the delivery vehicle or discharge line. Sampling of dry materials such as sugar and flour can be accomplished by inserting a long trier or specialized sampling equipment through the opened manhole at the top of the car or truck, and withdrawing and combining portions taken from several directions. There are also automatic sampling devices which divert a certain portion of the ingredient as it flows into the plant receiving system. Sampling problems resulting from stratification of either liquid or powder loads must be kept in mind constantly in order to avoid unpleasant surprises on the production line.

Verifying Delivered Amounts

The importance of having accurate receiving scales can hardly be exaggerated. Unless all ingredient receipts are checked for weight, and this includes bulk shipments, no control can be exerted over either shipper's mistakes or deliberate fraud. Certified public scales and railroad scales are usually accurate but do not allow positive and direct control of the weighing operations by the purchaser. Buyers must recognize that it is possible some of their suppliers do not even have shipping scales and do not weigh their shipments but only estimate them.

Weighing equipment used at the receiving point can include railroad track scales, motor truck scales, bulk handling scales, bag check weighers, and dock platform scales. By installing an automatic bulk materials receiving scale in the unloading line leading from the truck or rail car to the storage bins, and by installing automatic tank scales under the liquid storage tanks, shipping weights can be verified and inventories of bulk ingredients can be kept under close surveillance. In one type of bulk receiving scale, the material is air-conveyed to an overhead surge hopper, and then fed through a weigh hopper on a scale until a preset weight is reached, at which time the fill feeder stops and the weight is recorded. The weigh hopper then discharges the lot of material to a designated storage hopper. When the weigh hopper is empty, the residual weight is recorded, and the cycle begins again. The computer adds each batch weight and subtracts each residual weight so that the contents of the shipping vehicle are known exactly at the termination of the receiving process.

Amounts of material retained in tank trucks can be considerable if viscous materials are being delivered. In the case of liquid chocolate, several percent of the load can adhere to the inside surface of the tanker. Therefore, the weight of material loaded into the tanker by the supplier gives only an upper limit on the delivery and is by no means a satisfactory figure to use in determining payment. Weighing the vehicle on a public scale before and after delivery is a possible check on delivered amounts, but can be subject to all the usual errors and tricks that accompany such measurements.

LIQUID INGREDIENTS

Equipment and Principles

Since water was the first ingredient received and conveyed in bulk, it is to be expected that other liquid handling plants would be based to a considerable extent on engineering principles first established for water distribution systems. It was found necessary to introduce many new design principles in lines intended for some of the other common fluid ingredients, such as the viscous, relatively unstable, and mildly corrosive syrups used as sweeteners. Liquids delivered and stored in bulk include not only aqueous fluids such as sugar syrups, but oily materials as well. These oils and fats have requirements considerably different from those of the aqueous systems.

Product contacting surfaces in pumps, pipes, valves, and meters should be made of stainless steel, if possible. In practice, some piping may be made of galvanized iron because of the cost factor, and occasionally copper is used. Plastic piping is a possible alternative for some applications. Piping must be installed so as to allow complete drainage of the system for cleaning. The entire transfer system is normally kept full if liquid sugar is being handled. Pipes in corn syrup installations are sometimes traced with heat tapes or steam pipes in order to permit use of less powerful pumps. This may lead to overheating and the delivery of discolored and off-flavored syrup. Overheating can occur during a shutdown, when a static load of syrup in the heated pipe encounters high temperatures for many hours or days. The obvious corrective measure is to shut off, or reduce, the heat when these conditions occur.

The design of the most economical system for transferring a liquid food requires a determination of the pipe diameter which will give a minimum resistance to flow (and thus requires the smallest pumps) while still conforming to restrictions on pipe cost and space. The rheology of the food or ingredient, as well as the required rate of flow, must be taken into account. Various schemes have been suggested for these determinations (see Garcia and Steffe 1986).

Automated bulk handling systems obviously require electrically activated valves which are electronically coordinated with pumps, scales, etc. The valves must be of sanitary design and of materials compatible with the ingredient being handled. The newest types of "smart valves" will include, for example, a valve control unit mounted in a water-proof housing on top of the air actuator and consisting of a four-way air piloted solenoid valve, a microprocessor printed circuit board, and a photosensor printed circuit board. These will be connected to a switch/indicator control unit that indicates the status of the valves and combines output transmitters and input receivers, each with its own light to indicate valve position. All of these will be connected to a master control unit that maintains two-way communication between the plant computer, the switch/indicator control units, and the valves (Anon. 1986).

Sweeteners

Bakeries of medium to large size will often find it to be economical and profitable to obtain fluid sweeteners, including corn syrups, sucrose syrups, invert syrup, and various combinations of these, in rail cars or trucks. Molasses, refiners' syrups, and other specialty items as well as made-to-order mixtures can be obtained by bulk delivery in many sections of the country. The economic advantages of buying cane or beet sugar as a syrup rather than in granulated form vary in different geographical areas, and the financial aspects of this method of handling sucrose supplies should be carefully analyzed before a commitment is made to change from granular deliveries. As mentioned elsewhere, sugar syrup is not an acceptable substitute for granulated sugar in some formulas. Anyone purchasing large volumes of corn hydrolysates would, however, almost certainly save money by installing a bulk system for this material rather than continuing to buy it in drums. Invert sugar, being available only in syrup form, is also cheaper to receive in bulk if the use rate can justify installation of the necessary equipment.

Stainless steel is the preferred construction material for storage tanks. Many tanks for sweeteners are constructed of mild steel coated on the inside with a plastic lining meeting FDA specifications. Unlined aluminum tanks are sometimes used for liquid sugar tanks, but not for corn syrup or invert sugar tanks because their low pH makes them too corrosive. The top of the tank should be convex for strength and for good drainage inside and out. Outlets should be placed in the center of cone shaped bottoms.

Sugar syrup and invert sugar syrup—The following preparations for unloading shipments of sweetener syrups have been recommended by Pancoast and Junk (1980):

(1) After properly spotting the car or truck, check the vehicle's number against the invoice or shipping document and determine if the contents are appropriate for the loading port at which the vehicle is spotted.

(2) Break the seal, open the dust dome, and check the outlet valve to make sure it is tightly closed. An air filter is needed for the car dome during unloading. This may be made of fine-mesh metal or fabric on a round frame. Available air filters may be adapted to a plate with a rubber gasket so the plate can fit the dome flange.

(3) Remove the cap from the vehicle outlet and rinse the inside of the pipe or fitting with very hot water or, preferably, with a sodium hypochlorite solution having a concentration of about 100 ppm available chlorine. Assemble hose connection between car and piping system leading to storage tank. Open outlet valve.

(4) Keep dome cover closed and fastened when actual unloading is not in progress.

Receiving line connections must be designed to minimize the likelihood of contamination or damage. A standard type connector for outside tanks is a 3 inch butterfly valve equipped with a quick-coupling connection and a locking cap. If the storage tank is located inside a building, the butterfly valve may also be inside with a short pipe extension through a hole in the wall. The quick-coupling closure is, of course, located on the outside of the wall, and it should be protected from vandalism or inadvertent damage by a metal cover provided with a lock. A check valve should be placed in the receiving line to prevent loss of syrup from the storage tank in case the main valve is left open or fails to close for some reason.

Since cane or beet sugar syrup at the usual 66 to 67% solids concentration will support the growth of certain microorganisms (particularly osmophilic yeast), these bulk products are subject to spoilage. Careful attention to good sanitary procedures and occasional cleaning of the tank with sterilizing solutions are essential. It is recommended that ultraviolet light appliances be placed in the head space of the tank. Ultraviolet radiation inhibits microbiological activity above the liquid level, but it is effective only in the air space and in a thin top layer of the liquid because penetrating power of ultraviolet rays is very poor in aqueous solutions. Consequently, osmophilic yeasts may become established in lower levels of the liquid and begin a slow fermentation. Sucrose syrups of higher concentration are not feasible at normal temperatures, but combinations of sucrose and invert with solids up to about 76% are available and are definitely more stable. No matter how stable the syrup is expected to be, however, it is good practice to have plate counts run on the material at least every month and, preferably, weekly.

Air in the head space of the tank is saturated with water vapor in equilibrium with the water activity of the syrup, the latter factor being related to the temperature and solids content of the syrup. Some of this moisture can condense on the tank walls when the walls decrease significantly in temperature. If this condensate runs down the walls and forms a film on the surface of the syrup, an environment is created which is very conducive to rapid growth of yeasts, molds, and bacteria. To prevent condensate from forming, an air circulation system is regarded as an essential feature of tank design unless the whole tank is inclosed in a controlled temperature space. These systems include a fan to draw in the outside air and a filter to extract microorganisms from the air.

For sucrose-based syrups, a two-tank system is preferred, each tank being completely emptied before a new load is placed in it. This procedure minimizes cross-contamination. Figure 1.1 illustrates a receiving, storage, and handling facility for liquid sugar.

Liquid sugar, but not corn syrup, can be refrigerated down to about 45°F as a means for adjusting dough temperatures. When cooled, liquid sugar and invert sugar syrups do not exhibit unacceptably high viscosities. On the other hand, chilling corn syrup increases its viscosity to a point which makes the ingredient extraordinarily difficult to transfer and to incorporate into dough or batter.

Regular conversion corn syrup—Corn syrups are supplied at 76% solids or higher. These syrups will not be subject to microbiological attack unless diluted by condensate or water from other sources. Corn syrup can be heated, either for simplifying processing steps (as in manufacturing confectionery or icing) or for reducing the viscosity so as to facilitate pumping. In either case, it is desirable to accomplish the temperature change shortly before the ingredient goes into the mixer so as to minimize undesirable changes such as caramelization from temperatures which are too high or which have continued for too long. In fact, most corn syrups will have to be heated before they can be pumped, metered, and dispensed satisfactorily in a bulk system, since corn syrups of low DE and the high solids now being produced are extremely viscous at low temperatures.

Although it is possible to inclose the tank in a room where the temperature is maintained at, for example, 100°F, this treatment is not considered to be the method of choice because it has a tendency to cause browning of the syrup, especially if it takes a considerable time to use up the contents. A second alternative is to build inside the tank a chamber containing heating coils so as to raise the temperature only in the lower section of the storage tank. As syrup is removed from the tank, more syrup flows into the heated chamber where it rises in temperature sufficiently to make it readily pumpable. This system is not applicable to high fructose corn syrup, since the whole mass of this ingredient must be kept warm (85°F) in order to prevent crystallization of the dextrose which constitutes a large part of this material.

Still another alternative for improving flowability of regular corn syrup involves adding heating means to the outside of the tank. For example, electric strip heaters with thermostatic controls can be placed under the tank end nearest the discharge port. A temperature of about 105°F is suitable. The tank should have a screened vent, a circulating blower of 40 cfm capacity on top, and a manhole access port of at least

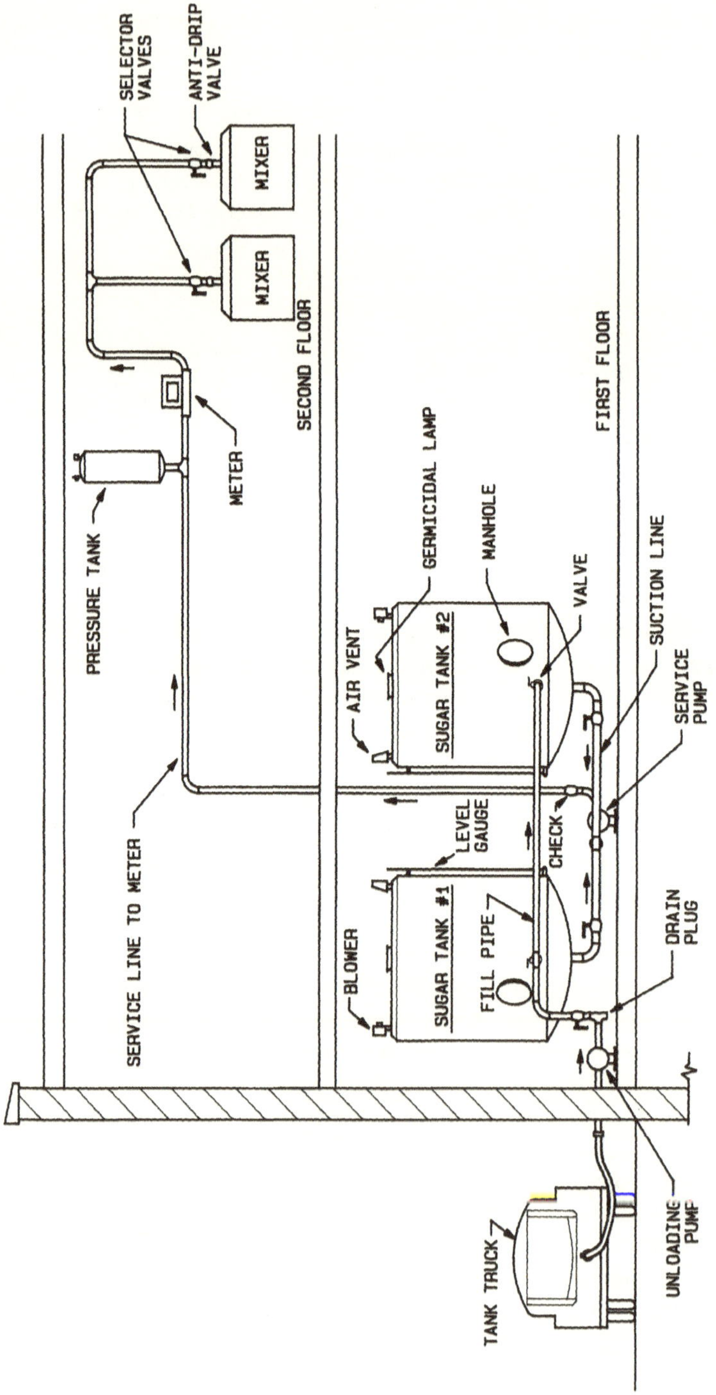

FIG. 1.1. A RECEIVING, STORAGE, AND HANDLING FACILITY FOR LIQUID SUGAR.

18 inches diameter to allow cleaning and inspecting of the interior.

Tank ventilating systems should be installed in order to keep the empty space free of condensate. Filtered air is drawn through the dome space by a fan, which operates either continuously or intermittently. Absolute filters—designed to remove microorganisms from the air passing through them—are used in present day installations. The filter element must be changed periodically.

A single tank for each type of corn syrup is the usual arrangement. Fresh loads are mixed with tank contents remaining from previous deliveries. Storage tanks are usually fabricated of mild steel with a baked-on lining inside. Aluminum tanks are not considered satisfactory because they are not lined and the metal would be unacceptably corroded by these low pH ingredients. It is now common practice to build tanks large enough to handle at least one jumbo car of 20,000 gal and, in the biggest installations, up to 50,000 gal. Tanks of 5,000 gal or less may be suitable for smaller factories which obtain their supply by truck.

Stainless steel piping is preferred for corn syrup handling systems, for the usual reasons. Galvanized pipe has been used because of its lower cost. Black iron and copper are not recommended because corrosion of these metals can cause contamination of the ingredient. Diameters of 4 to 6 inches are the most common sizes. It may be possible to use 3 inch lines if the pumping rate is low and the distance is short. Pipes should be heat-traced, that is, wound with small diameter copper pipe through which hot water or steam can flow. Or, they may be wrapped with a tape containing embedded resistance wire so that the pipe can be electrically heated. The pipes should also be insulated to retain heat and facilitate temperature control.

Positive displacement pumps of low rpm are preferred for transferring corn syrups. Rotary positive displacement pumps should have iron casings and bronze or hardened steel trim. Reciprocating piston pumps should have iron casings with bronze trim including metal snap rings on the piston. Rod and shaft packing should be of reasonably impervious and resistant material. Measurement of sugar syrups can be accomplished by volumetric meters or by a constant rate pump and time controller. The pumps should be equipped with a pressure release valve to protect the system. Sometimes insulated heat boxes of stainless steel are placed around the pumps and meters. These enclosures are heated by thermostatically controlled electric units. In the absence of heat boxes, the proper temperature must be maintained by heat tracing the meters or pumps with copper tubing containing circulating hot water, and surrounding them with insulation.

Rubber lined valves with stainless steel butterfly elements are recommended, but gate valves and lubricated plug types have been used satisfactorily. Butterfly valves are said to be less expensive and they minimize corrosion caused by to the low pH of the syrup. At the discharge, spring loaded anti-drain or no-drip valves of non-corroding construction should be installed. The spring loading feature insures the valve will open only when there is positive pressure in the piping, contributing to the accurate delivery of metered amounts of corn syrup. Check valves of non-corrosive materials should be included in the system to prevent back flow.

Corn syrup can be measured with meters which can be pre-set to deliver a given amount of syrup. These meters can be calibrated by volume or weight, and some of them include a totalizing register for inventory control.

High fructose corn syrup—High fructose corn syrup is transported in either insulated tank cars or stainless steel trailers. The requirements for unloading and handling of high fructose corn syrups are somewhat different from those for regular corn syrup and more like those of invert sugar syrups. This situation results from the solubility characteristics of dextrose and fructose. Fructose crystallization does not occur in these syrups, so the newer generations of HFCS, with high fructose:glucose ratios present fewer problems of crystallization, but the concentration of dextrose in the older and less expensive types of HFCS is high enough to cause precipitation if the temperature falls below about 80°F. HFCS 42 is loaded at a temperature of 90° to 95°F in the summer and 100° to 110F in cold weather. HFCS 42 should be stored at 80° to 90°F to minimize crystallization and color development. Because it contains less dextrose, HFCS 55 can be stored at temperatures between 75° and 85°F.

Glucose crystallization is reversible simply by heating the syrup. If the syrup is to be mixed with other aqueous solutions, there is probably no need to dissolve the crystals insofar as finished product quality is concerned. The real problem is the effect the crystals have on the physical properties of the syrup in the bulk handling system, where clogging of valves, pipes, and meters can occur. Generally, the crystals can be dissolved by taking the syrup to about 100°F, but heating must be done with care to prevent discoloration from taking place. It may take several hours for the syrup to become clear when heated at this temperature.

Dextrose—The limited solubility of dextrose necessitates the continuous application of heat to tanks containing solutions having concentrations greater than 50%. Therefore, if solutions of 65 to 67% are shipped considerable distances, the truck or railcar must be provided with heating facilities adequate to prevent crystallization. For short hauls in insulated vehicles, solutions which are filled hot may remain high enough in temperature to preclude crystallization without the use of heating units.

Solutions containing 65 to 67% dextrose are usually stored at about 130°F, often in tanks covered with insulation. The localized heating systems used in most corn syrup tanks are not suitable for dextrose storage units.

Other corn derivatives—Corn starch can be handled in bulk. Corn syrup solids, maltodextrins, and cereal solids are probably not being delivered or handled in bulk because of problems associated with flowability and humectancy.

Shortening

Equipment for storage and transfer—Bulk systems for edible fats and oils consist essentially of tanks, pumps, pipes, valves, and meters. Bulk oil shipments are usually made in standard tank cars containing 60,000 lb, jumbo cars containing 150,000 lb, or in tank trucks containing 20,000 or 30,000 to 45,000 lb. Smaller tank cars of 30,000 lb capacity are sometimes available, but at higher freight rates. Figure 1.2 is a diagram of a bulk oil system.

When oil or fat is delivered by truck, the driver connects a hose to the intake of the plant system and discharges the ingredient using a pump powered by the tractor engine. Plant personnel must take charge of emptying tank cars, using plant pumps and equipment. A basket strainer should be installed on the intake of the pump to prevent damage from scrap metal, and a finer filter on the pump discharge is desirable to remove extraneous material. Cartridge or bag-type filters with a porosity of 25 microns will remove visible particles. A bypass should be provided in case partially solidified fat must be pumped (Woerfel 1981).

Either positive displacement pumps or centrifugal pumps are suitable for moving oil. They may be made of stainless steel or cast iron. Mechanical seals are preferred over packing seals. Rotary positive displacement pumps are quite good because the lubrication of the oil gives them long life. They give constant delivery rates regardless of the head. High pressure relief valves should be provided.

Both round and rectangular tanks are used. A convenient arrangement consists of two tanks for each type of shortening being procured, each tank having a capacity of 65 to 70 thousand pounds. Since each

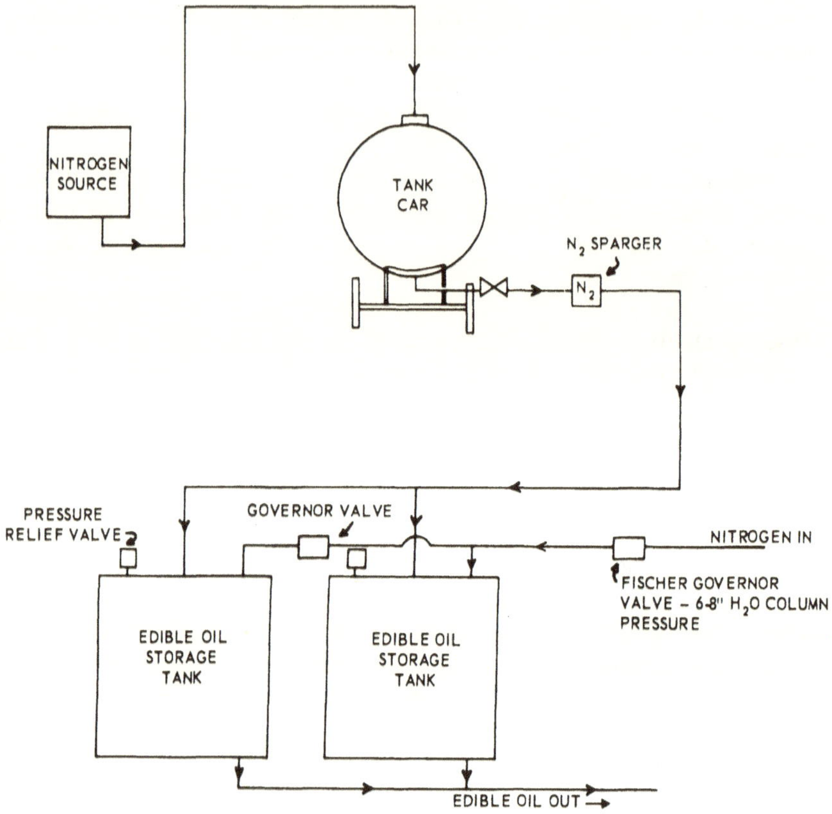

FIG. 1.2. RECEIVING, STORAGE, AND NITROGEN TREATING OF BULK EDIBLE OIL.

tank will hold the contents of a full tank car of oil, it will not be necessary to mix fresh oil with residues left in the tank from previous shipments. Tanks of 35,000 lb capacity are more convenient for truck deliveries.

Stainless steel, type 302 or 304, is the preferred material for constructing tanks for storing fats and oils. These metals have no adverse effects on shortenings and are not affected by common cleaning solutions, but they are expensive and difficult to fabricate. Consequently, most tanks are made of mild steel plate or even of black iron. These tanks are satisfactory as long as precautions are taken to prevent rust formation. Oil will readily pick up the thin film of iron oxide that rapidly forms on a thoroughly cleaned steel surface. Therefore, oil-contacting surface should be wiped with oil immediately after cleaning, before rust forms. It has been suggested that fiberglass reinforced plastics be used for storing oil, but it is believed there are not many of these installations (Woerfel 1981).

The configuration of the tank and the location of outlets should be such that complete drainage of the contents can be obtained. To achieve this result, vertically oriented tanks can be made with dished or cone-shaped bottoms and rectangular tanks can be slightly tilted toward the outlet. Tanks must be equipped with gauges, manholes, etc. The inlet pipe should be brought to within a few inches of the bottom to minimize splashing of the oil when a load is being delivered. A siphon breaker consisting of a 0.5 in pipe with a check valve should be installed slightly above the high point of the fill line.

Liquid level gauges in shortening tanks can be either sight tubes or float type indicators. The latter can be arranged to operate signals or controls. Sight tubes are not considered suitable for use with high temperature installations. The installer should calibrate the tank in terms of depth versus volume (or weight) of the contents.

Liquefied fats should be held at about 10°F above the AOCS capillary closed-tube melting point—not the Wiley melting point. Lard will be completely liquid at 120°F, but oleo and some vegetable shortenings will require slightly higher temperatures. Some suppliers recommend installation of stainless steel screens at the inlet of the tank, while others indicate this is not required. Welded construction is highly desirable since few pipe compounds or gaskets are compatible with edible fats and oils.

Transfer pipes for bulk shortening systems within the plant are almost always heated and insulated. A convenient method of heating is by means of an electrical heating tape wrapped around the pipe. Steam tracing or hot water tracing is also used. Unloading lines are not heated. They are freed of fat after use by draining or by blowing nitrogen or air through them. A packed type, ball check, reciprocating pump will give good accuracy with low viscosity liquids such as oils. Some varieties of gear-type single or double-lobe pumps have been used to dispense liquid shortening. However, these pumps have a high degree of slip and do not accurately measure low viscosity liquids if any fluctuation in head occurs. Positive displacement pumps can be used. There are pumps expressly designed for transferring shortenings. Lines and valves may be of standard iron or steel. No copper-containing alloys such as brass, Monel, or bronze should ever be allowed to contact the fat in shortening storage and transfer systems. Gate-type valves are recommended.

It is good practice to incorporate in a liquid fat system a temporary holding tank located close to the use points. A float switch in this tank is connected to a pump which maintains the liquid level within a narrow range. The temperature is thermostatically controlled. Maintaining the temperature and the head constant increases the accuracy of dispensing meters and confers other advantages.

Handling compounded fluid shortenings—Many products, such as certain kinds of cakes, require plastic shortening at some stage in their processing. These can be produced from liquid fats by chilling and working them to the proper texture in equipment such as the Votator. The product that emerges from the Votator is not fully equivalent to a plasticized shortening procured in that form, however, since an additional tempering period of about 48 hr is necessary to permit crystal growth and other changes leading to the preferred crystal form. Consequently, changes may be needed in processing or formula to accommodate these differences.

Development of "fluid" shortenings has simplified bulk handling of this ingredient, as compared to melted shortenings which are solid at room temperature and may have to be Votated before being incorporated into doughs and batters. A fluid shortening can be composed of fat, oil, stearine, emulsifier, and other additives. Some of these additives may be in the form of small solid (or semi-solid) particles floating in the liquid. Since these ingredients flow freely at normal room temperatures, they can be stored and transferred with ordinary liquid handling techniques.

Among the advantages claimed for fluid shortenings are (Petricca 1976): (1) They are easier to handle; (2) They are easier to incorporate into automated systems; (3) They reduce mixing problems and contain readily dispersed and more effective emulsifier systems; (4) Since they are, in effect, a kind of premix of emulsifiers and other additives, they can reduce scaling errors involved in adding separate ingredients and reduce labor required for separate scaling operations; and, (5) Because their principal constituent can be a polyunsaturated oil, they can perhaps have some nutritional advantage over solid (generally saturated) fats. Fluid shortenings consisting of oils with added stearine flakes and emulsifiers do not need to be heated, and, in fact, heating above the melting point of the stearine can lead to separation on subsequent cooling with very bad effects on the ingredient's quality. These fluids should be kept under continuous agitation with a slow moving blade that circulates all the way around the inner surface of the tank. At temperatures within the range of 70° to 87°F, the shortening should retain good odor and flavor for at least 3 to 4 weeks.

Stability problems—Solid shortenings are more stable than liquid shortenings. Unless a solid shortening (hydrogenated soybean oil, for example) has been aerated with a gas containing oxygen (rarely done), it will exhibit signs of oxidative rancidity only on the surfaces—particularly surfaces in contact with porous materials such as cardboard—and then only after several weeks. The situation is somewhat different with bulk shortenings because there is much greater opportunity for them to absorb oxygen. Recommended turnover is at least a tank every 2 to 3 weeks. Many operators take four weeks, or more, to use a full tank, however.

The tanks should be thoroughly cleaned at least every six months or at the first sign of sediment or of off-odors suggestive of rancidity. When off-odors appear, it is too late to save the contents of the tank and the oil must be sold for feed or soap use or for some other non-food application. A new shipment of shortening should never be dumped on top of the residue from a preceding shipment. Antioxidants should be used up to the legal limit and bulk oils are often supplied with these additives.

As stated previously, liquid shortening should be stored at temperatures about 10 degrees above the melting point (note the exceptions given in a preceding paragraph for compounded fluid shortenings containing stearine, etc.). In most climates, and for most shortenings, this requires a heated tank or a tank which is inclosed in a heated room. The fat can be heated by internal coils of pipe carrying hot water or steam, or externally by pipes or a jacket similarly heated. Hot water heating is preferred since steam may cause localized overheating with damage to the fat. Nonaerating agitators should be installed if steam is to be used for heating. Side-entering propeller-type agitators or recirculating pumps are satisfactory. They should be run at speeds which do not create a vortex; this would draw air into the shortening and accelerate rancidity development. Tanks may also be installed in a thermostatically-controlled room or some other type of heated enclosure, in which case no additional temperature-adjusting equipment will be necessary.

Controlled Atmosphere Storage—Nitrogen blanketing has been recommended as a means of increasing the storage life of shortening. Melted shortening will absorb oxygen while it is being manufactured and during transfer to the user's plant. Blanketing the shortening under a layer of nitrogen while it is in the user's tank will retard further uptake of oxygen, but the amount already present can cause sufficient deterioration to make the shortening unusable after a short time. Oxidation at the surface is small compared to that resulting from entrained and dissolved oxygen. Therefore, the nitrogen blanket is not a sure preventive of rancidity, but it is a further precaution which, in combination with other good practices including prompt turnover of stock, will assure a constant supply of shortening having excellent organoleptic qualities and which goes into the product with enough shelf-life remaining to permit the ultimate consumer to receive a finished baked product having no detectable rancidity.

Equipment for nitrogen blanketing is relatively cheap, while the expenditure for gas should be less than $10 per month per tank. To hold down condensation in the head space of a tank held in a room where the temperature fluctuates, a worthwhile precaution is to blow air across the surface of the liquid with a fan of about 90 cfm capacity.

Bubbling nitrogen through the fat in short bursts is a good method for preventing stratification or settling of higher melting point fractions. This is probably not needed for vegetable oils, but may be helpful for maintaining uniformity in hydrogenated shortening, oleo, or lard held near the melting point, or if the required temperature is not being maintained in all regions of the tank.

Chocolate Products

In many parts of the country, chocolate, milk chocolate, sweet chocolate, and imitation chocolate coatings as well as other kinds of fat-based enrobing products can be procured in bulk. The liquid material replaces the usual paper-wrapped 10 lb blocks and eliminates the need to have a special unit to melt the chocolate at point of use. It is necessary to transport and store the bulk coatings at elevated temperatures in order to keep them in liquid form. If they are allowed to solidify at any point in the transport system, many problems result. Then, slow remelting at moderate temperatures is absolutely essential if heat damage to flavor and other properties is to be avoided. The usual tendency is to apply excessive heat to speed up the melting of the congealed mass, with the result that the entire content of the tank is spoiled. What is even worse, is that the damage to flavor is sometimes ignored and the defective ingredient is used to produce many tons of inferior products.

Chocolate and similar enrobing materials are transported in insulated tanks, which may have a duct system in the walls to permit delivery of the ingredient at nearly the same temperature at which it was loaded, i.e., in the range of 100° to 130°F. It is important that the tanks and fittings be clean and dry when the coating is loaded, since even traces of moisture can increase the viscosity of the product considerably and may be impossible to remove by any method available to the user. The tankers may have their own delivery pumps or it may be necessary for the customer to supply his own unloading system. Pipes 3 or 4 inches in diameter and jacketed with hot water or electrically traced are used. Heavy duty positive displacement pumps are recommended for rapid unloading.

Chocolate and sweet chocolate are more resistant to oxidative rancidity than are purified fats and oils because the former materials contain considerable quantities of natural antioxidants. It is rather unusual to come across a rancid sample of pure chocolate. White and pastel colored enrobing materials are also fairly resistant to rancidity because they consist largely of saturated fats; of course, they probably contain added antioxidants as well. As a result, nitrogen blanketing of chocolate and similar coatings is seldom, if ever, done. Milk chocolate is a somewhat different matter since it contains substantial amounts of milk fat, which is very susceptible to oxidation.

These coatings are not only much more viscous than liquid fats and oils but they behave in a non-Newtonian manner so that selection of the proper horsepower and type of pump is largely based on empirical methods. Coatings are also more abrasive since they contain particles of crystalline sugar and fibrous cocoa, both of which can cause fairly rapid wear of pumps, valves, and meters. As stated above, viscosity of the coating is greatly increased by fractional percentages of water, so every possible means should be used to prevent contamination of the material with condensate or residual cleaning water.

Tanks for chocolate coating are generally smaller than those used for liquid fats because the usage rate of coating is nearly always lower, and management is understandably reluctant to tie up the capital required to purchase large inventories of these expensive ingredients. Either vertical or horizontal tanks can be satisfactory, though a vertical tank with a cone bottom and a 4 inch discharge valve is commonly recommended. The tank can either be jacketed or contain stainless steel heating coils. Automatic temperature control is recommended in all cases.

Constant or frequent agitation of chocolate coatings is essential to the maintenance of a uniform product. A slow sweeping agitator is normally used, since rapid stirring is neither necessary nor desirable. Failure to agitate the coating allows the suspended insoluble material to settle, resulting in differences in viscosity and composition at different levels, a condition which will cause severe difficulties in production and unacceptable variations in product quality.

Pipes must be heated to prevent solidification of the coating during transfer. Methods which have been used for heating the pipes include: (1) Hot water tracing by means of copper tubing, (2) Electric heating blankets, (3) Use of fully jacketed pipes, (4) Welding a structural angle to the pipeline to form a channel for steam or hot water, and (5) Using an electric tracing cable.

GRANULAR AND POWDERED INGREDIENTS

General Considerations

Most granular or powdered dry ingredients can be received, transported, stored, and dispensed in bulk by either mechanical or pneumatic systems. Of course, combinations of the two methods can also be used. Which of the two systems is better for a particular purpose depends on the type of ingredient being handled, the rate of use of the material, the availability of suppliers who can conform to the needs of the particular plant, acceptability of the cost parameters, and other factors peculiar to a specific situation.

The suitability of a particular dry granular ingredient for bulk handling will depend to a large extent on its flow properties. These affect its movement out of bins and through chutes and are also related to its adaptability to pneumatic transfer methods. Procedures for measuring the pertinent physical properties of a material have been discussed in an earlier section of this chapter. Readers who are interested in more details about this aspect of bulk handling can begin their search with Bureau of Mines Information Circular IC-8552 (Pariseau and Fowkes 1972) and an article by Peleg *et al.* (1973).

Improving the gravity flow of powdered materials out of bins and hoppers has been a problem ever since these ingredients have been handled in bulk form. Some ancient impromptu methods such as pounding on the side of the bin with a mallet or poking and prodding the material with a stick are still in use. More sophisticated equipment is available, however. Magnetic and pneumatic vibrators attached to the walls provide constant or intermittent movement to dislodge packed or clinging powders and to neutralize the avalanching of floodable powders. A variety of internal agitating devices have been developed. These include double walled bins with the inner wall flexed by compressed air, injection of air directly into the powder at critical points, pulsating air pads, etc. Attempts to overcome the limitations of manual prodding methods have led to the development of powered internal agitators such as screws, sweep arms, revolving scrapers, and the like. These devices function by pulling material through the bin

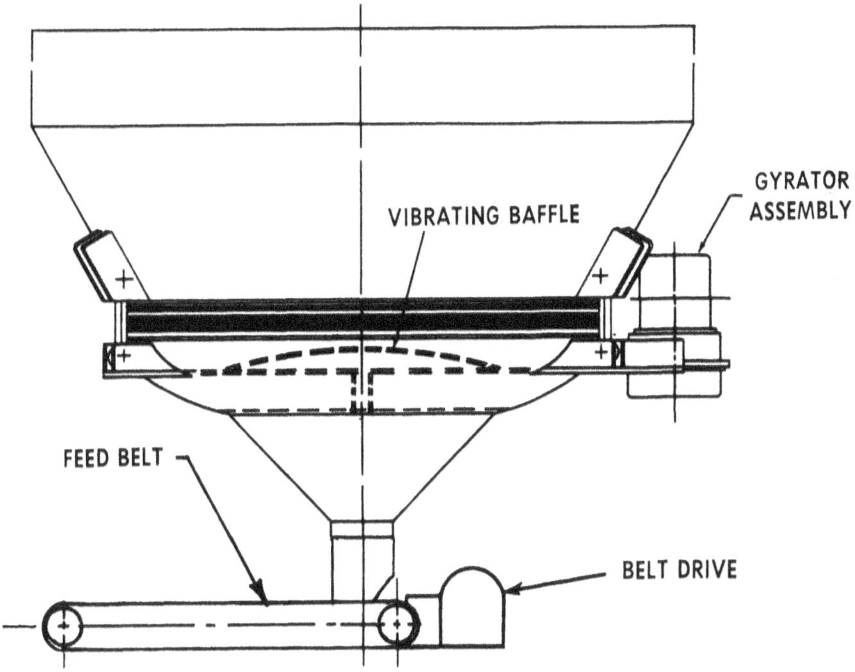

FIG. 1.3. A BIN ACTIVATOR ASSEMBLY FOR IMPROVING THE DISCHARGE OF POWDERED INGREDIENTS.

SOURCE: VIBRASCREW, INC.

opening. They work well with some powders, but do not function satisfactorily if the material tends to arch over the bottom or to tunnel. One of the most successful designs combines a bin activator (vibrating hopper) with a curved vibrating baffle affixed several inches above the outlet (see Figure 1.3).

The organization of this section of the chapter will be to first discuss those pneumatic handling principles generally applicable to bakery ingredients and then to discuss the specific problems involved in mechanical and pneumatic handling of some of the more important ingredients.

Pneumatic Transfer Principles

The essential feature of pneumatic handling, insofar as it refers to bakery ingredients, is the movement of particles through tubes by streams of air. Dense phase pneumatic conveying methods involve relatively low air-to-solids ratios in which plugs of granular materials are pushed through the lines. Dilute phase pneumatic conveying methods involve high air-to-solids ratios in which the particles are surrounded by air.

Although low pressure (3 to 5 psig) systems transfer material at relatively low rates, the equipment is considerably less costly than high presure handling units. The fans used in low pressure lines are less expensive than positive displacement pumps and generally require less maintenance and last longer. Venturi injectors can be used to introduce the ingredient into the air stream, and simple cyclone-type separators can be used to remove the air from the ingredient at discharge points. Such systems are used for moving low tonnages of material over short distances at low product-to-air ratios. They may also be used where a trickle feed is required or when it is necessary to cool the ingredient.

Dilute phase systems can be classified as (1) negative pressure or vacuum systems, (2) positive pressure systems, and (3) combination systems. The simplest possible arrangement is shown in Figure 1.4. In many respects, vacuum systems are simpler and can be built from less expensive components, especially when they must convey ingredients from several locations to one discharge point. Pressure systems usually require a rotary valve and manifold at each pickup point, whereas only a manifold and cutoff valve may be needed for vacuum systems. Product is not blown out at leaks in negative systems and, if bags or drums are dumped into the system, the conveying air controls the dust, thus eliminating the need for a separate air collector. Vacuum systems permit conveying directly from a rail car or truck by simple suction nozzles, making it possible to use simpler unloading equipment. Less expensive diverter valves can be used because special seals are not needed. Diverter valves with sliding-type seals (available with multiple outlets) work well because product build-up in dead areas is minimized. When conveying from two pick-up points into a single line, a Y-branch with cutoff valves in each leg can be used.

Positive pressure systems are best suited for conveying material from one pickup location to several discharge points. Cost is often the decisive factor, since savings may result from not needing a rotary valve at each discharge point (though such valves are needed at the intakes). The discharge can often be a cyclone collector with a vent to the atmosphere on top and a simple spout connection at the bottom. If it is necessary to convey to several hoppers, a simple bag-type filter may suffice. Pressure systems require smaller pipelines because they operate with about 1.5 times the differential pressure of vacuum lines, and as a result the product-to-air ratio is much higher. Maximum operating pressures are usually 10 to 12 psig when rotary airlock seals are

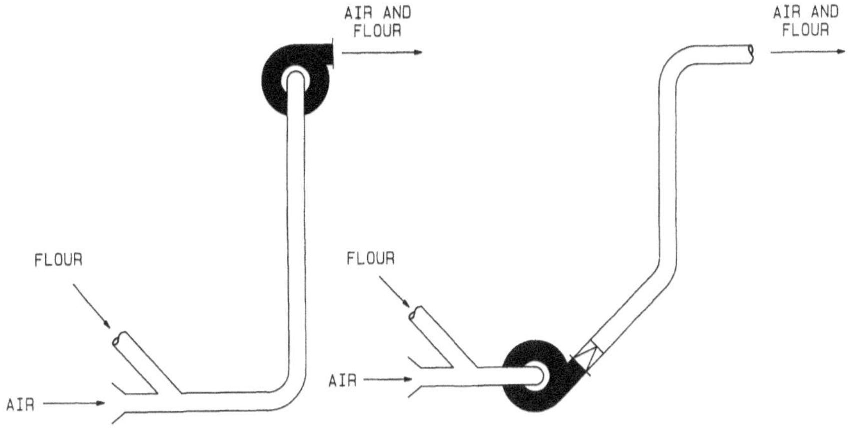

FIG. 1.4. THE BASIC PATTERN OF NEGATIVE AND POSITIVE PRESSURE SYSTEMS FOR PNEUMATIC TRANSFER OF POWDERED INGREDIENTS.

placed at the inlet. Most rotary valves can't tolerate pressure differentials much higher than this. Negative systems usually operate on pressure differentials (vacuum) of about 7.5 psig. In dilute phase systems, such as we are discussing here, about 20 lb of material can be conveyed per pound of air, equal to about one cubic foot of product per 20 cu ft of standard air. If a separate venting system is needed to take care of blowback air or dust generated at a dump hopper,the pressure system becomes more expensive.

Since the same amount of product is moved the same distance, the horsepower needed to operate the two types of systems is about the same, except for small differences in the efficiencies of the blowers.

Combination systems can have the advantages of both the vacuum and the pressure procedures. They are particularly well suited for conveying ingredients from several pickup points to several discharge points. The usual arrangement is to bring the material from the bulk carrier by vacuum, and collect it in a tank fitted with a cyclone or filter receiver. Subsequent movement is by positive pressure. Generally two blowers are required, although some plants have been designed with only one blower. With a single blower, air velocity on the vacuum leg will be higher than the velocity on the pressure leg because of the compression ratios involved. Low velocity on the presure side tends to cause plugs, while speeding up the blower to increase velocity on the pressure side causes the velocity on the vacuum side to become excessive with resulting product breakdown and other problems.

Flour and sugar are the principal particulate ingredients handled in pneumatic systems by bakeries. The following operations are most often involved in these systems:

(1) Unloading of bulk railroad cars and trailer trucks.
(2) Transfer from unloading point to bulk bins.
(3) Transfer from sack blender to bins.
(4) Transfer from bulk bins to use bins and recirculation between bins.
(5) Transfer with in-line sifting.
(6) Delivery to scales with return line.

Flour

Bulk delivery vehicles—Bakeries receive bulk deliveries from vehicles which may be either specially equipped freight cars or trucks. Mechanical equipment, primarily screw conveyors, can be used to unload the delivery vehicle, but pneumatic equipment is employed more frequently. If a truck is used, unloading equipment is often carried as part of the vehicle. Pneumatic unloading equipment for freight cars may be either the pressure or the suction type, with the pressure type probably the most common. Figure 1.5 is a diagram of a combined pneumatic and mechanical system for unloading, storing, and in-plant transfer of flour.

There is a self-contained wheeled unloader for GATX Airslide cars (Hallman 1987) which provides air for activating the slides of the car and also transfers flour at a rate of 600 to 1000 lb per hr. In the pressure system, flour is fluidized by injecting air which transports it to the top of the flour bin. Capacity of railroad cars is 90,000 lb while most trucks will carry 40,000 lb. Unloading rates will vary, but may reach 40,000 lb in 15 minutes. Unloading points may be located as much as 500 to 600 ft from the storage silos. Power and equipment costs usually become prohibitive for longer distances.

Pneumatic conveying—Although it is possible to design a bulk handling system for flour which is based entirely on mechanical conveying, nearly all modern large-scale equipment uses pneumatic transfer principles. Mechanical conveying of flour is inherently less sanitary, requires greater capital expenditures (for complete large-scale installations), and is not as reliable or as versatile as pneumatic conveying. Many pneumatic conveying systems do include one or more mechanical conveying steps, however. Movement of flour by gravity, as in some milling operations, is not generally applicable to bakeries.

There are no product changes of importance caused by pneumatic

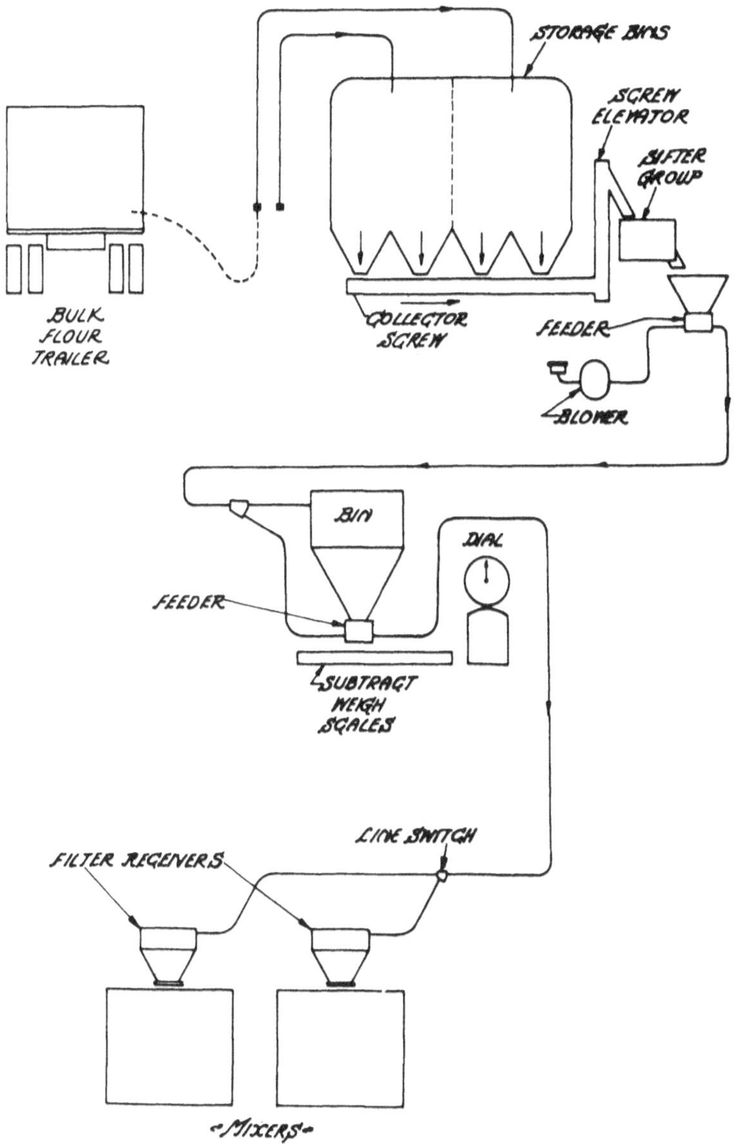

FIG. 1.5. TYPICAL ARRANGEMENT FOR UNLOADING BULK FLOUR TRAILERS, SHOWING A SCREW CONVEYOR UNDER STORAGE BINS.
SOURCE: FRED D. PFENING CO.

conveying of flour. Moisture loss may be somewhat greater in bulk flour as opposed to sacked flour, but this involves no reduction in the active ingredient. There have been no verified reports of stratification by particle size, and particle breakdown is not a problem in flour, contrary to the situation with sugar. Oxidation or aging of the ingredient reaches an equilibrium condition much sooner in flour handled by these systems than in sacked flour.

Pneumatic handling can be applied within the bakery to flour received in bags, but it is more efficient to receive the material from bulk cars or trucks. Pneumatic flour handling systems are classified as either negative pressure systems or positive pressure systems. Although there are fundamental design differences between the two types, in either case their operation depends upon fluidizing the flour, i.e., causing each particle to be surrounded by turbulent gases. In positive systems, the conveying air stream is at a pressure higher than atmospheric and the pumped air pushes the flour through the line. Normal operating pressures do not exceed 10 to 11 psig and the required air volume seldom exceeds 100 cfm. Negative pressure systems can be regarded as vacuum systems, where a low pressure is created by pumping air out of the last storage or surge tank downstream from the receiving port. That is, air entering the system at atmospheric pressure pushes flour through the lines. It is possible to design systems combining these two basic types of air conveying methods.

There are at least seven groups of equipment in pneumatic flour handling installations. These are:

(1) A mill-to-bakery transporting vehicle with unloading means compatible with the bakery receiving system.

(2) Equipment for receiving flour from the vehicle and transferring it to bins.

(3) Storage bins sufficient in number and capacity to hold the different types and required amounts of flour.

(4) Sifters and conveying means leading from the storage bins to sifters and from sifters to subsequent points.

(5) Surge or service bins feeding the scale hoppers.

(6) Scale hoppers and conveying means to and from them.

(7) Filter receivers discharging to mixers.

Connections between the different pieces of equipment are conveying tubes made of metal (frequently aluminum) or plastic and with diameters calculated to efficiently carry the required volumes. The interior walls of these tubes should be very smooth to reduce friction, facilitate air flow, and minimize abrasion of the particles. Bends, turns, and elbows should be made on a wide radius. Grounding is essential to prevent the buildup of static electricity. The system should include quick-disconnect joints at critical points so that blockage may be corrected easily.

Provision should be made for withdrawing relatively small amounts of flour for dusting purposes. Two methods are (1) provide an additional flour scale weigh hopper in the air-line loop, or (2) return the residue from the flour side air line and divert it to a flour dust receptacle. A third alternative is to bypass the bulk handling system entirely and receive dusting flour in bags. This allows the procurement of a special and perhaps less expensive flour for dusting, and is probably not much more inconvenient than withdrawing small amounts of flour from the bulk tanks.

In positive pressure systems, there will be blower-feeder groups at the car or trailer discharge point, a storage bin discharge, a central scaling unit, and use or service bins. The blowers may be of the positive displacement type, having two close-fitting impellers rotating against each other like gears. Air is discharged through filters and it is important that plenty of filter area be provided to prevent pressure build-up in the bins. Figure 1.6 shows how many small bag filters can be applied to individual collecting vessels for venting air from pneumatic delivery systems. An arrangement which has been suggested for eliminating the necessity for an expensive dust collector is to have all storage bins vented into a common system so that the the combined individual dust bags will relieve pressure. Although the passive exhaust system is satisfactory, a more efficient means of exhausting air is by suction blower and filter.

Bins and silos for flour can be horizontally or vertically aligned, and, in either case, are usually of circular cross-section. In some older installations, particularly where space limitations were severe, there will be found horizontal bins with angular—approximately rectangular—configurations. The availability of floor space in relation to ceiling height usually dictates the style of bin construction. Bins must be air tight and dust tight. In horizontal bins, leveling of the contents so as to get a complete fill becomes more difficult as the length and width of the bin increases. The bottoms are always slanted, but even so, flow of the material to the outlet is not as good as in vertical silos.

It is often necessary to install flour storage bins outside the main building in order to avoid infringing on valuable processing areas. Some experts (Haile 1973) recommend placing flour storage silos outside the plant in all cases where that is possible, to save on cost. No doubt this is perfectly satisfactory in most cases, but there are a few negative factors to consider. When outside, the bins are exposed to rapid changes in temperature, so there is a possibility of condensate

FIG. 1.6. PNEUMATIC FLOUR HANDLING SYSTEM FOR FLOUR RECEIVED IN BAGS. MANY SMALL FILTER BAGS USED TO RELIEVE PRESSURE ON RECEIVING BINS.
SOURCE: APV (BAKER PERKINS)

forming on the inner wall and running down into the flour if the temperature drops rapidly outside the bin. Insulation can be applied to the exterior to slow down the rate of heat exchange in the empty portion, but some suppliers have expressed doubts about the efficacy of this measure. A better arrangement in extreme climates is to enclose the bins in an inexpensive shelter (see Figure 1.7). Condensation can be prevented by installing an air circulation system to exchange the air above the flour so that its relative humidity is always in equilibrium with the outside atmosphere. When the silos are placed outside the building, longer supply lines are often needed, and this is also a negative factor.

Coatings are usually not recommended for the flour-contacting interior surfaces of bins, clean polished metal surfaces being preferred. Flour has an abrasive effect which tends to keep the surfaces polished, but it does remove coatings.

Various systems have been developed to expedite flour flow through exit ports, as all powdered materials have a tendency to hang up at low angles. These systems may include vibrators on the outside of the

FIG. 1.7. SHELTER FOR FLOUR STORAGE BINS.

hopper, agitators on the inside, or conveyors moving along the bottom. Hopper slopes should be as steep as the bin height allows, and not less than a 60° angle unless some mechanical flow assistance is provided. Even at this angle, vibrators may be required to expedite flow. Both horizontal and vertical bins can be unloaded by screw conveyors or inclined air chutes. Inflatable membrane silo liners which fit against the inner surface of the bin wall and floor are said to allow complete emptying even of flat bottom silos. When the control system receives a signal for product discharge, the membrane is inflated under very low pressure. As the membrane inflates, the product's angle of repose changes, rolling and pushing the powder to the discharge opening (Anon. 1985).

"Use bins," intermediate receiving containers into which the flour is transferred from the silo and stored for a brief period before use, are present in some systems but are generally considered unnecessary in present practice. Where blending of flours is required, the bulk systems may include storage bins, blending screws, and daily usage bins. When two different types of flour are called for by a formula, consideration should be given to blending them in the mixer immediately prior to adding the other ingredients rather than blending them in a separate set of equipment.

Flour should be sifted before use. The sifter should be placed as close to the mixers as possible. There are two types of sifters, the older design of rectangular shape containing a stack of wire or cloth screens moved in a horizontally reciprocating pattern by a motor and transmission located under it, and the newer design or "in-line" type which can withstand pressure so that powders can be conveyed directly through it (see Figure 1.8). Both types perform their intended function satisfactorily.

Explosion hazards—In any space where small combustible particles are suspended in air, the possibility of an explosion must be considered. It is true that dust explosions which have occurred in bakeries seldom have had catastrophic effects equal to some recorded explosions in grain elevators, where deaths of several workers and damage amounting to hundreds of thousands of dollars have occurred, but the consequences can be tragic, nonetheless. Bakery explosions are said to have resulted from the bursting of a breather bag on top of a mixer, ignition of flour inside a large flour silo, welding a funnel at the bottom of a sifter while the sifter was running, etc.

Pneumatic flour, sugar, and starch handling systems contain many spaces where explosions can occur, and they must be designed with full recognition of the hazard. See, for example, National Fire Protection Standard No. 66, entitled *Pneumatic Conveying of Agricultural Dusts*. The safety measures to be considered include, but are not limited to, prevention of dust emissions into the atmosphere and elimination of every possible source of sparking (e.g., static electricity and metal-to-metal abrasion). Interlocks should be provided to shut down the process

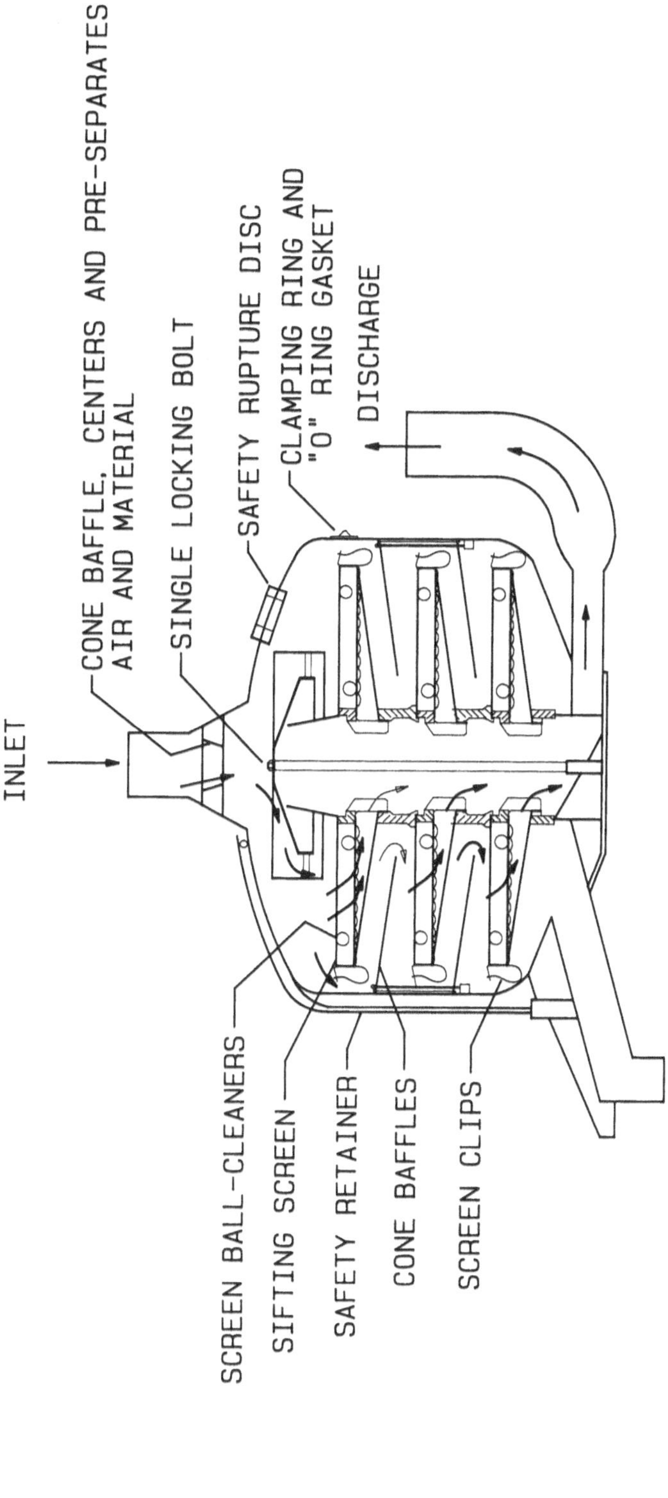

FIG. 1.8. An in-line pressurized sifter for powdered ingredients.

Source: Fred D. Pfening Co.

in the event of overfilling or overflowing of a storage or use bin or hopper. If an explosion should occur, damage can be limited by providing for explosion venting on the air-material separators and on the room housing the air-material separators (Coons 1979).

Sugar

When filled with granulated sugar, the usual railcar of 5,000 to 6,000 cu ft capacity will hold 200,000 to 220,000 lbs. Two discharge systems are practical—pressure differential and gravity. Air slide discharge is said to be of limited value because sugar cannot be fluidized with any significant degree of success. The pressure differential railcar is suitable for sugar and it is designed on the principle of pressurizing the storage compartment at about 14 psig, which allows relatively fast movement to a remote receiving hopper. Rates of 50,000 lb per hr can be easily achieved. The advantages of pressure differential discharge are said to be reduced loss of conveying air as a result of eliminating rotary valves, less breakdown of particles, and faster discharge due to higher product-to-air ratios (Reimelt 1987).

Gravity discharge railcars are sometimes designed to include discharge-assisting means, including some described in the above paragraph as being of limited value. Outlets of these railcars are connected by screw feeders and rotary valves to a pneumatic dilute phase conveying system. The most cost-efficient discharge system for sugar is said to be a combination of vacuum and pressure conveying, sometimes called the push-pull method.

Trucks are very commonly used to convey granulated sugar. They obviously have more flexibility in routing as compared to railcars. The smaller load is frequently an advantage. If the source of sugar is more than 300 miles distant, truck delivery could very well be more costly than railcar delivery, however. The discharge method used for trucks is basically the same as for railcars. The average unloading rate for a truck is approximately 40,000 lb per hr.

Bulk sugar can also be delivered to the food factory in portable bins. These containers are received and moved by forklift trucks. The most common portable bins have capacities of 3,000 to 4,000 lb. About 8 to 12 bins would be carried on each truck. One of the disadvantages of using portable bins is that transportation costs for the bins have to be paid both ways.

Either screw conveyors or pneumatic units can be utilized for unloading purposes. In mechanical systems, the delivery vehicle is put into position above a hydraulically activated "boot" to which its discharge outlet is connected. After the outlet valve is opened, sugar falls to a screw or belt conveyor for removal to a bucket elevator. The latter unit moves sugar to the top of a silo. If the system is pneumatic, hose connections are made between the sugar delivery truck and the receiving pipes. Two pipes are used, one for conveying sugar and air from the truck to the storage bin and the other for returning pressurized air and sugar dust to the truck. A blower and a dust collector are mounted on the truck so that no additional receiving equipment is required at the processor's plant. When shipments are received by rail car, the user must install additional equipment. The remainder of the system is somewhat simpler than for flour because no sifters are required. More air volume is required for moving sugar, however, and larger filtering equipment is needed to exhaust the air. Reuse of the air (closed system) helps to maintain a constant relative humidity in the storage and conveying circuit and may reduce caking problems.

Pneumatic conveying—The main operational features in a typical pneumatic system for storing, conveying, weighing, and dispensing have been summarized by Pancoast and Junk (1980):

(1) Air flow is developed by a blower located near the storage bin or under the bin skirt.

(2) Sugar is fed from the storage bin through a rotary valve airlock. It is picked up and conveyed by pipes to a receiver equipped with a dust collector. These pipes are designed with long sweeping curves to reduce the attrition of sugar crystals.

(3) Sugar is fed by a rotary valve from the receiver into the weigh hopper. When the requisite weight is reached, a rotary valve above the hopper shuts off. Simultaneously, a rotary valve below the storage bin also stops. The blower continues to run for a short period to clean out the line, then it shuts off.

(4) The weighed sugar is discharged through a clam valve to a process kettle or a mix tank.

The devices making up a blower unit assembly are usually combined on one chassis and normally consist of a blower, filter, silencer, check valve, pressure gauges, motor, controls, drives, and guard.

The principal problem in handling bulk sugar is caking due to moisture absorption. This can lead to bridging in the silo. Suppliers should load sugar at a maximum moisture content of 0.2% and a maximum temperature of 95°F (Hagedorn 1965). In the bins, an equilibrium is established between the water content of the sugar and the relative humidity of the ambient air. Flowability remains unchanged in the range of of 20 to 60% relative humidity, since in this range the equilibrium moisture content changes very little. At relative humidities higher than 60%, however, caking may occur.

Bins should be kept as small as possible and contents should be recirculated during plant shutdowns. Use of multiple screws at the bin discharge will facilitate movement of sugar agglomerates. The bins should be completely emptied at convenient intervals so that any sugar caked on the sides can be removed.

The tubes through which sugar is transferred to the silos, and through which the air with its entrained sugar is returned for filtering and reuse, are ordinarily six inches in diameter. Straight sections can be made of thin-walled aluminum tubing, but the elbows should be made of stainless steel curved with a radius of at least eight feet. The inside of the tubing should be smooth and without projections or cracks. Design of the transfer system should have as one of its priorities the minimizing of sugar crystal breakdown.

Bins should be of sanitary construction with a minimum number of cracks and crevices. They should be of steel, cylindrical in shape, and vertically oriented. The minimum practical capacity is around 50,000 lb. The bin can be placed at an elevation such that gravity feed can be used to deliver sugar to the process area, since the delivery truck can elevate and convey sugar for a considerable distance.

Powdered sugar—Some bakers have complained of caking and sticking of sugar conveyed with high humidity air. This seems to be a particularly serious problem when powdered sugar is conveyed from the mill to holding tanks. The problem can be alleviated by dehumidifying the air. Refrigeration-type dehumidifiers seem to be preferred, but silica gel dehydrators have also been mentioned.

Powdered sugar is seldom if ever procured in bulk because of its poor flow properties and its extreme tendency to consolidate into large lumps. It is either procured in bags or totes or (usually) made in the plant by grinding granulated sugar.

Pneumatic conveying of granulated sugar to the pulverizer and of powdered sugar from the pulverizer is common practice (see Figure 1.9). Granulated sugar is delivered from a silo or use bin to a receiving hopper above the mill by a system that is automatically controlled by sensors in the hopper. When it is necessary to replenish the stock of powdered sugar, an airlock at the bottom of the hopper meters granulated sugar into the mill. About 3% starch is added at this point to improve the flow properties of the powdered sugar. The pulverized material is discharged from the underside of the mill into a pipeline through which it is carried by cooled conveying air until it reaches a

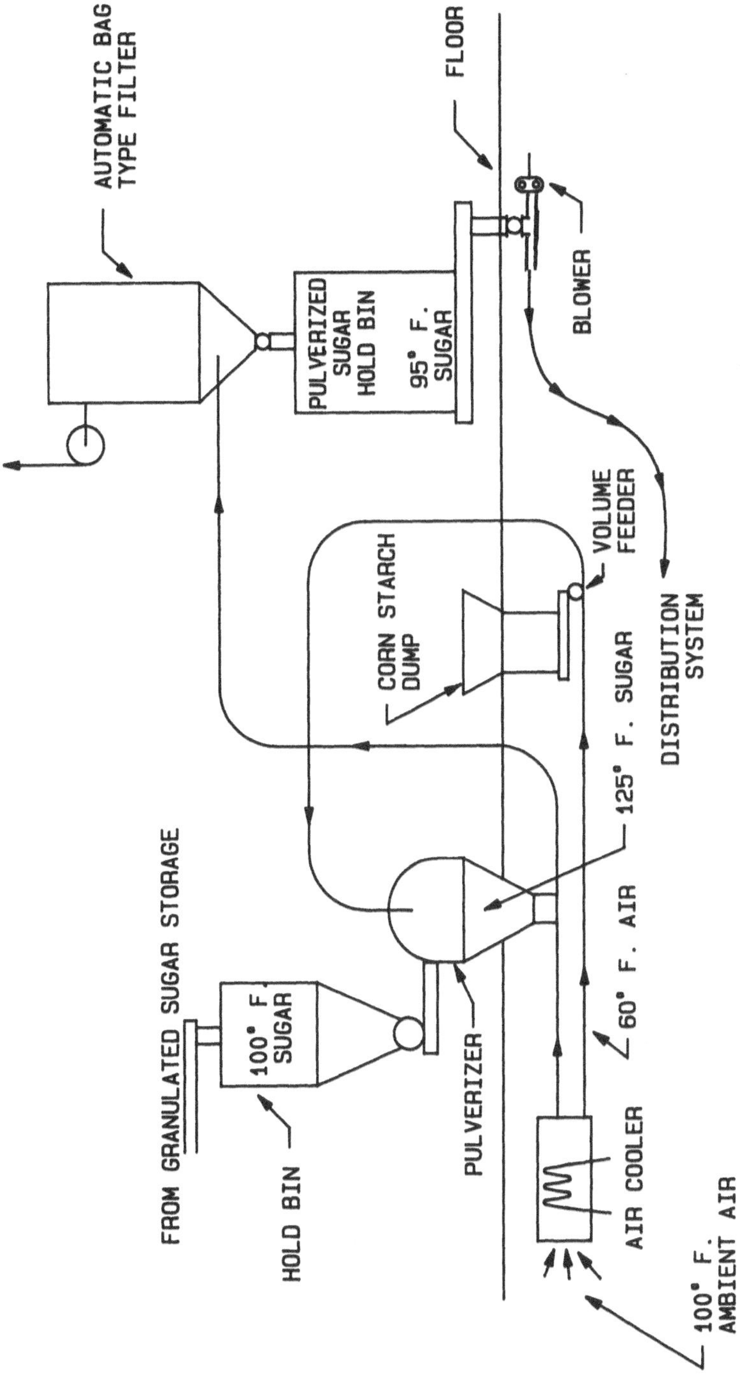

FIG. 1.9. Pneumatic transfer of material to and from a powdered sugar mill.

dust collector, where the air is separated from the product. Pre-cooled air is used in the system because it removes heat put into the sugar by the grinding action. Since moisture is removed from the air by the cooler, it also helps sugar to resist caking in bins. The sugar temperature can be lowered from 125°F to about 95°F during transit if 60°F air is used.

Mechanical conveying—Sugar in granular form is generally moved by pneumatic means in bulk handling systems, though some older systems are entirely mechanical. Mechanical conveying of sugar has some advantages because there is generally less attrition, or particle breakdown, in these systems than in pneumatic handling. Attrition occurs more readily in sugar and has more of an effect on the performance of this ingredient than is the case with flour.

The main operational steps in the transfer of sugar in a typical mechanical system are:

(1) Unloading from the delivery vehicle.
(2) Transferring to the elevator.
(3) Elevating to the top of the storage silo.
(4) Distributing the sugar evenly across the silo.
(5) Conveying from the silo bottom to elevator.
(6) Elevating to the top of a surge hopper.
(7) Transferring from surge hopper to weigh hopper.
(8) Dumping from weigh hopper to mixer or receiving bin.

There may be additional steps, but seldom fewer. Figure 1.10 shows some of the devices used in mechanical and gravity handling systems.

Mechanical systems are based on combinations of conventional transfer devices. The receiving unit usually consists of a receiving hopper and a standard screw conveyer feeding a bucket or chain elevator that terminates slightly above the level of the top of a storage bin. Sugar may also be discharged directly from the delivery vehicle into a receiving pit under the roadway or railroad tracks. There will be a conveyor leading from the pit for transferring the sugar to the elevating means. At the terminus of the elevator, another screw conveyor carries the sugar to the entrance part of the bin. The sugar is reclaimed from the bottom of the storage bin by a screw, belt, or other type of mechanical conveyor and then discharged into a surge bin feeding an automatic weigh scale that delivers the selected amount of sugar into the mixer.

The break up of sugar crystals during transfer in either mechanical or pneumatic handling systems can be a serious problem in cookie production and some other processes. Cookie spread during baking and the texture of finished cookies are affected by the particle size of sugar

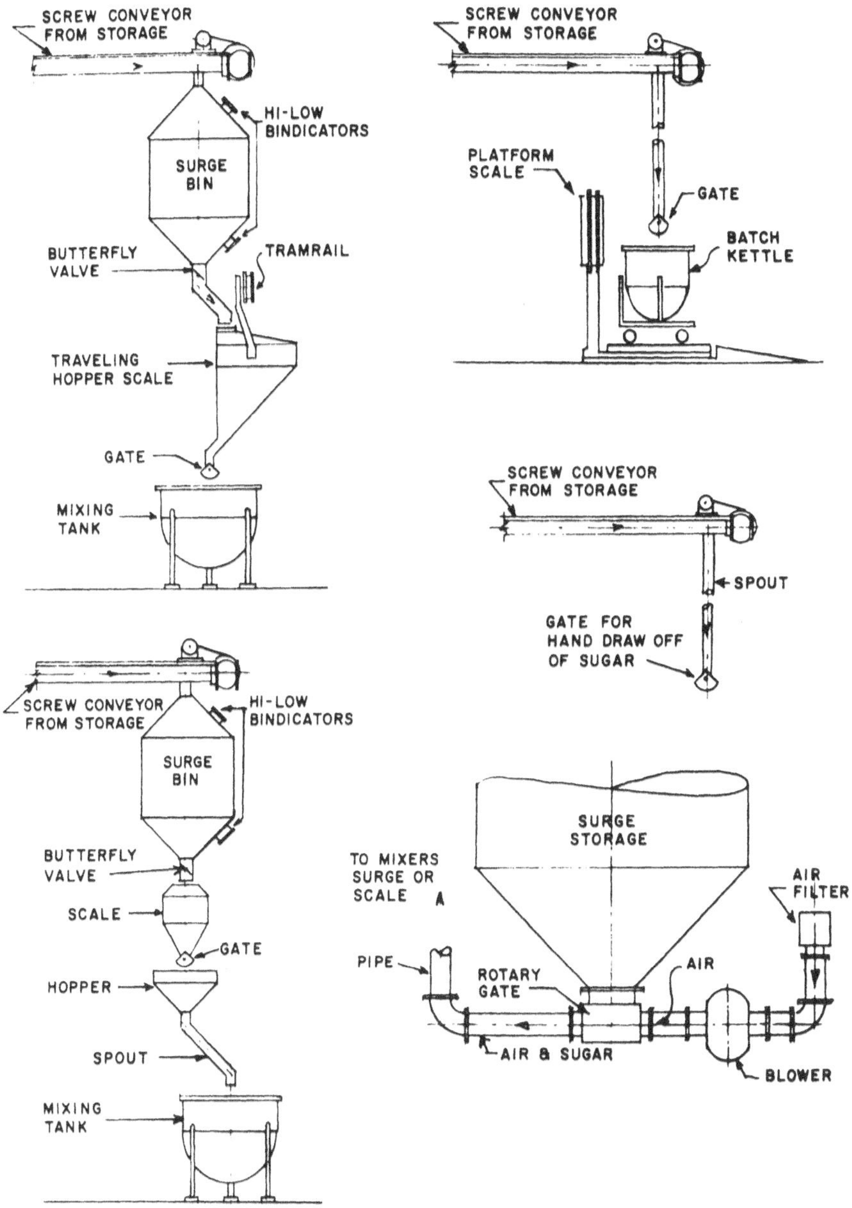

FIG. 1.10. Inplant distribution systems for bulk sugar.
Source: C and H Sugar Refining Corp.

used in the dough. If the particle size distribution remained the same at all times, attrition could be compensated for by a one-time adjustment of processing conditions or of formula proportions, and only a slight amount of inconvenience would result from granule fracture during transport. In some systems, however, there appears to be a fluctuation in the amount of fines which are generated, related to variations in air velocity or ambient temperature and humidity, as well as to unknown factors. This unpredictable variation makes it difficult or impossible to establish a standard corrective factor. Some of the preceding information was based on a discussion by Reimelt (1987).

Gravity feed—Of all the major ingredients, sugar is the one for which gravity feed assumes the most importance. Of course, all ingredients move under the influence of gravity at some point, even if only during the movement from top to bottom of a bin or tank, but we are concerned here with transfer from one major unit to another, as from a storage silo to a weigh hopper. Granulated white sugar flows relatively freely and can, in certain circumstances, be moved efficiently by gravity transfer provided it has been held under appropriate storage conditions.

In gravity transfer operations, it is important to remember that sugar, like all granular and powdered products, has a tendency to separate according to size when it is moved about or subjected to vibration. If sugar is allowed to fall for several feet, as is the case when a silo is being filled, a conical mound is built up. When additional sugar falls on the apex of the mound, coarse grains tend to roll down the sides and concentrate on the periphery while finer grains concentrate on the middle of the enlarging mound. If material is then withdrawn from the bottom center of the mound, the first sugar will have a different average particle size than that which is withdrawn later.

Controlled temperature storage—It is always desirable, but sometimes not practical, to control the temperature at which ingredients are stored. This is particularly important in the case of bulk-stored ingredients. Wide fluctuations in ambient temperatures can lead to moisture condensation on the interior of the bin, tank, or silo. Water droplets can coalesce and run down to ingredient level, leading to an increase in moisture content at the top rim of the ingredient. Although the increase is always a small percentage if measured on the basis of the total ingredient content of the container, it can be a much higher percentage at the rim. As a result, microorganisms (molds in flour and sugar, yeasts in corn syrup, etc.) can flourish, causing unacceptable changes in the ingredient. Water can cause hydrolytic rancidity and other undesirable changes in shortening. Clumping can occur in any particulate ingredient exposed to liquid water, and chemical reactions can occur in soda, leavening acids, etc.

Condensation can be minimized and even prevented by holding the ambient temperature within a narrow range. This can best be accomplished by locating the bulk tank within a temperature-controlled space—either within the plant itself or in a separate building which has air conditioning facilities. If this alternative is not feasible, the tank can be insulated. If there is rapid and continuous use of the tank's contents, the effects of condensation are minimized, but stasis over a weekend (for example) is enough to cause trouble.

Other Ingredients

Brown sugar in granulated form is not a good candidate for bulk handling because of its strong propensity for caking; possibly, some of the lower moisture forms could be handled by mechanical conveying systems. A better strategy is to procure this ingredient in liquid form, or replace it by a combination of molasses and either granulated or dissolved sugar. Liquid sugar (i.e., sugar syrup) can be made from granulated sugar in the plant, or procured, received, stored, and transferred in the liquid form.

Bulk granular dextrose can be shipped in trucks or Airslide cars. The handling systems and storage bins are much like those used for bulk sucrose. Pancoast and Junk (1980) say that mechanical handling of dextrose is preferred over pneumatic because of the crystal attrition that occurs in the latter systems. Crystalline dextrose tends to cake in storage silos and clog handling systems. This problem can be alleviated by using a bin with a cone-shaped bottom and equipped with a vibrator. Other discharge-assisting means, such as live bottoms consisting of multiple screw conveyors placed under a large bottom opening, may be effective.

Corn starch is stored in silos with airslide type bottoms and handled similarly to flour. Railcars can transport typically 40 tons of starch. Starch sometimes contains very fine particles, in the 10 to 20 micron range, which can escape through the pores of regular bin filter relief bags. Starch has a high shear friction when tightly packed, and this affects the internal design of feeders and diverters. Because of the explosive potential of starch aerosols, explosion proof gearmotors are used instead of standard enclosed motors.

OTHER SYSTEMS FOR BULK INGREDIENTS

It is possible to make bins out of flexible materials, by hanging a cylindrical fabric container from a circular frame at the top and attaching some sort of hopper to the bottom. In Europe, a woven plastic material called Trevira has been heavily promoted for this purpose. There are two basic types of these bins: (1) Hanging silos, shaped much like a conventional silo but made of Trevira and suspended from a suitable steel framework, and (2) Lifting type silos ("variable geometry" silos) which can be partially collapsed to assist in removing the contents. Manufacturers of these fabric containers say that air can pass through the walls and tops of the silo but all flour particles are retained. Accordingly, neither explosion relief panels nor top dust control panels are necessary. Discharge of, for example, flour is assisted by passing pressurized air through a device called an "Alivac feeder" in the hopper area.

Brined cherries are sometimes transported in bulk from growing areas to the plants where they are made into maraschino cherries. Stainless steel tank trucks hold 100 barrels, or 25,000 lb of cherries, replacing individual drums. The trailers are insulated to keep the fruit cool during the trip. Unloading takes about 75 min with a gentle centrifugal pump that prevents damage to cherries.

Caramel color is transported in both 8,000 gal tank trucks and 16,000 gal railroad cars, replacing the usual 55 gal steel drums. The tank cars have special food grade lining and, for winter shipments, have steam coils under the outside of the car.

Large users can buy liquid eggs in stainless steel tank trailers. Smaller sealed tanks can also be used. Payload of a bulk liquid trailer is the equivalent of 442,000 eggs, an increase of almost 50% over a trailerload of shell eggs. Trailers can be divided into two separate compartments—2,500 and 3,000 gal—to permit combination shipments of whole eggs, yolks, whites, or special mixtures. Insulated construction prevents temperature rise.

BULK HANDLING OF PRODUCTS

Ingredients are much more adaptable to bulk handling than are most bakery products. The large, fragile, and often non-uniform units—such as loaves of bread—seem particularly unsuited for such treatment, and yet, significant strides have been made in automated transfer, order assembly, and the like (Rader 1984). Of course, there comes a point where the cost of designing a system to transfer and handle a few units each of many different types of products becomes uneconomical compared to the wages of a small crew of experienced workers who can adapt quickly to changes in production schedules.

Bulk handling can sometimes be employed in the transfer, packing, and assembly of small-sized units which are made in large quantities. Confectionery and snack foods are already handled in bulk (see Figure

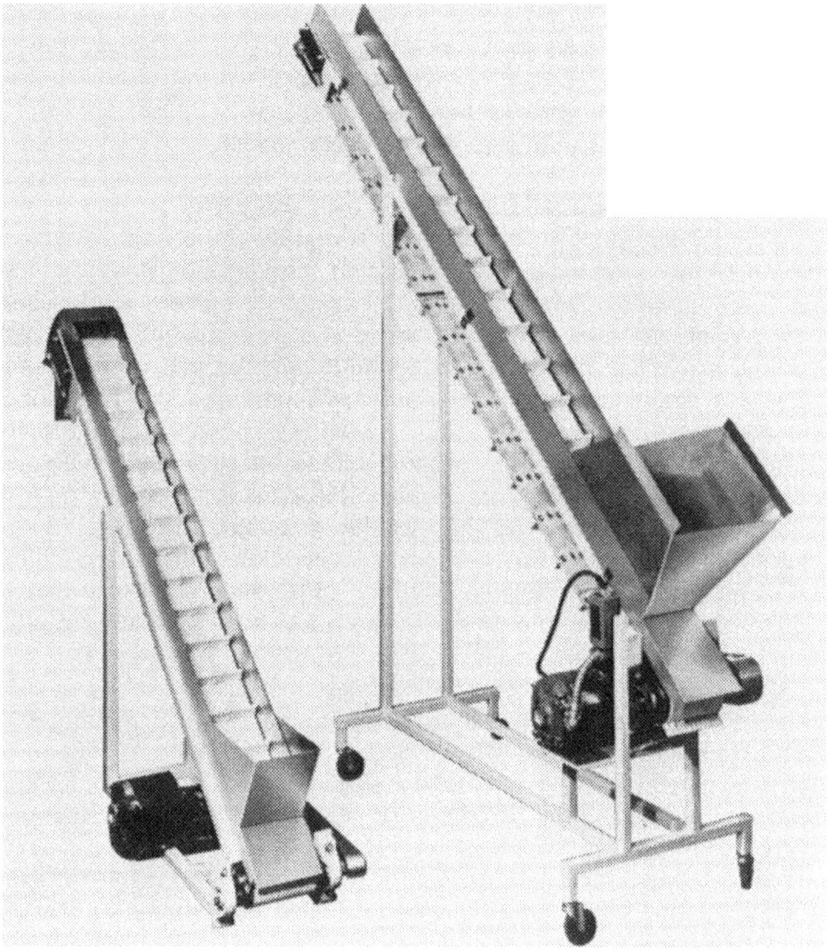

FIG. 1.11. CONVEYORS FOR TRANSFERRING FRAGILE SNACK PRODUCTS.
SOURCE: DORSEY-MCCOMB, INC.

1.11), sometimes pneumatically, and they overlap in size and quantity such bakery products as cookies and doughnuts. Of course, the volume must be large in order to justify the application of true bulk handling systems to such products. Popcorn, expanded corn snacks, and the like are good subjects for this type of treatment and can even be transferred by pneumatic conveying and stored in small silos.

Breading, a manufactured ingredient, has been stored and transported in bulk; tank trucks holding about 40,000 can be loaded or unloaded in about 30 min using positive pressure pneumatic handling throuigh stainless steel pipes. Shelled peanuts can be transported in sealed hopper cars holding 130,000 lb of nuts. The rail car is divided into three sections and unloaded pneumatically. After loading, a fumigant is placed in the cars and they are sealed.

COMPUTERIZED INVENTORYING

A big advantage of bulk handling of ingredients is that it permits computerized inventorying. Signals from load cells or strain gauges relay information on bin weights to the computer memory bank. These data can be processed almost instantly to give a readout of the total stock of any given ingredient. Ingredients metered out of the bins or tanks are deducted from the previous totals, giving a double-check of the contents. Received materials, measured by bin weighing or by meters on the intake pumps, provide a check on the supplier's invoice and the current bin contents. Among the types of inventory reports which are possible are, (1) the complete extended inventory, (2) the perpetual inventory, (3) the comparative inventory, and (4) ingredient and packaging requirements (Koehler 1981). Re-order levels can be entered for each item and provision made for an automatic printout of the needed materials each day or other convenient period.

A computer plus the appropriate software permits the ready calculation and provision of: (1) recap of all products ordered, (2) bake shop production orders having breakdowns for racks, pans, and pieces, (3) formula sheets for the mixing room, (4) raw material usage, (5) packaging material usage, and (6) cost of the day's bake with labor included (Cooper 1984).

A weak point in some bulk handling systems for flour is the difficulty of adequate inventory control. None of the methods currently used for measuring bin contents is completely satisfactory. Both volumetric and gravimetric devices are being used. Among the gravimetric types are strain gauges operating as part of an electronic indicating circuit and liquid load cells based on a combination of hydraulic and mechanical principles.

BIBLIOGRAPHY

ANON. 1985. Bulk storage system ensures 99% cleanout and uninterrupted flow. Food Engineering *57,* No 11, 110.

ANON. 1986. Smart valves save money at Palm Dairies. Prepared Foods *155,* No. 4, 82.

BORDEN, B. 1958. Bulk liquid sweeteners—engineering and economics. Proc. Am. Soc. Bakery Engineers *1958,* 79-84.

COONS, J. 1979. Dust explosion hazards in bakeries. Proc. Am. Soc. Bakery Engineers *1979*, 206-211.

COOPER, M. R. 1984. Computers for the baking industry. Proc. Am. Soc. Bakery Engineers *1984*, 99-104.

DAVIS, L. B. 1953. Pneumatic conveying systems. Proc. Am. Soc. Bakery Engineers *1953*, 123-128.

DIVER, J. J. 1982. Conveyors. Proc. Am. Soc. Bakery Engineers *1982*, 142-150.

FARRAND, W. J. 1962. Flour packing, warehousing, and bulk transport. Biscuit Maker and Plant Baker, May 1962, 407-418.

FISCHER, J. 1959. Conveying flour by air. Food Processing *20*, No. 9, 37-39, 43.

GARCIA, E. J. and STEFFE, J. E. 1986. Optimum economic pipe diameter for pumping Herschel-Bulkley fluids in laminar flow. J. Food Process Engineering *8*, 117-136.

GERCHOW, F. J. 1975. How to select a pneumatic conveying system. Chem. Eng. *82*, Feb. 17.

HAGEDORN, H. G. 1965. New practices in bulk handling of materials with special emphasis on instrumentation. Proc. Am. Soc. Bakery Engineers *1965*, 148-152.

HAILE, F. 1973. Bulk material handling equipment. Proc. Am. Soc. Bakery Engineers *1973*, 178-184.

HALLMAN, R. B. 1987. Personal communication. General American Transportation Co., Chicago, IL

HOWARD, R. M. 1956. Bulk flour. Proc. Am. Soc. Bakery Engineers *1956*, 78-84.

IRWIN, J. G. 1976. Basic principles of pneumatic conveying. Bakers Digest *50*, No. 5, 31-34.

KICE, J. 1985. Skilled Air Manual. Kice Metal Products, Wichita, KS

KOEHLER, W. H., JR. 1981. Computers in the bakery. Proc. Am. Soc. Bakery Engineers *1981*, 179-184.

KOLLMAN, W. C. 1961. Some engineering and economic aspects of bulk systems. Proc. Am. Soc. Bakery Engineers *1961*, 199-206.

LONG, J. W. 1984. Minor ingredient systems. Proc. Am. Soc. Bakery Engineers *1984*, 92-98.

MORRIS, G. C. 1971. Pneumatic bulk handling systems in the bakery. Bakers Digest *45*, 49-53, 66.

PANCOAST, H. M., and JUNK, W. R. 1980. Handbook of Sugars, Second Ed. AVI Publishing Co., Westport, CT

PARISEAU, W. G. and FOWKES, R. S. 1972. Bin hopper engineering and bulk materials flow: A state-of-the-art report of empirical and theoretical analyses. U. S. Bur. Mines Inf. Circ. *8552*.

PELEG, M., MANNHEIM, C. H., and PASSY, N. 1973. Flow properties of some food powders. J. Food Sci. *38*, 959-964.

PETRICCA, T. 1976. Fluid bakery shortenings. Bakers Digest *50*, No. 5, 39-41.

RADER, J. R. 1984. Automated packaged product system. Proc. Am. Soc. Bakery Engineers *1984*, 75-85.

REIMELT, S. 1987. Personal communication. Sept. 21.

ROTHFUS, P. R. 1968. Working concepts of fluid flow. V. Flow measurement. Instrum. Control Syst. 41, No. 7, 105-108.

SCHROEDER, W. F. 1956. Bulk fat handling. Proc. Am. Soc. Bakery Engineers *1956,* 85-90.

WAHL, R. C. 1978. Handling bulk materials—simple tests predict handling problems. Plastics World, Nov. 1978.

WOERFEL, J. B. 1981. Bulk handling of fats and oils. Cereal Foods World *26,* 446-448.

CHAPTER TWO

WEIGHING AND METERING EQUIPMENT

INTRODUCTION

The measurement of ingredients is obviously one of the most critical operations in the bakery. Not only is the accuracy of measurement directly related to the quality of finished products, but it also affects costs and, therefore, the profitability and continued existence of the company. Since composition is regulated to some extent by Federal, state, and local regulations in the case of standardized items such as bread and must, in any case, conform to the label declaration of ingredients, bakery management must be certain that the amount of each component found in the finished product will fall within acceptable tolerances if they expect to avoid legal entanglements.

The hand scaling of individual ingredients has been replaced at an ever increasing rate in recent years by automatic weighing and metering. Some of the advantages which result from these changes are faster batch preparation, reduction of human errors, lower manpower requirements, and improved sanitation. In practice, accuracy is usually improved. Automatic measuring is almost essential when bulk handling systems are used.

Designers of automatic measuring equipment for ingredients, doughs, and finished products have borrowed techniques freely from other industries. Water meters of various kinds have proved to be adaptable, with changes of varying degrees of complexity, to the measuring of other liquid ingredients. Figure 2.1 is an electronic water batching controller for bakeries which, in combination with a solenoid valve and a turbine meter having electrical pulse output, allows the automatic dispensing of accurately measured amounts of water into the mixer or other batch container. Weighing devices originally used in other food industries or in nonfood applications have been modified for bakery use. A great deal of ingenuity and expertise has gone into some of these adaptations and modifications.

Powders and granular materials are generally measured by gravimetric techniques, while liquids are more often measured volumetrically. Volumetric equipment is the least expensive, as a rule, but it is also inherently less accurate and dependable since the feed rate necessarily depends upon the density and (usually) the flow characteristics of the material being measured, and these factors cannot be expected to be perfectly uniform, especially in powdered materials. As a practical matter, however, volumetric feeders function quite satisfactorily in many bakery applications for sugar and other relatively free-flowing ingredients.

Gravimetric measurement is usually somewhat more difficult to automate, but is theoretically capable of greater accuracy, at least for powdered and granular materials which can vary considerably in density. The subsequent sections of this chapter describe equipment which has been developed for the automatic weighing and the automatic metering (i.e., volumetric measuring) of ingredients. This chapter will not include discussions of dough measuring devices, which are described in another chapter in the section on dough dividers.

DEVICES FOR MEASURING MASS

Basic Principles

The two major categories of weighing devices are balances and force-deflection systems. The simplest weight, or force, measuring system is the ordinary equal-arm balance. It operates on the principle of moment (tendency to produce motion about a point) comparison. The moment produced by an unknown weight is compared with that produced by a known weight. If the two arms of the balance are equal in length, a null balance will be obtained when the two weights are equal; for all practical purposes, a null balance is indicated when the beam forms a perfectly horizontal line. Usually, a pointer is attached to the beam at the pivot point so that the pointer is exactly perpendicular when the null balance is achieved. When the two arms are unequal, the product of the moment and length of the arm it will be equal to the product of the weight and length on the other side. The device known as the beam scale results when the unknown weight is attached to a very short arm, the long arm is a beam with graduations along its length, and a known weight is slid along the graduated arm until a null balance condition is obtained. The ancient and still familiar steelyard is an example of a beam scale.

Some industrial equipment is based on the principle of the beam scale. An unknown weight on a short arm is balanced by moving a poise of known weight along a calibrated long arm. The farther it is necessary to move the known weight ("poise") away from the fulcrum or pivot before the indicator becomes vertical, the heavier is the unknown weight. That is, the ratio of the arm lengths is varied to obtain a null reading. Multiple beam industrial scales introduce additional levers between the unknown weight and the beam to increase the ratio of unknown weight to poise weight, so that hundreds of pounds can be balanced by a much smaller poise. This strategem permits reductions

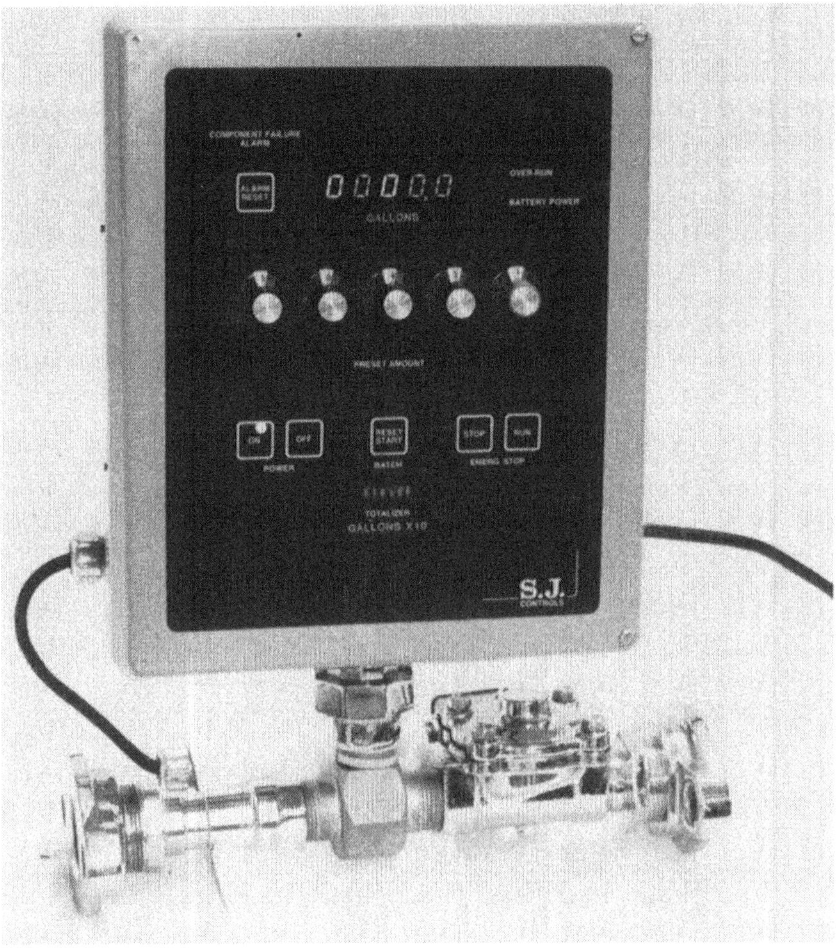

FIG. 2.1. ELECTRONIC BATCHING UNIT FOR LIQUID FLOWMETERS.
SOURCE: S. J. CONTROLS, INC.

in size of the equipment and, usually, improves accuracy. Small capacity bench balances are examples of the uneven arm scale while many platform scales and suspended hopper scales make use of the multiple beam design. Improvements which have been made in the design of these scales consist of electronic or electric readouts, automatic balancing systems, damping mechanisms, and recording devices.

Pendulum balanced scales use one or more "pendulums" (i.e., weights mounted on one end of a rigid rod which rotates about a pivot at its other end) to balance increasing loads as the pendulums move from vertical to horizontal. Because the weight which the pendulum will balance varies as the sine of the displacement angle, equal increments of weight will not cause equal rotation of the pivot and so cams are inserted into the system to give linear movement to the scale pointer. Metal tapes transmit the motion from platform to pendulum. These scales are commonly used where a dial indication of weight is required. They are available in recording and printing models, and with electric cutoff for batching.

Because the deflection of a spring or any other elastic element is directly proportional to the force applied to it (within its elastic limit) a calibrated spring, tube, rod, or plate can serve as a weight-measuring device. This principle is applied in the spring scale and the torsion balance, as well as other kinds of weighing equipment. The spring, rod, or other elastic element is used to support a platform on which the unknown weight is placed. The larger the weight, the larger the deformation of the elastic element. The latter is measured mechanically, electrically, or electronically. These devices are subject to hysteresis, metal fatigue, and temperature errors, but can still be effective weighing devices when properly designed and used. Figure 2.2 is a diagram of a platform scale based on flexure plates, the deformation of which is transmitted through a series of arms to an indicator beam.

Other force-sensing elements adaptable to weight measurement include strain-gauge load cells and pneumatic pressure cells. In a sense, strain gauges are deflection plates, but the deformation is measured as a change in an electrical signal rather than directly as a change in dimensions.

The electrical resistance of an electronic load cell varies with the tension or strain on the cell. These cells are the basic measuring instruments in many types of automatic weighing systems. The principle of such systems is that one or more of the supports for a tank, for example, is fitted with one of these elements. The load cell is distorted by an amount which is proportional to the weight resting on the support. An electronic circuit incorporating the load cell actuates a device that will transform the electrical charges into indications of weight. Equipment of this type is normally guaranteed to have an overall accuracy of 0.05% of the indicator's total capacity. In combination with the hydraulic load cell system, electronic load cells can be used in batching or other weighing operations both to indicate the weight on the load cell and to automatically activate a valve or other control when a preset weight is reached.

The hydraulic or pneumatic load cell can also be used as part of the

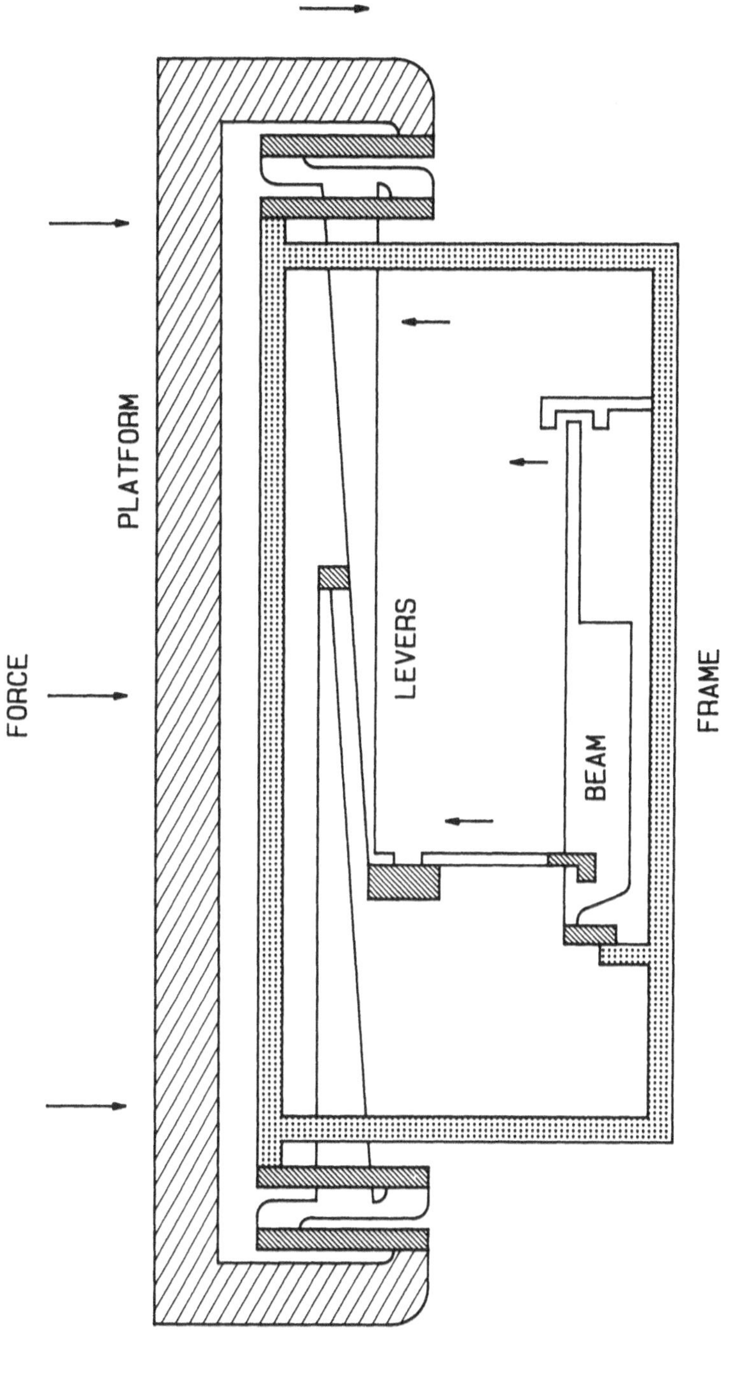

FIG. 2.2. PLATFORM SCALE BASED ON FLEXURE PLATES.
SOURCE: THAYER DIV., CUTLER-HAMMER

supporting member for an ingredient container. These cells generate hydraulic or pneumatic forces that vary with the weight they sustain, and transfer them through the medium of liquid contained in rigid tubes to an indicator or recorder. Accuracy of these units is said to be within 1% of load range.

Industrial Scales

The manual scaling of minor ingredients (sometimes major ingredients, too) is common practice in small bakeries. The scales may be stationary or mobile, mechanical or electronic, or vary in other characteristics. Inexpensive dial type mechanical scales with manual taring are common pieces of equipment. They often have dial pointers and circular graduations where multiple revolution readings are necessary, and they may have either an adjustable or a fixed scale. Additive weighing is done with a fixed scale and absolute weighing with an adjustable scale. The mechanical scales are relatively inaccurate, difficult to read, and have no capacity for recording weights or other data (Long 1984). Their resolution, or the smallest difference which can be read, is limited by the type of scale; customarily a dial with 1,000 lb total can be read to the nearest pound. In addition to their relative low cost, their advantages are ease of maintenance and calibration, and the ability to cancel out the tare weight so the ingredient weight can cover the entire dial (or beam).

Mechanical scales are gradually being replaced by electronic scales based on load cells and having digital readout and automatic taring. They can be obtained with metric or avoirdupois readouts, often interchangeable by pressing a button, and are available for all weight ranges, from micrograms to tons. For some models, readouts can be located at any distance from the scale, and recording and analysis programs of almost any degree of complexity can be put into use. Although these scales were initially too expensive for use in the smaller bakeries, many are now cheaper than equivalent mechanical scales. Sensors are usually electromechanical load cells of the compression-tension or deflector rod type. Advantages of electronic scales are their potentially better accuracy and resolution and their ability to be tared over their entire weighing range. Additionally, their output is suitable for transmission to a printer or for computer processing. In theory, they are not necessarily more accurate than mechanical scales but in practice they are less subject to reading errors and to other operator's mistakes. They occupy less space than mechanical scales, but also may be less sturdy and more sensitive to environmental problems.

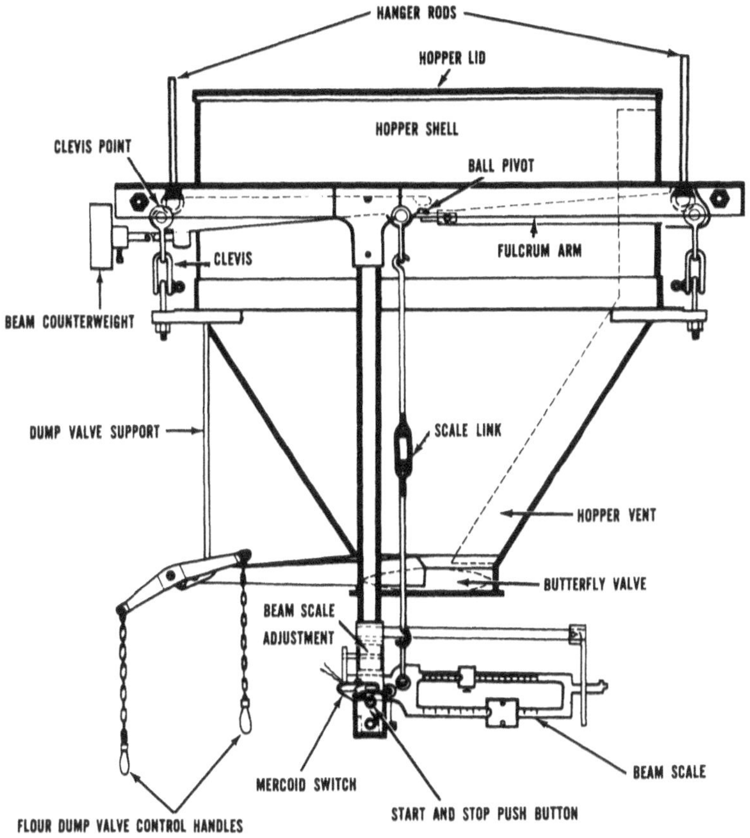

FIG. 2.3. FLOUR HOPPER AND SCALE.

Continuous and Automatic Scales

Continuous and automatic operating scales for batch weighing can be based on the same weight-comparison principles found in manually operated unequal arm balances. Sensing means are used to energize the mechanisms that start and stop the flow of material into the weighing hopper. For example, weighing of flour in most modern bakeries is accomplished by an automatic flour scale located immediately above the dough mixer. Figure 2.3 is a schematic diagram of a flour hopper scale. In some cases, the hopper is on a trolley, and can be moved over several mixers, as required. The usual model consists of a conical steel hopper mounted on four point-suspension bearings of knife-edged pivots made from case-hardened steel. Once the scale beam is set for the necessary amount of flour and the switch is activated, operation is completely automatic. A mercury switch activated by the scale beam shuts off the flour input when a preset weight has been reached. At the proper time, the operator discharges the contents of the hopper directly into the mixer. Capacities range from 200 to 1,000 lb of flour. Transfer operations are dustless, the outlet of the hopper being connected to the mixer by a sliding sleeve. Air displaced as the flour goes into the mixer is conducted to the upper section of the hopper by a venting tube. The scale must be adjusted by the operator or automatically compensated for conveyor overrun to obtain satisfactory accuracy.

In another type of automatic scale, a balancing weight is positioned by a reversible automatic motor. Deflection of the beam makes an electrical contact which causes the motor to move the weight in the direction necessary to restore balance, and the final balance position is translated into a signal by means of a potentiometer or digital encoding disc. The signal can be used for recording or control purposes, or both.

Automatic batch scales are adaptable to continuous flows of liquids, granular materials, or powders. If dry materials are being measured, the ingredient flows from a feed hopper through an adjustable gate into the scale hopper until the preselected weight is reached. Then, a trip mechanism closes the gate and opens the outlet. When the scale hopper is empty, the weight of the tare forces the door closed, resets the trip, and opens the gate for another cycle. A dribble feed, resulting from a partial closing of the gate as the weight target is approached, reduces the rate of inflow so that the extent of overfill is reduced. A counter can be used to record the number of cycles and the total amount of material which has been dumped.

Weighing Minor Ingredients

Many different types of scales and feeders are being used to dispense into batching bins, hoppers, and mixers those ingredients which are required in small amounts. There seem to be more manufacturers offering equipment for measuring these ingredients, probably because they are sold in greater numbers and the investment in plant is less than for making the bigger equipment. Operating principles are for the most part the same as for larger metering devices, but, of course, greater sensitivity and accuracy is necessary, as compared to multi-hundredweight dispensers.

Counter-rotating augers, sometimes combined with live bins or vi-

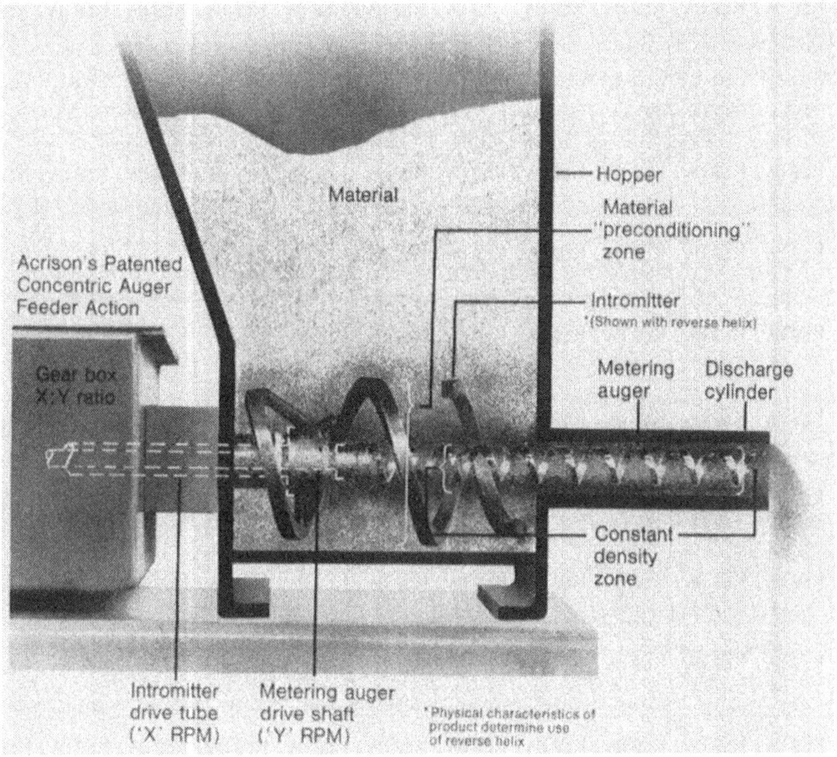

FIG. 2.4. CUTAWAY VIEW OF DRY INGREDIENT FEEDER.
SOURCE: ACRISON, INC.

brated hoppers, are frequently used to assist flow and to make the density of powders more uniform as they reach the measuring region of the device (see Figure 2.4).

In one unit suitable for measuring amounts as low as 0.15 lb per hr, a rotating and vibrating screw feeder is combined with a vibrating hopper or "live bin." These devices are said to be capable of handling many kinds of powders, pellets, flakes, and agglomerates which cause difficulties in other types of of equipment. The hopper is subjected to continuous gyratory vibrations that are transmitted throughout the contents. This permits feeding from the hopper without bridging or flooding, producing a steady, positive flow of material into the screw chamber below the hopper. The vibrating action also tends to condition the material so that the bulk density when it enters the screw chamber is uniform (Anon. 1987). Units which can measure as much as 8,000 lb per hr are available.

Even in very large bakeries, there are usually facilities for weighing minor ingredients which require a certain degree of manual attention. In one arrangement for manually weighing minor ingredients, a carousel conveyor carries small tubs past a stationary scale. As each bin stops on the scaling platform, a worker deposits in it ingredients taken from bags or drums.

Small ingredient storage bins can be mounted in a row along a wall. They are filled either by pneumatically conveying material from a dumping station or by dumping the ingredient manually into the top of the bin from a platform above it. An auger fed dispenser at the bottom of the hopper leads into a mechanical scale or electronically controlled weighing mechanism (Figure 2.5). Although the small silos are generally of the usual rigidly-attached metal construction, lift-type bins are also available; these are lowered so that the top is near floor level for filling, then raised to a level convenient for dispensing ingredient from their bottoms. Such bins are available in sizes from 1 to 2 ft in diameter and 6 to 15 ft high, with capacities of 600 to 1,500 lb of material. To assist in dispensing ingredients which are not free flowing, bins can be fitted with mechanical cone wipers, discharge screws, or air fluidizing means.

In an automated version of minor ingredient scaling, a mobile weighing device holding a batch receptacle moves on a track or roller conveyor beneath a series of storage bins. A weighed portion of ingredient is discharged from each bin when the mobile scale is positioned directly below its outlet. The scale signal controls the shut-off point for ingredient delivery. A trailing cable reaching along the entire length of the track connects the computer's formula storage unit to the mobile scale.

In medium- and large-size bakeries, automated minor ingredient scaling involves programmed and pre-set conveying to the scale, performing the weighing, and discharging weighed ingredients. The formula weights and the sequence of weighing are programmed for those products requiring minor ingredients.

The rate at which minor ingredients are conveyed from the storage bins to scales must be adjustable to give coarse, fine, or dribble feed into the scale hopper. Choices of equipment combinations include pneumatic conveying, mechanical conveying by metering screws, vibratory feeders, air slide feeders, conveying belts, and rotary feeding valves.

Design and construction of the scale hopper and associated equipment must meet certain design criteria. These units should be dust-

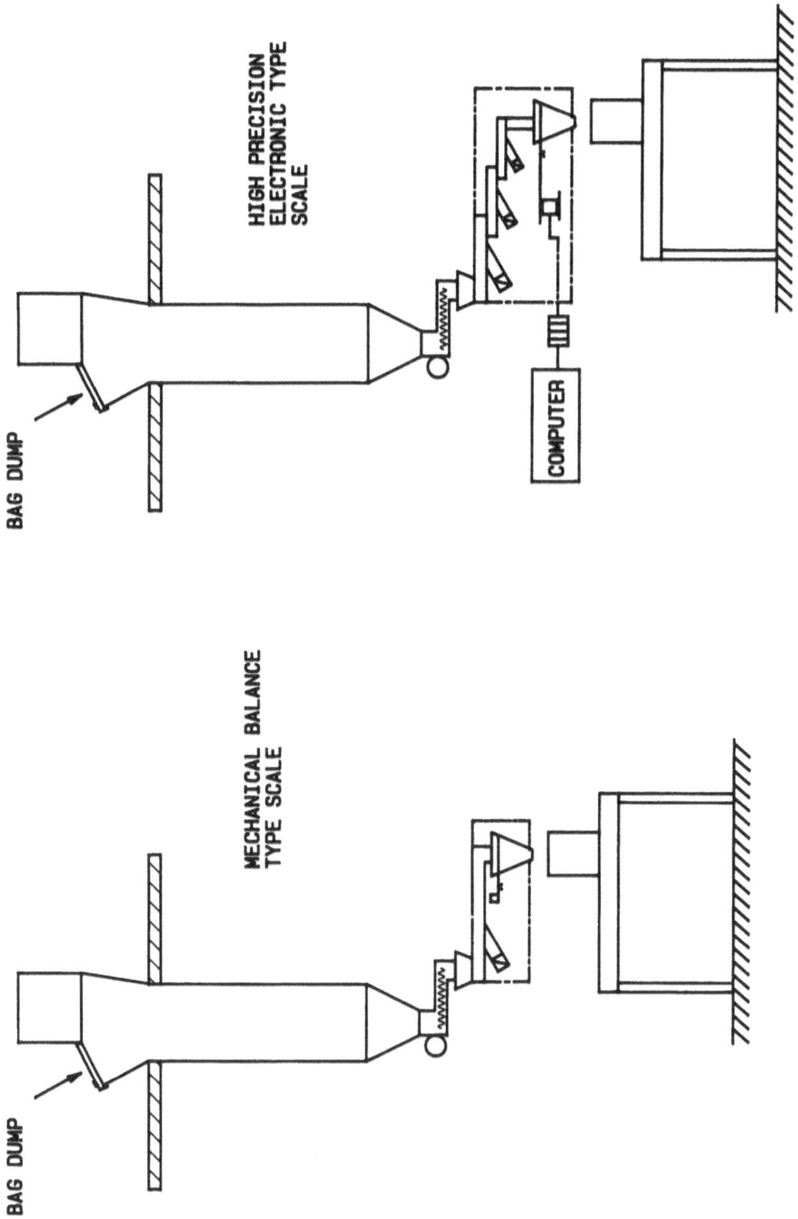

FIG. 2.5. BINS FOR MINOR INGREDIENTS WITH (LEFT) MECHANICAL SCALE AND (RIGHT) ELECTRONIC SCALE.

tight but vented for charging and discharging, and they must be designed to eliminate the formation of condensation. Capacity of the hopper must be adequate for the maximum charge it must carry. Sanitation and maintenance features must meet the usual standards for food plant equipment.

The single ingredient scale is utilized in applications where a large number of batches must be weighed within a short period of time, or where several ingredients have to be weighed simultaneously, or where the application requires that no trace of one ingredient can be allowed to contaminate another (such as with strongly flavored or colored ingredients), or where very accurate weighing is required over a small range. Figure 2.6 illustrates a moderately complex system for combined simultaneous and successive batch weighing. Single ingredient scales employed in these lines can be either the mechanical or electronic type. They are generally stationary and are used exclusively with a vertical silo or bin containing only a single ingredient (Long 1984). The choice between mechanical and electronic scales is determined by the characteristics of the ingredient to be weighed and how it is to be weighed. The electronic scale with data storage and retrieval capability should be chosen for non-free-flowing ingredients which must be weighed with high accuracy and low tolerances from several pre-set formulas. Mechanical scales are said to be adequate only for free-flowing ingredients which are weighed to a single, pre-set weight and for which relatively large tolerances are acceptable. The electronic and mechanical systems can be combined by joining a single tension-type load cell to one end of the pull rod of the precision linkage of the mechanical scale.

Weighing Materials on Moving Belts

Some scales are designed to determine the amount of material on a belt which is passing over the weighing mechanism. Although it might seem that movement of ingredient and interference from the supporting belt would seriously hamper accurate weighing, properly designed equipment can, in fact, deliver measurements sufficiently accurate for many purposes.

The simplest forms of this type of device weigh the ingredient but do not adjust the delivery rate so as to maintain a constant flow, i.e., they have no feedback control. They are useful in automatic batching, but are not readily adaptable to continuous processing where the rate of delivery and not the total weight is important. In one semicontinuous model, the scale belt runs continuously but is fed intermittently with loads that do not extend its full length so that a isolated load is being

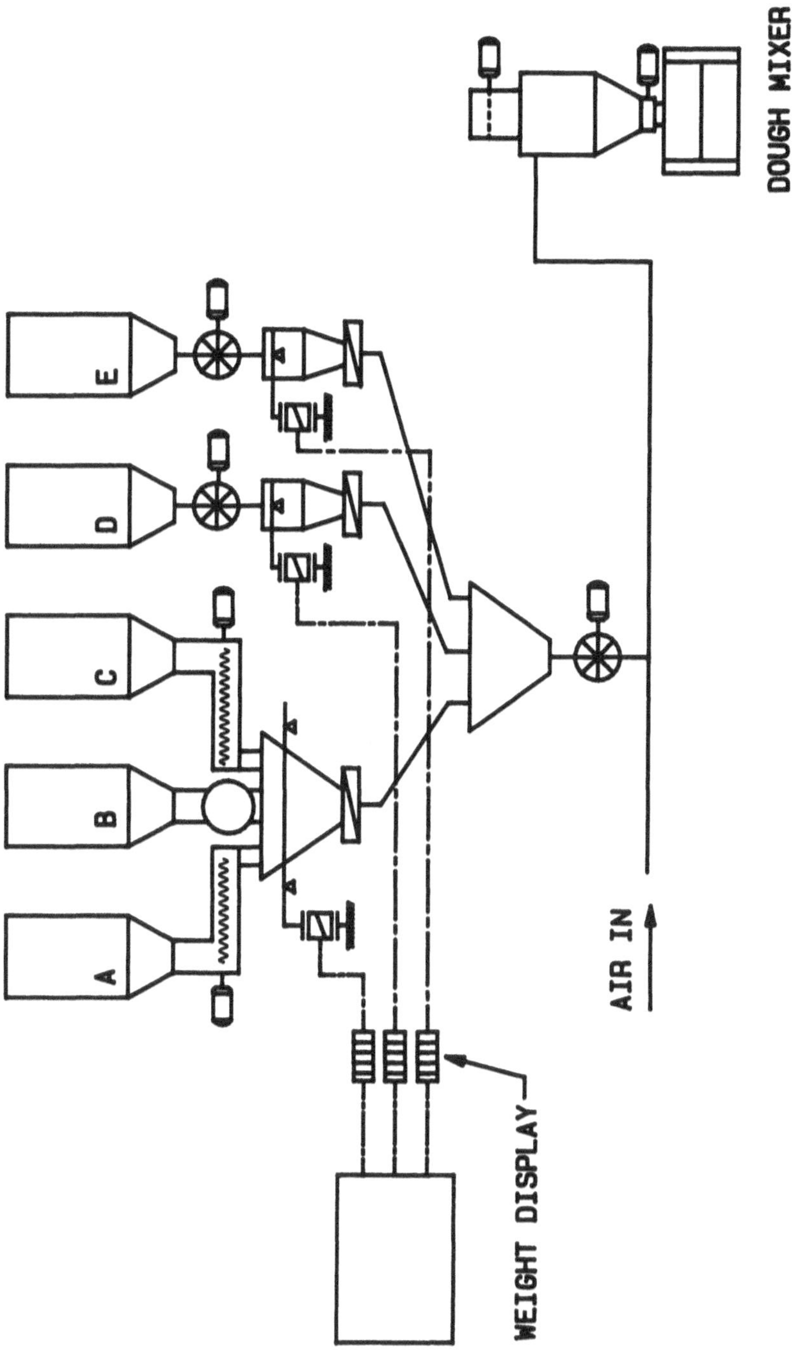

FIG. 2.6. TYPICAL ARRANGEMENT FOR COMBINED SIMULTANEOUS AND SUCCESSIVE BATCH WEIGHING FROM MINOR INGREDIENT BINS.

weighed at any given time. Since the increments are deposited at short intervals, they approximate continuous flow. The weigh belt is mounted on a scale mechanism which records and sums the amounts transported.

A second version of the belt weigher, the continuous conveyor scale, uses a scale-supported section of a belt conveyor to totalize the load which is being constantly deposited on it. The difference between this and the former type, is that the entire conveyor (not just the belt) is supported by the weighing mechanism. The forms of weighing devices used in such equipment include spring-balanced beams, strain gauges, and pneumatic load cells.

Fully automated belt weighers with feedback to control the rate of delivery are useful in continuous or intermittent weighing. The most common type utilizes a conveyor belt balanced on a weigh beam. When the belt is driven at a constant speed and the total weight of the belt, material, and associated mechanisms is held constant, the rate at which the material comes off the end of the scale is also constant and the total weight for any known time interval can be computed. Imbalance of the beam actuates a change of the rate of material deposit onto the belt in the direction of restoring balance, by a mechanical adjustment of the feed gate or by varying the speed of a belt or screw feeding the weighing conveyor. Accuracies are said to be as high as 0.0001% (!) for certain products and equipment, but a more likely figure is about plus or minus 0.5% for continuous measuring scales used in the bakery.

Following is a description of one brand of weigh belt feeder (see Figure 2.7). Material is fed to a continuous weigh belt by a vibrating nozzle having an adjustable gate. The complete belt assembly, as well as the material on it, are weighed to eliminate errors due to variations in belt tension. The fulcrum is located directly under the infeed point, eliminating errors due to momentum of material discharging onto the belt. In addition, the input end is supported by Inconel flexures, providing a sturdy pivot that can withstand accidental overloads or shock loading yet is highly sensitive to small weight changes. Actual weighing is done by a temperature compensated linear voltage differential transformer located at the discharge end of the belt. Such units can withstand overloads of 500 lb or more without damage. If there is a difference between the preset feed rate and the delivered feed rate, an equalizer circuit supplies instantaneous correction commands to the belt drive. A sophisticated digital setpoint, digital readout controller provides easier, more accurate setting, monitoring, and calibration. The controller furnishes direct readout of the feed rate in a lb per min

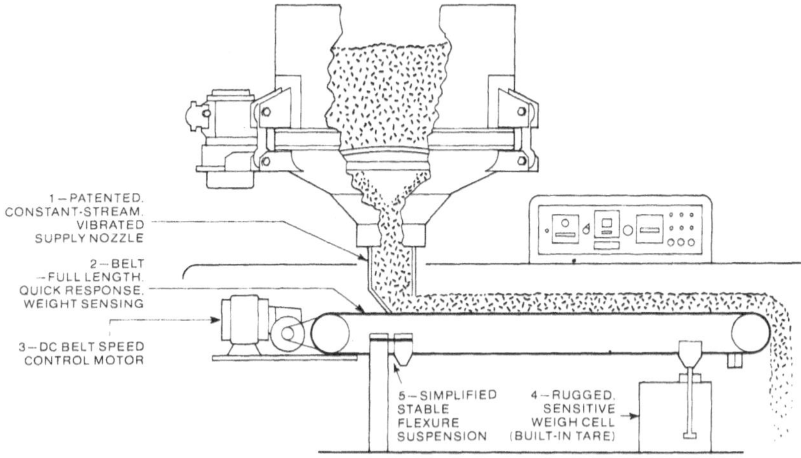

FIG. 2.7. PRINCIPLES OF OPERATION OF A WEIGHBELT SYSTEM RECEIVING MATERIAL FROM A VIBRATING BIN AND NOZZLE.

SOURCE: VIBRASCREW, INC.

display. Control units may be rack mounted or remotely located up to 100 ft away from the feeder. Models are available to feed as little as 1 lb per min and as much as 700 lb per min. In addition to digital readout of the feed rate, there is available an optional totalizer, which logs the amount of material crossing the belt from startup to shutdown and provides a readout of the amount. Inventory records may be logged from this unit for production control, material reordering, or supply hopper recharging. A pre-set totalizer for selecting desired batch weights is also available (Anon. 1987).

In another design of automatic belt gravimetric feeder, variation in amplitude of vibration of a feed tray is used as a means of controlling the rate of delivery to the belt. A cam-operated mechanism driven off the belt transmission oscillates a driving plate at constant frequency in the direction of an opposing receiving plate connected to the feed tray. A rubber control wedge suspended from the scale beam transmits the vibration from the driving plate to the receiving plate. The amplitude of vibration is thus regulated by the position of the wedge between the plates. If the weight on the belt is excessive, the wedge is raised, the vibration is diminished and less material is transferred, and vice versa.

Loss-in-weight Feeders

Instead of measuring the amount of ingredient deposited in a batching receptacle by weighing that container and its contents, the loss in weight of the storage bin or hopper from which the ingredient has been withdrawn can be determined. In a typical loss-in-weight feeder, the entire feed hopper is mounted on a scale which controls the rate of removal of material from the hopper. Material may be released from the hopper by a rotating valve powered with a variable speed motor or by some other mechanism. If the counterpoise on the scale beam is retracted continuously by a constant speed drive, the rate of delivery will be constant and the equipment will be suitable for continuous processes.

Several modifications of loss-in-weight feeders are available for bakers. Vibratory, screw, pneumatic, and belt methods may be used to remove material from the hopper which is mounted on a multiple-beam scale mechanism. The hopper is filled and the scale beam balanced by manually adjusting the poise. The control dial on the rate setter is set to the desired rate of feed in pounds per hour. The rate setter, operated by a synchronous motor, retracts the poise by a lead screw at the exact feed rate desired. As long as the feeder delivers material from the hopper in a manner such that the hopper loses weight at the same rate the poise is being retracted, the scale beam will stay in balance. If the feeder delivers too much or too little ingredient per unit of time, causing the hopper to lose weight faster or slower than the poise is being retracted, the beam will tip and operate controls which cause the feeder to correct its rate. Only a very small beam movement is required for control, and the scales are so sensitive that the beam is essentially in balance at all times.

When the loss-in-weight feeder is electronically controlled, it can be coupled to the flow from another loss-in-weight feeder or from liquid flowing through a Venturi tube, orifice plate, or other flow measuring device so that two or more ingredients are delivered in correct proportions. The metering can be continuous or intermittent, and in the latter case, the sensing of a predetermined amount of one ingredient leads to the operation of the feeder for a fixed number of seconds.

Control Methods for Gravimetric Feeders

Belt gravimetric feeders can be mechanically, pneumatically, or electronically controlled. Operation of belt gravimetric feeders is based on the principle of maintaining a constant weight of material on the moving belt through activation of a positive-acting gate on the feed hopper. The rate of feed may be changed by varying the belt load and/or changing belt speed. Net weight of belt load may be sensed directly by a force balance pneumatic transmitter and load cell or by a differential transformer which will transmit any variation from set load to the solid-state electronic controller.

The feedback system actuates the control gate to maintain belt load at the set point and maintain weigh platforms in a null position. Electronically controlled belt gravimetric feeders are available with capacities from about 0.1 lb per minute to over 60,000 lb per hr and accuracies of plus or minus 1%. The belt type with preset cutoff uses a scale and tachometer signal to provide an instantaneous indication of the rate of material flow by weight per unit of time. The resulting analog signal is converted into digital form by an integrator. A pulse counter records the digital information to provide visual indications of the total flow of material. The pulse counter can also be provided with a stop-point setting which will cause a contact to close and thus halt the conveyor when the prescribed amount of material has been transferred.

The proportional belt feeder also relies on a scale and tachometer output to provide instantaneous rate of flow measurement. In this case, however, the resultant signal is electronically compared with that from a percent-setter to regulate conveyor speeds. The system can be designed to have a stop-point similar to the arrangement described above. Two ingredients can be measured at proportional rates and blended on a conveyor. In belt type systems, one of the chief factors affecting overall accuracy is the location of the sensing device in relation to the discharge point; the closer they are together, the greater the accuracy.

Hopper weighing systems can be provided with either analog or digital controls. The analog system employs a scale and a conventional sensing arrangement. Other types of weighing devices such as load cells can be used as sensors, provided their output is electrical. The amplified signal is compared with the preset input through the balance detector. When the signals balance, indicating that hopper weight is at the desired level, the solenoid cutoff closes the feeder. The analog indicator (e.g., a large scale head with a thousand graduations) continually supplies progressive weight information.

The digital system employs a mechanical weighbeam and lever arrangement and a conventional scale head. The basic sensor can assume a variety of configurations and use several kinds of components, but a pulse signal giving a number of pulses proportioned to the material weight must be obtained. The remaining portion of the system features a preset counter with digital weight indication.

VOLUMETRIC MEASURING DEVICES

A volumetric measuring device in its simplest form is a container of known volume, such as a tank, barrel, pipette, or volumetric flask. The container may be calibrated at various points—as in burettes, graduated cylinders, or some storage tanks. Volumetric measurements can be applied to solid and pulverulent materials as well as liquids; for example, in home cooking procedures where nearly all proportioning is done volumetrically in spoons and cups. Large scale manufacturing processes do not customarily include volumetric measuring systems for powders and the like because of the difficulty in maintaining uniform density in these materials.

Metering Liquids

Location of the liquid surface, and hence the amount, of liquid in a tank can be determined by several methods. A tape or chain connected at one end to an indicator and at the other end to a float on the liquid surface represents about the simplest possible method. As the float moves up, the indicator moves down along a calibrated scale, and vice versa. A calibrated or plain glass tube attached to outlets near the top and bottom of the tank can also be used for directly indicating liquid level since the level of liquid in the tube corresponds, at least approximately, to the surface of the liquid in the tank. The dielectric, conducting, or absorption properties of the liquid can also be the basis of this measurement. If the liquid rises between two vertically oriented plates of a condensor, a capacitance change will be produced proportional to the length of the conductor which is wetted and thus proportional to the depth. Absorption of radiation by the liquid lying between a radioactive source and a sensor will be proportional to the depth of the liquid. The electrical signals can, of course, be converted into analog or digital indications of volume and can, if desired, activate an alarm, pump, valve, etc. These volume measuring—or, rather, depth measuring—devices are much more suitable for rough determinations of inventory than for controlling delivery of ingredients to processing areas. Their precision is seldom satisfactory for the latter task.

In modern bakery practice, the amounts of liquid ingredients delivered to processing equipment are measured by totalizing flow meters wherever possible. This type of equipment is almost indispensable in bulk-handling systems. A typical mechanical totalizing flow meter installation for dispensing water and shortening is shown in Figure 2.8.

Much engineering expertise has gone into the design of water meters, and virtually all fluid control and measuring systems are based on instruments originally designed for water. Although we rarely think of it in this way, water is a classical example of an ingredient received, transported, and dispensed by bulk transfer methods.

FIG. 2.8. TOTALIZING FLOW METERS FOR WATER AND SHORTENING.
SOURCE: APV (BAKER PERKINS)

Water is a rather difficult liquid to meter because of its lack of lubricity, its promoting of electrolytic erosion, and its corrosiveness (Hesley 1988). Other liquid ingredients add different or additional parameters to the measuring problem, and meters designed to meet the challenges of water dispensing may have to be substantially modified so they can handle hot, viscous, abrasive, and otherwise difficult materials.

Several versions of flow meters sufficiently accurate for bakery use are commercially available. These have been classified as (1) inferential meters, in which the liquid actuates a screw, a vane, or some other inertia-dependent mechanism, and (2) positive displacement devices in which a defined volume of water is allowed to pass during each complete cycle of the mechanism. Both types are accurate to a few percentage points of the total reading at high rates of flow. At low rates of flow, the displacement meters are generally more reliable.

Headmeters measure the loss in pressure between two points in a pipe containing a flowing liquid. The three most widely used headmeters are pitot tubes, orifice plates, and Venturi meters. Only brief discussions on the principles of these instruments will be included here; for more details on design and theory, consult Cheremisinoff (1979, 1984) and Anon. (1959).

The pitot tube is used for measuring local fluid velocities and consists of a tube inserted into the main pipe with its inlet turned upstream to receive the full impact of the flowing liquid. The impact is completely converted into pressure head and superimposed on the existing static pressure of the fluid. In practice, pitot tube instruments consist of both an impact tube and a piezometer tube (to measure static pressure) combined into an S-shaped tube, the combination allowing the static pressure to be subtracted from the total pressure—the difference being the velocity head.

An orifice meter consists of a thin plate having a hole of accurate size and placed across a pipe through which a liquid is flowing. In accordance with Bernoulli's equation, reduction of the cross section of the flowing stream as it passes through the orifice increases the velocity head at the expense of pressure head. Gauges or manometers are used to measure the pressure upstream and downstream of the constriction to give the data on which calculation of the flow rate can be made.

A Venturi meter consists of a pipe of narrower diameter inserted in the main pipe. The measurement principle is based on determinations of the reduction of flow pressure and increase in velocity. The pressure drop in the upstream connecting section is used to determine the rate of flow through the Venturi. On the discharge side, the fluid velocity is decreased and the original pressure is recovered. Practical applications of these devices involve some method of automatically sensing the pressures, comparing them, and converting the differences to amounts of liquid passing through the pipe in each unit of time. When the rate is known, the duration of flow is a measure of the total amount delivered. Different kinds of head meters are shown in diagrammatic form in Figure 2.9.

It is also possible to relate the flow rate to the area change needed to obtain a constant pressure drop. This principle is the basis of the rotameter, an inferential meter which is the most common type of variable area meter used in the food industry. Rotameters consist of a vertical tube (usually of glass) having its bore gradually tapered (becoming larger toward the top), with the fluid moving upward through it, and a float capable of unrestricted movement up and down the tube. The float will assume an equilibrium position such that fluid drag and buoyancy just equal the downward force of gravity. Fluid drag is related to the area of flow between the float and the sides of the tube. If the flow rate increases, the bob moves upward in the tapered tube so that the separation between the inner surface of the tube and the surface of the bob is increased. This increases the flow area and keeps the drag constant. The pressure drop through the instrument remains almost independent of the flow rate. The early versions of rotameters were designed so that the bob rotated, hence the name. A spinning float has

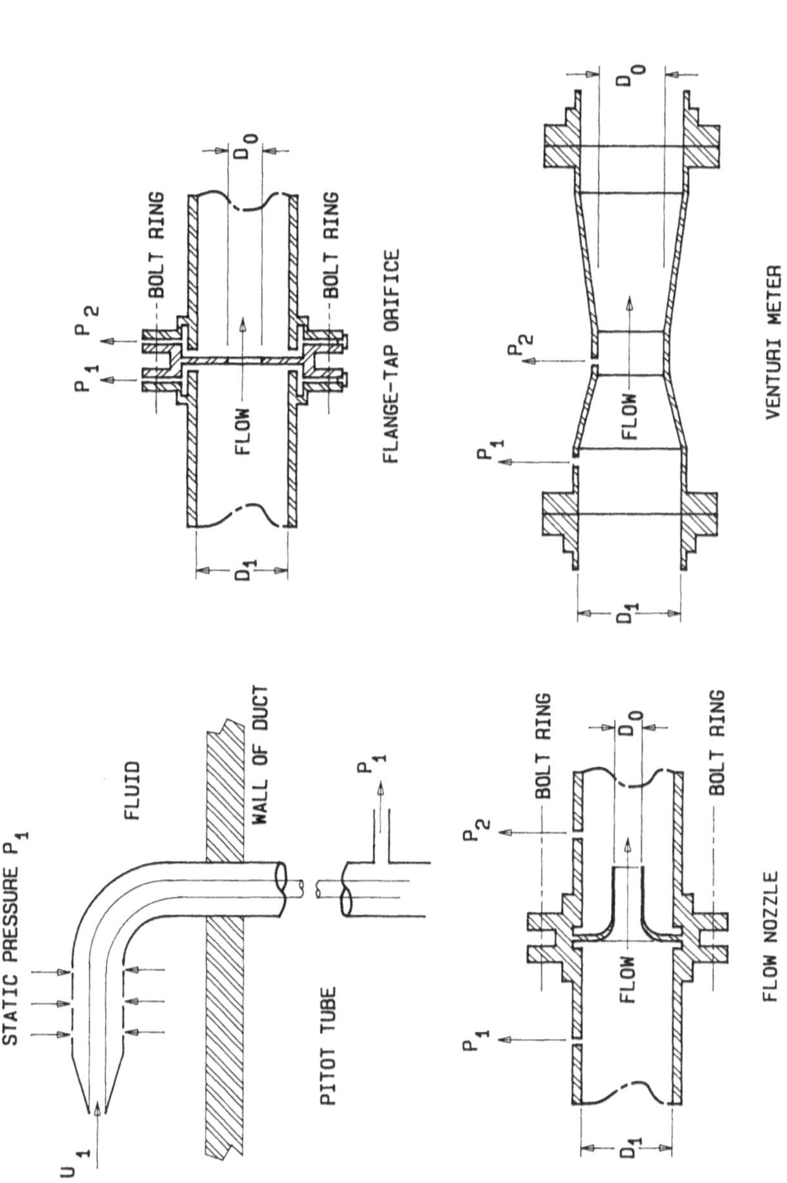

FIG. 2.9. VARIABLE HEAD METERS.

more stability, is easier to read, and keeps itself cleaner. Most modern rotameters perform satisfactorily with non-rotating bobs.

The calibration of rotameters will vary with float dimensions, tube taper, and fluid properties such as viscosity and density. Special float designs are available which are relatively insensitive to viscosity effects, but in the bakery, each meter will normally be used for one ingredient only, so viscosity changes should not be a complication that needs to be considered. Means for compensating for fluid density changes can be obtained if needed, however. The mathematical treatment of fluid flow measurement is detailed in many texts on chemical engineering and hydraulics, but a good summary of this aspect of metering has been given by Rothfus (1968). By affixing a magnet or armature to the float and placing a sensing device outside the tube, rotameter readings can be transmitted to other instruments for recording and control purposes.

Turbine meters (another type of inferential meter) have come into use for measuring such liquids as water, invert syrup, and ammonium bicarbonate solution. Their small size and weight permit installation at almost any convenient point. These meters consist essentially of a a sensing device which measures the turning of a rotor located in a tubular housing inserted as part of the pipe line. The flowing liquid impinges upon the turbine blades which freely rotate about an axis along the center line of the surrounding tube. The angular velocity of the turbine rotor is directly proportional to the fluid velocity through the turbine. The angle of the rotor blades to the stream affects the rotor velocity. Blade angles are usually between 20° and 40° because greater angles result in excessive end thrust and bearing friction while smaller angles cause undesirably low angular velocity and loss of repeatability. Models are available in which the rotor spins freely in a low pressure chamber, no thrust bearings being needed because there is established a hydraulically balanced rotor position which keeps it free of the chamber walls. Both the K factor and linearity of turbine meters are seriously affected by changes in viscosity when the viscosity of the fluid they are measuring exceeds 100 cp. For such liquids, positive displacement meters are recommended for determining volume. Viscous fluids can also be measured satisfactorily by coriolis mass flow meters (Hesley 1988).

Sensing of the rotation of the turbine is done by magnetic interaction between the blade tip and a pickup coil located outside the tubing. The rotor blades are made of paramagnetic material and the pickup coil contains a permanent alnico magnet. The surrounding pipe is made of stainless steel, so that it does not interfere with the magnetic interaction. Frequency of the magnetic pulses will be proportional to the flow rate. Pulses per unit volume may be reduced in meters of small capacity by making some of the blades nonmagnetic. In large diameter meters, resolution is increased by installing a large number of small magnetic buttons on a ring rotated by the turbine. An alternating current output proportional to the flow rate is created and sensed by calibrated instruments which amplify it and use it for recording flow rate, totalizing the delivered amount, or controlling other instruments. Turbine flowmeter systems have, in addition to the flow sensing element, a frequency converter and an electric potentiometer. The frequency converter delivers a DC output directly proportional to the frequency of the AC input from the sensing element. A digital counter of the roller type registers a proportion of the pulses.

In addition to the wide range through which accurate measurements can be made (on the order of 10:1), a major advantage of the turbine meter is that each electrical pulse is proportional to a small incremental volume of flow. This incremental output is digital in nature, and so can be totalized with a maximum error of one pulse regardless of the volume measured. Accuracies of plus or minus 0.5% of the actual flow rate or within 0.25% over selected flow ranges have been claimed for these systems.

Continuously flowing streams of liquid material can be measured volumetrically by displacement meters. There are several types based on nutating discs, reciprocating pistons, rotating vanes, etc. Figure 2.10 shows the measuring elements in several commercial positive displacement meters. The piston meter is like a piston pump operating backwards and it is capable of accuracy to 0.1%. For precise volume measurements, corrections for temperature must be made because of its effect on the density of the fluid being measured and on the dimensions of the volumetric device. Pressure is generally not a factor since common liquids are noncompressible under the conditions of measurement and pressures normally encountered in bakery equipment do not have a significant effect on meter dimensions.

For many years, the most common method for measuring water delivered to customers has been the nutating meter. In meters of this type, a disc piston fits approximately horizontally in a chamber defined top and bottom by truncated cones (the apexes facing each other). A vertical diaphragm also separates the chamber. Liquid flows into the chamber on one side of the vertical diaphragm and pushes the disc up and down with a nodding (or nutating) motion. The disc does not rotate. At each complete cycle, the piston discharges a volume of liquid equal to the capacity of the measuring chamber. A spindle affixed to the large

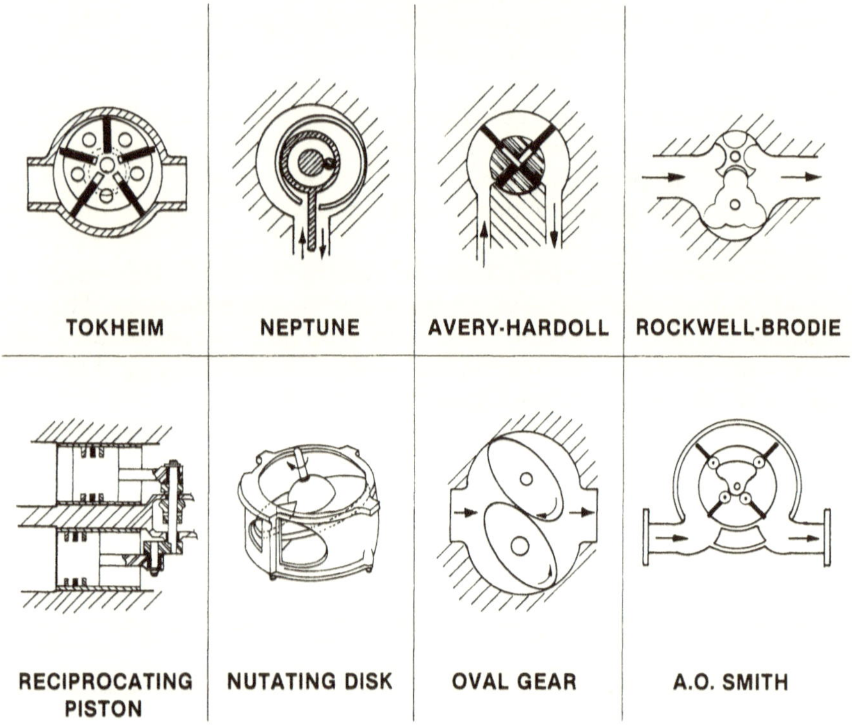

FIG. 2.10. WORKING PRINCIPLES OF COMMERCIAL POSITIVE DISPLACEMENT METERS.

spherical bearing which supports the piston is made to describe a circle as a result of the disc's nutation and this motion is transmitted through a gear train to a register. The only moving part in the measuring chamber is the piston (Anon. 1963).

Another common device for delivering precisely measured quantities of fluid, the positive displacement metering pump, is electrically actuated by a controller which has a synchronous motor drive. The motor moves the hands on an indicator which can be preset to shut off the flow when a desired quantity of liquid has been delivered. The metering cycle continues until another synchronous motor in the controller has counted the predetermined number of revolutions of the metering pump, after which the controller stops the two motors at the same time. Since both the metering pump and the controller are driven by synchronous motors energized by current of the same frequency, their motions will always be proportional. Each cycle of the pump will empty and fill a cavity of known volume. Control of metering can also be accomplished by micro-switches activated by digital counters operated by cams on the pump shaft. When the shut-off point is reached, the meter automatically resets itself to the original amount, ready to dispense the ingredient for another batch. Accuracy of 0.1% has been claimed for positive displacement flow meters utilizing a system of oval gear wheels to measure the liquid.

Any positive displacement pump in good condition can be calibrated for delivery at constant infeed and discharge pressures and used as a metering device. Revolutions may be counted by inserting a magnet in the pump coupling and using a sensor which detects the movement of the magnet. By counting the revolutions, the delivered quantity can be determined. If the pump is run at a constant rate, the delivery per unit of time can be used to make the pump serve as a meter. Accuracy is not of the best in such installations, but is sufficient for some purposes. For somewhat better accuracy, gear or lobe type pumps can be used with constant speed drives for positive displacement metering of liquids of uniform density by operating them under constant suction and discharge pressures. An automatic timer or automatic revolution counter can make the pump turn a selected number of revolutions equivalent to a desired quantity of material (Anderson 1976).

Metering pumps must be fed at a pressure sufficient to avoid cavitation, and piping must be also be assembled so that the meter does not "pump air." If these conditions are not met, satisfactory accuracy is not likely to be achieved.

The electromagnetic type of liquid measuring device has not been used much in bakeries, but has great potential for metering viscous fluids and does not require a constriction in the pipe (Anderson 1976). It requires a primary element or flow transmitter which is connected by electric cable to a secondary instrument for indicating, recording, controlling, or converting to other compatible signals. It is based on Faraday's law of electromagnetic induction which states that the voltage induced in a conductor of fixed length moving through a magnetic field is proportional to the velocity of the conductor. The voltage generated by liquid moving through the magnetic field in the pipe is sensed by two point-type electrodes located diametrically opposite and flush with the inside of the pipe.

A system of this type consists of a flow transmitter, an electrically insulated liner on the inside of the product tube, and the electrodes previously mentioned. Teflon, Neoprene, Kel-F, or glass are the insulators commonly used. In order to measure product flow with these flowmeters, it is necessary that the ingredient be a conductor of electric current. Many liquid food ingredients are sufficiently good conductors for this purpose. The output of the flow transmitter is linear and directly proportional to the average velocity of the product flowing through it. Converting the small voltage generated by the fluid to a digital readout or other function results in a certain amount of error, but the accuracy will normally be within a range of plus or minus one percent. Advantages of the electromagnetic measuring system are that there are no moving parts or adjustments in the transmitter, it can be used with liquids of any viscosity, it measures different types of liquid without need for correcting for conductivity changes, it can be mounted in any position, and it has a low installation cost.

Metering Solids

As mentioned earlier, difficulties in accounting for variations in the density of the ingredient create problems whenever volumetric measurements are to be made on powdered or granular materials such as flour and sugar. Volumetric feeders for pulverulent materials can be classified on the basis of their action as : (1) a belt, disc, roller, or screw pushes or pulls the ingredient through a gate which usually can be moved so as to change the size of the opening, (2) a helix moves the material through a tube at a rate governed by the speed of the screw or the duration of its rotation, or (3) pockets are first filled with powder as they pass under a hopper and then are rotated to dump their contents, the speed of movement of the pockets past the release point governing the rate of delivery.

Volumetric measurers relying on helical feed devices have been used in the baking industry for many years, as discussed briefly in an earlier section of this chapter. In one version of this type of feeder, two concentrically mounted and independently driven augers rotate in the same direction but at dissimilar speeds. The larger outer helix tends by its slower rotation to create constant motion in a zone of material surrounding the faster speed smaller inner auger (or metering auger). An optimum ratio of the two speeds is selected to give a constant uniform density in the material surrounding the metering screw. If product characteristics require it, a reverse helix can be added to further stabilize the flow pattern. Since the metering screw operates in an environment of stabilized density, it can deliver a constant rate of ingredient at any given speed. Metering accuracy of plus or minus 1% of the set rate is claimed to be achievable for most materials.

Still another type of volumetric feeder is frequently used in the food industry to dispense "micro" ingredients such as vitamins and oxidizers at low rates. The feeding range can be varied from 4 oz to 60 lb per hr. The principle is that a vertically-movable slide or gate controls the depth of powder on a horizontally rotating feed disc which is drawing the material from a hopper. The powder, after it is removed from the disc by a screw, falls down a chute to the mixing vessel. Addition rates are controlled by varying the speed of the disc or the height of the opening in the hopper. Disc speed can be varied by changing the gears or the speed of the motor. The gate is adjusted by a micrometer screw to give a 1:20 variation in the height of the opening. As in all equipment of this type, feed rate is not necessarily directly related to the dimensions of the opening or the speed of the disc, and it is important to calibrate the settings by weighing the material discharged during a given time interval.

Another approach to auger feeding of small ingredients from loss-in-weight bins utilizes a flexible vinyl hopper against which two paddles press in alternating cycles. The undulating action of the paddles creates fault lines in the material and usually overcomes the common problems of bridging and rat-holing. It also conditions the ingredient into a uniform bulk density so that each flight of the feeding screw receives the same amount of material. The amplitude and frequency of paddle movement can be adjusted to meet the requirements of individual powders. Feeders of this type can be purchased with a microprocessor-based controller and a counterbalanced scale engineered for the unit. Loss-in-weight accuracies of 0.25 to 0.5% and volumetric accuracies of 0.5 to 2% are claimed by the manufacturer. Units sized to give flow rates of 0.0001 to 1,000 cu ft per hr are stocked.

OTHER METHODS OF MEASUREMENT

The coriolus mass flowmeter uses a different principle of measurement from the devices described above. The following discussion is mostly based on information provided by Smith (1988).

A mass flowmeter of the type to be described consists of one or two flow tubes enclosed in a sensor housing. The meter uses Newton's Second Law of Motion (Force = Mass X Acceleration) to determine the precise amount of matter flowing through the tubes. Inside the sensor housing, the horizontally positioned U-shaped tube is vibrated at its natural frequency by a magnetic drive coil located at the center of the bend in the tube. The vibration is small in amplitude and rapid, similar to that of a tuning fork. As fluid flows through the tube, it is forced to take on the vertical momentum of the vibrating tube. When the tube is moving upward (during half of its vibration cycle), the fluid flowing into the meter resists, thereby exerting a downward force on the tube. The fluid moving toward the outlet of the tube, having by that time been forced upward by the tube walls, resists the downward movement of the tube. These differently oriented forces cause the tube to twist—in one direction during the upward motion and in the opposite direction during the downward motion. Due to the tube's elasticity, the amount of twist imparted to it is directly proportional to the mass flow rate of the fluid passing through the tube. Magnetic sensors located on each side of the tube measure its up-and-down velocity and send this information to an electronics unit where the data is processed and converted into an output signal proportional to the mass flow rate. Figure 2.11 illustrates the operating principles of a mass flow meter.

Coriolus flow meters can measure mass flow rate of gases, liquids, and slurries. Because they measure mass directly, problems associated with variations in fluid density, viscosity, temperature, and pressure are less important than in some other designs. For example, entrained air bubbles in a viscous liquid affect the readings very little. Also, viscosity has no effect, contrary to the situation with turbine meters. The precision of mass flow meters may be somewhat less than that of turbine meters—one estimate of the repeatability of mass flow meters is about 0.4% as compared to 0.1% for turbine meters. Linearity is about the same in the two types (Hesley 1988). Temperature can have an effect on the readings, however, as a result of its influence on the flow tube's modulus of rigidity. It appears these devices are being used mostly in the chemical process industries, but commercial installations in wineries and fructose syrup factories have been reported. Recently, some bakery applications have been investigated, including the measurement of both ingredients and liquid sponges. Since the latter process intermediate tends to be nonuniform in density, it would seem that mass flow meters would be particularly suitable for controlling the weight of material to be dispensed (Anon. 1986, Kocher 1987).

AUTOMATIC BATCHING SYSTEMS

Weighing operations for flour, sugar, and some other ingredients are integral parts of bulk handling systems. Ingredients which, for one reason or another, are not being handled in bulk transfer systems, can be dumped into receiving hoppers. Bag unloaders, sifters, and conveying systems (mechanical or pneumatic) will precede the receiving or surge hopper. Automatic conveying and metering or weighing devices are often used to eliminate manual transfer thereafter.

Automated minor ingredient scaling, as found in medium and large bakeries, involves programmed and pre-set (1) conveying to the scale, (2) weighing, and (3) subsequent discharging of weighed ingredients. One or more scales can be used. The sequence of weighing, as well as the weights for the individual ingredients in a formula, is pre-set and

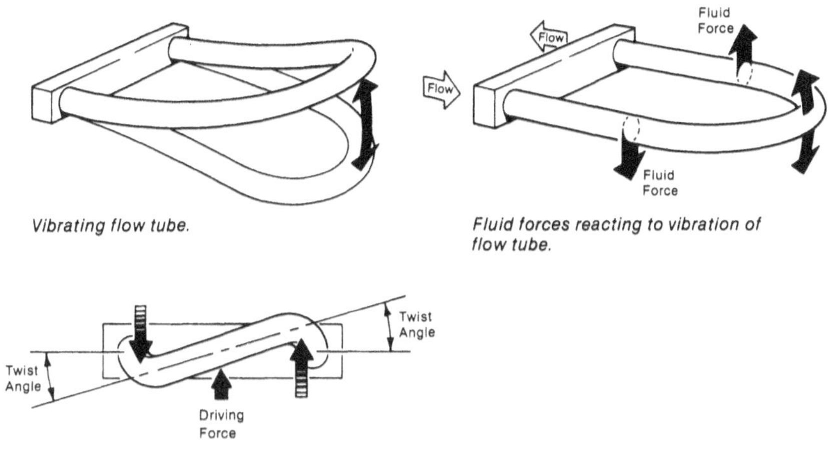

Vibrating flow tube.

Fluid forces reacting to vibration of flow tube.

End view of flow tube showing twist.

FIG. 2.11. THE MEASURING ELEMENT IN A MASS FLOWMETER.
SOURCE: MICRO MOTION, INC.

dispensing proceeds automatically without any need for manual interference (Long 1984).

Central batch weighing systems have one or more centrally located scales which deliver to several use points in a defined order. Hoppers can be located over each mixer, not as weighing hoppers, but as receving or holding bins for ingredients weighed elsewhere. As a result, a weighed and fully assembled formula of dry ingredients is always ready to drop into each mixer as soon as the mixer is empty and ready to receive the next batch. This system is said to be particularly economical where a large number of use points is involved as well as where a large number of different ingredients is used. Such systems require computer control of the weighing process because several different formulas for several different use points must be stored in the computer's memory to be available on demand. There can be fully automatic checkout of the system, covering such things as verification of the condition of the scale, the conveyors, and tolerance, relieving the operator at the use point of responsibility for these checks. Central batch weighing can also include pneumatic mixing, providing the benefit of a homogeneous mixture of minor ingredients to be combined later with the major ingredients (Schraps 1982).

Automated bakeries generally use a central control panel registering and controlling the operation of many remote scales. In one form of the automated weighing system, formula weights are set in advance of each day's schedule by a supervisor who adjusts the weight selector dials behind the console. When the mixer operator signals readiness to begin a new batch, the central control operator starts the material flow by pressing a button. A further advance is the computerized system, in which recipes for several kinds of products are recorded on tape, on circuit boards, on punched cards, or in the memory bank of the computer. After the mixer operator signals for his ingredients, the computer takes over all other functions. Changes, as for example to compensate for differing flour moisture contents, can be made at will by typing them on input keyboard which is part of the computer assemblage.

The advantages claimed for automatic batching include: (1) elimination of human errors, (2) more consistent weights, (3) better sanitation, (4) less labor, and (5) reduction in loss of costly ingredients.

The automatic scale or meter is a system, or part of a system, made up of material handling devices, the weighing or metering equipment, data readouts, and the controls which program the entire series of functions for automatic operation. All automatic scales transfer matrials from some type of storage to a scale and from the scale to some destination. From the scale comes information in the form of electrical signals which controls the weights of ingredients and which can also be used to print out weight data for quality control, inventory control, and other management requirements.

The simultaneous coordination of all feeders can be achieved by powering each of the critical drives with an induction motor receiving its electrical power from a central adjustable frequency source, such as a Varidyne power unit on the blender. The higher the frequency, the faster an induction motor revolves.

A typical digital blending system will include: (1) A master unit with integral controls for setting system demand rate, batch size, valve ramp rates (up and down), and preshutdown point. Automatic shutdown will be initiated by either measured or demand total. (2) Ratio unit (one or more) for setting individual component ratios by manual thumb switches. Multicomponent ratios are available, and 3 or 4 digit settings are optional. (3) Individual component controllers, either pacing or memory, to provide control of the addition rates of the separate ingredients. Standard features include integrated total flow indication, manual valve control, and a low flow rate alarm to warn the operator when the measuring device is functioning below the linear flow range.

Computers can be made part of the system to control any or all of the functions and to coordinate their operation. Automatic temperature compensation units can be used to continuously adjust quantity measurements for variations in density due to temperature fluctuations of liquid ingredients.

BIBLIOGRAPHY

ANDERSON, R. C. 1976. Weighing and metering systems for liquids. Proc. Am. Soc. Bakery Engineers *1976*, 63-70.

ANON. 1956. Bread Baking. U. S. Dept. Army Tech. Manual *TM* 10-410.

ANON. 1959. Fluid Meters—Their Theory and Application. The American Society of Mechanical Engineers, New York.

ANON. 1965. Conveyorized weighers. Food Engineering *37*, No. 12, 67-71.

ANON. 1968. Omega gravimetric dry materials feeders and weighers. BIF, Providence, RI

ANON. 1985. Continuous Weighing and Feeding. Schenck Weighing Systems, Totowa, NJ

ANON. 1986. Dry Materials Feeding Handbook. AccuRate, Inc., Whitewater, WI

ANON. 1987. Vibra Screw Volumetric Live Bin Belt Feeder. Bulletin BF-2A. Vibra Screw, Inc., Totowa, NJ

ANON. 1988. Neptune meters. Neptune Meter Co., Long Island City, NY

BEATTIE, D. W. 1988. Personal communication. Liquid Controls Corp., North Chicago, IL

BENIER, J. 1983. Automated checkweighing. Proc. Am. Soc. Bakery Engineers *1983*, 113-119.

CHEREMISINOFF, N. P. 1979. Applied Fluid Flow Measurement. Marcel Dekker, New York.

CHEREMISINOFF, N. P. 1984. Fluid Flow Pocket Handbook. Gulf Publishing Co., Houston.

FULLER, W. S. 1970. Automatic weighing and dispensing of wet and dry sundry ingredients. Proc. Am. Soc. Bakery Engineers *1970*, 145-151.

HAGEDORN, H. G. 1965. New practices in bulk handling of materials with special emphasis on instrumentation. Proc. Am. Soc. Bakery Engineers *1965*, 148-152.

HESLEY, F. B., Jr. 1988. Personal communication. S. J. Controls, Inc., Long Beach, CA

KOCHER, J. M. 1987. Personal communication. AccuRate, Whitewater, WI

LONG, J. W. 1984. Minor ingredient systems. Proc. Am. Soc. Bakery Engineers *1984*, 92-98.

ROTHFUS, R. R. 1968. Working concepts of fluid flow. V. Flow measurement. Instr. Control Systems *41*, No. 7, 105-108.

SCHRAPS, S. 1982. Automatic batching of major and minor ingredients in the baking industry. Bakers Digest *56*, No. 3, 12-18.

SMITH, S. J. 1988. Personal communication. Micro Motion, Inc., Boulder, CO

SPANGLER, E. G. 1958. Storage and automatic dispensing of bulk lard and shortening. Proc. Am. Soc. Bakery Engineers *1958*, 71-78.

WILLIAMS, J. C., Jr. 1955. Instrumental weighing and control in the baking industry. Baking Ind. *104*, No. 3, 62-65.

ZIEMBA, J. V. 1965. Conveyorized weighers. Food Engineering *37*, No. 12, 67-71.

CHAPTER THREE

MIXERS AND MIXING

INTRODUCTION

Most of the operations in the bakery are specific to the manufacture of baked products. Mixing is an exception, and has been said to be the only bakery processing step classifiable as a chemical engineering unit operation. As such, it has been thoroughly studied in its simpler applications by many investigators. Mixing has been defined as a process designed to put a plurality of materials, originally existing separately or in a nonuniform combination, into such a condition that each particle of any one material lies as nearly adjacent as possible to a particle of each of the other materials. The reader will understand that mixing, as applied to doughs and batters, has a broader purpose, encompassing aeration and development as well as other functions, some of them poorly understood.

It is probably safe to say that the theoretical aspects of the unit operation of mixing have been of little help to bakery equipment developers, except possibly in the scale-up of equipment to larger models of the same design. Dough was being mixed successfully and equipment designers evolved an impressive family of efficient dough mixers, long before there were college courses on dough rheology. Still, it is to be hoped that basic theoretical concepts can be applied to the art to develop relationships helpful in designing even more efficient machines to mix complex materials such as dough.

A fairly consistent body of data has been accumulated relating power input to mixer design and operating variables, but this has mostly involved simple equipment mixing nearly ideal fluids. On the other hand, there is little to rely upon in the design and development of new mixers for highly viscous, pseudoplastic materials, except for intelligent trial and error in an experimental plan guided by practical experience. The main reasons for this situation are the numerous poorly understood factors affecting the response of doughs to mixing. Also, in any mixer operating on a dough of given formulation, viscosity and mixing velocity (two of the most important variables affecting power input) are interdependent. The apparent viscosity decreases with increasing rate of shear. It is almost impossible to define this relationship mathematically for purposes of equipment design, since most mixers must process several different formulas, each representing a different relationship between viscosity and mixing velocity and, hence, power input. Perry (1950) gives a good discussion of the fundamentals of mixing, including clear definitions of the various types of fluids: ideal, plastic, pseudoplastic, thixotropic, dilatant, and rheopectic. Doughs and batters are principally pseudoplastic.

Blending or mixing can be accomplished by many different types of equipment, but they all rely on the following types of action: (1) Devices using blades, paddles, helical metal ribbons, etc., to push portions of the mix through other portions, (2) Devices relying on the elevating and dropping of all or a portion of the mass so that the random rebounding of individual particles results in a redistribution of the particles, and (3) Devices creating turbulent movement by injecting currents of gas or liquids. There are changes occurring during the mixing of doughs and batters which are not encompassed by the preceding definition. For example, dough development is a separate phenomenon which happens to occur simultaneously with mixing when certain conditions are present. On the other hand, creaming and whipping fit the definition, since they involve the entrapment and reduction in size of bubbles of gas within a mass of other material.

There are significant advantages, in many cases, to separating the mixing process into two or more stages. In a preliminary step, all or part of the ingredients can be roughly blended into a dry mix or slurry which is subsequently divided into batch size, possibly mixed with other components, and then subjected to intensive action to insure uniformity and facilitate the physical and chemical changes needed to yield a finished dough or batter. Some of the implications of these procedures are discussed in the following section.

PREMIXING

In bakery processes there are some situations in which it is desirable to have a premix of part or all of the ingredients, and other situations in which a premix is indispensable. A bakery "premix" is a combination of ingredients that must undergo additional mixing—usually with other ingredients or premixes—before processing is completed. Although a bakery premix often contains only a few of the total ingredients required to complete a product, it may contain all of them, or all except the gaseous ingredient, air. Premixing is used to accomplish several different objectives, some of which will be reviewed below.

Advantages of Premixes in Batch Operations

Obtaining uniform distribution of micro-ingredients—When it is desired to distribute uniformly a very small quantity of one ingredient (salt, vitamins, yeast, etc.) throughout a much larger viscous mass of other materials (such as bread dough), it is advantageous to premix the minor ingredients with the flour and other major dry components prior to adding water and starting the final mixing operation. This saves time and energy, minimizes the temperature rise which normally accompanies the mixing of doughs, and helps prevent localized concentrations of minor ingredients.

Reducing multiple scaling operations—When a formula includes small quantities of several materials, and many batches of the formula are to be made, it is often helpful to prepare a large batch of premix containing the minor ingredients in the proper ratios. A calculated quantity of this premix is then added to each final batch. This saves a large number of weighing operations and reduces the expected frequency of errors in ingredient scaling. It does not necessarily increase the accuracy of scaling, however. For this process to be efficient, the premix must be easy to prepare, store, convey, and meter. A pumpable aqueous premix containing dissolved or dispersed minor ingredients is a good example of a premix.

The greater the number of minor ingredients which are included in the premix and the larger the premix batch, the greater the saving of weighing time. For example, a premix containing 10 minor ingredients which is made in batches large enough for 100 additions to the final dough mix would save 890 weighing operations (10 weighings for premix preparation and 100 weighings of the premix into the final batches versus 100 times 10 different weighings when the minor ingredients are added separately to each final batch). If only 2 ingredients are present in the premix, and it is made in batches suitable for only 10 additions, then only 8 weighing operations would be saved for each 10 complete doughs, i.e., a saving of 80 per 100 doughs.

Improving mechanical efficiency—When some of the ingredients require a strong shearing action to be dispersed but the final mix is of low viscosity, it may be difficult to obtain uniform distribution of all components if a single stage mixing operation is specified. For example, it is difficult to disperse lumps of nonfat dry milk in water. A possible solution is to use a high shear mixer to make a concentrated premix of all solid ingredients plus a small portion of the liquids, then add the rest of the liquids in a final mixing stage. Although the solid materials may not actually dissolve in the thick premix, lumps tend to be broken up and the particles uniformly dispersed by the high internal friction. As more fluid is added, the dispersed particles dissolve (or are wetted) more readily. Gluten and gelatin are two other ingredients which may benefit from this type of premixing. Successful utilization of the strategy depends upon the interaction of several factors which can't be predicted in advance. There are specialized mixing devices which create localized regions of very high shear and turbulence; if one of these is available, premixing with the full amount of liquid to yield a premix of fairly low viscosity may be feasible, even with these troublesome ingredients.

Premixing for convenience in shipment and ultimate use—For various industrial, military, foodservice, and domestic applications, prepared bakery mixes are customarily made up in large batches to save time and labor at the point of use. These can be packaged in batch size containers or in sacks or drums from which the baker weighs out enough mix for one batch. Although complete doughs and batters (including all liquid ingredients) are available for some products, it is usually not possible to supply a complete mix, since these tend to be unstable, difficult to handle, and lack versatility and flexibility. Consequently, dry premixes to which the baker adds water or some other liquid are more common than the finished doughs and batters. Complete doughs and batters in frozen or refrigerated form are widely distributed to both retail and wholesale buyers, however.

Advantages of Premixes in Continuous Operations

Many of the advantages of premixes for batch operations which have been described above are also found in continuous operations. There are, however, special problems in continuous mixing of doughs and batter which are greatly ameliorated by premix methods. Some of these will be discussed below.

Premixing for economies in equipment—It might be possible to feed a continuous mixer by metering each individual ingredient stream separately through its own calibrated pump or gravimetric feeder. Very accurate metering would be necessary since the continuous mixer must always operate on a dough or batter of uniform composition. Fluctuations of ingredient proportions could lead to malfunction of the equipment as well as to substandard products. It appears that direct feeding of ingredients to the mixer would require more equipment and be more expensive than an alternative pre-batching procedure.

Greater economy, and usually greater accuracy, results if several ingredients can be combined into a premix which is fed as one stream into the dough mixer. For instance, it is usually feasible to disperse in the water fraction all of the solid water-dispersible ingredients in a bread dough formula and meter this blend through a pump or other device. The premix can be prepared in batches (manually or automatically) or continuously.

Premixing to improve handling properties—It is rather difficult to feed dry materials directly into a continuous mixer operating at pressures exceeding atmospheric pressure. Special valving or a separate auger feeding section is required. A simple and convenient alternative is to assemble the ingredients in an open, atmospheric premixer, where batches in the form of a pumpable slurry are formed. Often, a storage tank or surge tank with agitators is provided between the blending unit and the continuous mixer. The liquid premix can be easily forced into the pressurized continuous mixer by any of several types of metering pumps.

Premixing Procedures

Premixes of solids—Premixing of flour, sugar, milk powder, dried eggs, chemical leavening agents, etc., is often necessary in the processing of prepared bakery mixes. If no shortening is to be added, satisfactory blending can be easily accomplished in blenders of the tumbling barrel type. Standard equipment in the industry is, however, the ribbon blender (see Figure 3.1). Such a mixer will have a horizontal trough-shaped (U-shaped) bowl with a horizontal agitator shaft passing through the flat ends. Agitator elements are steel ribbons in an approximately helical design affixed to the shaft by radial supports. As the shaft rotates, these agitators lift and move the trough contents axially and radially. Other types of dispersing elements may be added to break up the mass as it is conveyed throughout the trough.

These mixers are nearly always filled from the top and emptied from a discharge port at the bottom center of the trough. It is necessary to run the agitator during emptying if nearly complete discharge of the contents is to be obtained. Covers are needed for sanitation purposes

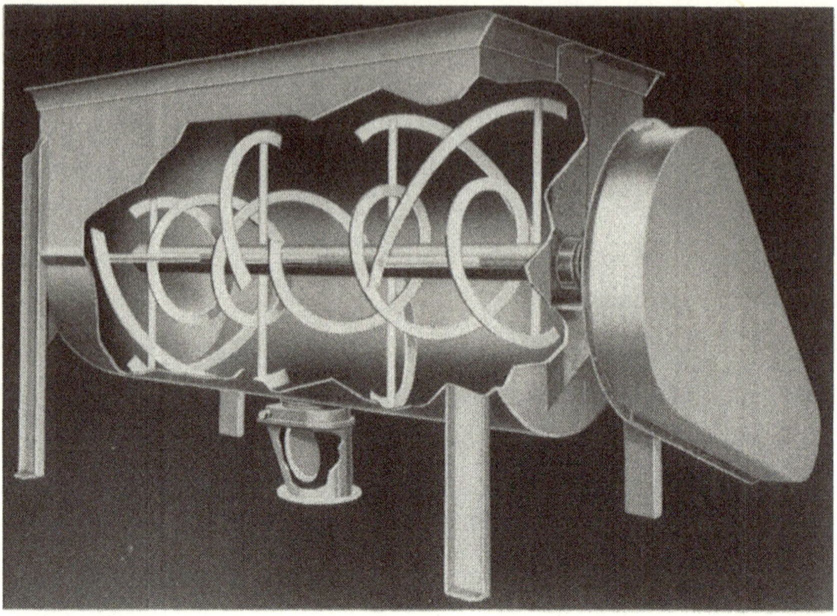

FIG. 3.1. RIBBON BLENDER.

and to prevent the mixing elements from tossing the contents out of the mixer.

One advantage of the ribbon blender is that it is available in a very wide range of sizes, from a one cubic foot, fractional horsepower unit to 500 cubic foot, 50 horsepower units. Larger mixers can be designed and built to special order. Maximum loading of the mixer should be the amount of the mix which just covers the top of the agitator blades when the mixer is operating. This will generally be a lesser amount than required to fill the mixer when it is at rest, since some air is incorporated even in granular materials during mixing and can raise the level of the contents to a point where the top layer experiences very little agitation.

When shortening, especially plastic shortening, is to be blended into the dry mix, requirements are somewhat more exacting, but mixing can be accomplished in conventional horizontal double-arm mixers, in ribbon blenders, or in vertical planetary type mixers. Ribbon blenders are commonly used for this application, and they represent the highest ratio of capacity to capital investment.

To make up a premix containing dry ingredients and shortening, the dry ingredients are first charged to the mixer and blended briefly. Then, oil or liquid shortening is pumped to a manifold feeding a row of atomizing spray heads mounted in the cover but above the contents so that the spray impinges on the dry solids as they are turned over by the agitator. The mixing action continuously exposes a fresh surface to the shortening mist, and the dry particles eventually become coated with the oil or crystallized shortening.

With some cake mixes, plastic shortening must be used. It is often pumped through a T-manifold and extruded in spaghetti-like strings onto the dry ingredients. The agitating action breaks up the shortening into discrete particles and eventually disperses it uniformly in the mix. In a typical installation, a 3000-lb batch is prepared in a ribbon blender of at least 75 cubic foot capacity. The mixing cycle extends typically seven minutes past the time the last shortening is added.

Premixes of fluids with solids—In the conventional batch mixing process for bread dough. minor ingredients are often pre-blended into a portion of the water. Among the types of equipment used for this purpose is the Readco Ingrediator, which consists of a small vertical stainless steel tank with a direct-driven high speed agitator. When the tank is charged with water, the agitator creates a vortex that draws in the powdered materials, then wets and dissolves (or disperses) them. Emulsification of the minor ingredients in the water fraction assists in distributing them evenly in the sponge or dough mass.

The shortening may be incorporated in a liquid phase pre-blend if the mixture is subjected to a sufficiently intense emulsifying action. Anderson and Mullen (1951) have reported work in which such premixes were made up in 1500 gal stainless-steel tanks and apportioned to a large number of bread dough batches. Use of this large batch eliminated many repetitive weighings of minor ingredients. A Manton-Gaulin homogenizer reduced the undispersed particle size in the premix enough to make the emulsion so stable it could be held throughout an operating day. This method appears to have considerable merit but has not come into general use, posssibly due to the increased use of automatic remote-controlled scaling systems in the most recent bakery plant installations. It should not be confused with the liquid ferment process, in which a large batch of refrigerated prefermented "brew" is metered to conventional horizontal mixers to make straight doughs.

It may safely be said that premixing is of definite benefit to the conventional batch dough mixing process, but it is not essential to the process. On the other hand, premixing is indispensable (or nearly so) to the continuous bread making process as that method has been developed. It would be impractical to introduce the flour into the pressurized developer head without first mixing it with some of the liquid ferment to form a pumpable slurry. The slurry is put into a continuous screw mixer-feeder and forced into the positive displacement dough metering pump.

In the batch cake mixing process for shortening-type cakes, premixing is very common, though it has normally been thought of as stage mixing. It will be discussed later in this chapter.

Premixing is essential to the continuous cake mixing process, and may be done in one or two stages, depending on the nature of the process and on the type of cake being mixed. The purpose of the premix is to provide to the final mixer a continuous flow of material containing the proper proportion of each ingredient except air. Air also could be incorporated in a premix of sufficiently high viscosity, but it would be difficult to control the specific gravity and hence the metering rate of the premix.

It is not necessary that the cake batter premix be completely homogeneous as it is fed to the final mixer. It may in fact contain fairly large shortening lumps. These would not, of themselves, affect the character of the final cake product. If the continuous mixer provides sufficiently intense hydraulic shear, the lumps will be completely dispersed, and it is only necessary that the feed from the premixer provide uniform overall composition in aliquots approximately the size of the hold-up in the final mixing head.

In most continuous cake mixing systems, the premixer is a large planetary vertical mixer (Figure 3.2). Many operators find a 500-lb premix can be formed in as little as three minutes, since absolute uniformity need not be obtained. This mixture is then pumped to a holding tank before being metered through the final system. Use of a planetary mixer may seem like over-design, since the premix could be made in much less expensive equipment, and the fine dispersion characteristic of the planetary mixer is not essential. This type of batch mixer has become standard, however, and most cake shops have had one or more units in operation before graduating to a continuous system.

If a vertical mixer is not available, a double-acting agitated kettle may be used for most cake premixes. Or, for the highly aerated cakes such as sponge cake and angel food cake, all that is required is a propeller agitated tank of suitable size and shape.

For heavy, shortening type cakes, a two-stage premix works well. Provided easily emulsifiable shortening is used, all ingredients except flour and air can be batch premixed in a vertical tank having a high speed propeller agitator. This liquid premix is then metered with a

FIG. 3.2. A CONTINUOUS MIXING SYSTEM FOR CAKE BATTER. RIGHT TO LEFT, PLANETARY MIXER FOR PREMIXING, SANITARY PUMP, SURGE TANK, AND CONTINUOUS MIXER FOR FINAL PROCESSING.

SOURCE: AMF

Waukesha-type pump to a Disc Blender, along with a stream of flour from a gravimetric feeder. The Disc Blender is a low hold-up, highly efficient premixer containing a twin-cone agitator that maintains an open vortex and incorporates dry ingredients much more effectively than does a propeller type agitator. The premix is pumped out of the bottom of the Disc Blender bowl into the final mixer.

Oakes slurry mixers have been designed specifically to furnish a total ingredient preblend to Oakes Continuous Automatic Mixers, but they may have other applications as well. Each unit consists of an electric motor and power train, a sealed mixing chamber, a sanitary pump, and a control console mounted on one frame. The slurry mixer is provided with both stationary and rotating mixing blades which are capable of blending all ingredients into a uniform slurry in approximately one minute without incorporating air. A high capacity pump then transfers the slurry into a holding tank to await further processing in the continuous mixer. The slurry mixer is said to be useful in mixing cake batters, pie fillings, pastry fillings, cream toppings, whipped toppings, and cheese cakes. Three sizes are available: 90 lb, 450 lb, and 1,000 lb nominal capacity.

BATCH MIXERS FOR DOUGH

Special Problems in Dough Mixing

When classified on the basis of the mixing action which is required, most bakery products fall into three categories: (1) Extensible doughs, generally but not always yeast leavened, including those for such products as bread, rolls, sweet doughs, saltines, pretzels, and puff pastry; (2) Flowable mixtures such as preferments or liquid sponges, cake batters, some cookie doughs, most muffin doughs, pancake batters, most icings, marshmallow, and meringue; and (3) Mixtures which are neither flowable or extensible, such as powdered premixes, certain types of cookie doughs, pie doughs, and streusels. The mixing of extensible doughs presents the most complex problem, and will be discussed in some detail.

The two most important functions accomplished by bread dough mixers are blending the ingredients and developing the gluten. Hydration of the flour, or more correctly, hydration of the gluten proteins, is an essential precursor of the development process. It is generally agreed that hydration of gluten proteins occurs rapidly once they come in contact with water, and the rate of hydration does not vary a great deal between flours. On the other hand, dough development is relatively slow and requires the input of a considerable amount of ordered force. The mixing response of a flour is determined to a large extent by the amount and quality of the gluten.

The kind of mixing action which appears to be the most effective in promoting gluten development is a repeated stretching and folding action. If the stretching and folding are always performed in the same direction, development will be particularly efficient. Although no completely satisfactory explanation of the changes occurring at the molecular level has been published, the chief result of the folding and stretching actions may be an orienting of many of the gluten molecules so they become extended and lie side by side, rather than randomly coiled or compacted in a more or less spherical pattern with most of their hydrogen bonds and disulfide bonds being formed intramolecularly. When positioned in the extended mode, there should be ample opportunity for disulfide bonds and hydrogen bonds to become established between adjacent protein molecules, leading to maximum strength of the gluten network and a configuration which is conducive to the formation of thin gluten films. It is a reasonable assumption that the intermolecular bonds are constantly breaking and reforming under the conditions existing in a normal dough.

When the dough approaches its state of optimum development, it will begin to enfold considerable amounts of air. One result of the gas entrapment is a marked reduction in density just before the maximum power requirement occurs. This phenomenon apparently marks a major change in status of the gluten. There are also measurable changes in plasticity, viscosity, adhesiveness, cohesiveness, elasticity, and extensibility throughout the mixing cycle.

The effectiveness of the mixing operation in forming a continuous gluten network which will have the maximum gas holding capacity has a large influence on the quality of the finished product. Well developed doughs usually result in loaves with high specific volume, soft, silky, and uniform grain and texture, and good shelf life. If the dough has not been mixed to its optimum state, it is difficult to compensate for this deficiency by changes in subsequent processing conditions.

Factors other than quality and amount of gluten which tend to increase the mixing requirements are: (1) Short fermentation of sponges, (2) High salt level, (3) Fast acting oxidizers, and (4) Low dough temperatures. Factors tending to decrease mixing requirements are: (1) Long fermentation of sponges, (2) Higher dough temperatures, (3) Proteolytic enzymes, (4) Reducing substances, (5) Alcohol derived from fermentation, and (6) Inactive dry yeast.

Types of Dough Mixers

The mixers to be considered in this section are bread dough mixers, which are also, of course, generally useful for sweet doughs and many other kinds of doughs although the type of agitator and the horsepower required of the power train may have to be different if optimum results are to be obtained. There are also many kinds of mixers which can be made to work with bread doughs even though they were designed for other types of bakery products. It is the former type which will be discussed here. Mixers for products such as cookies, crackers, pizzas, and pie doughs will be discussed elsewhere in this chapter or in other chapters devoted to specialized lines of machinery for making these products.

There are several possible ways to categorize dough mixers. It is convenient for the purpose of this section to classify them as horizontal dough mixers, planetary vertical mixers, continuous mixers, and others. Each of these types has sub-types.

Horizontal dough mixers—Many types of mixers can be, and are, used for batch mixing bread dough, but nearly all large wholesale bakeries use high speed horizontal dough mixers for this purpose. Figure 3.3 illustrates an installation for a medium-sized bakery in which a bag

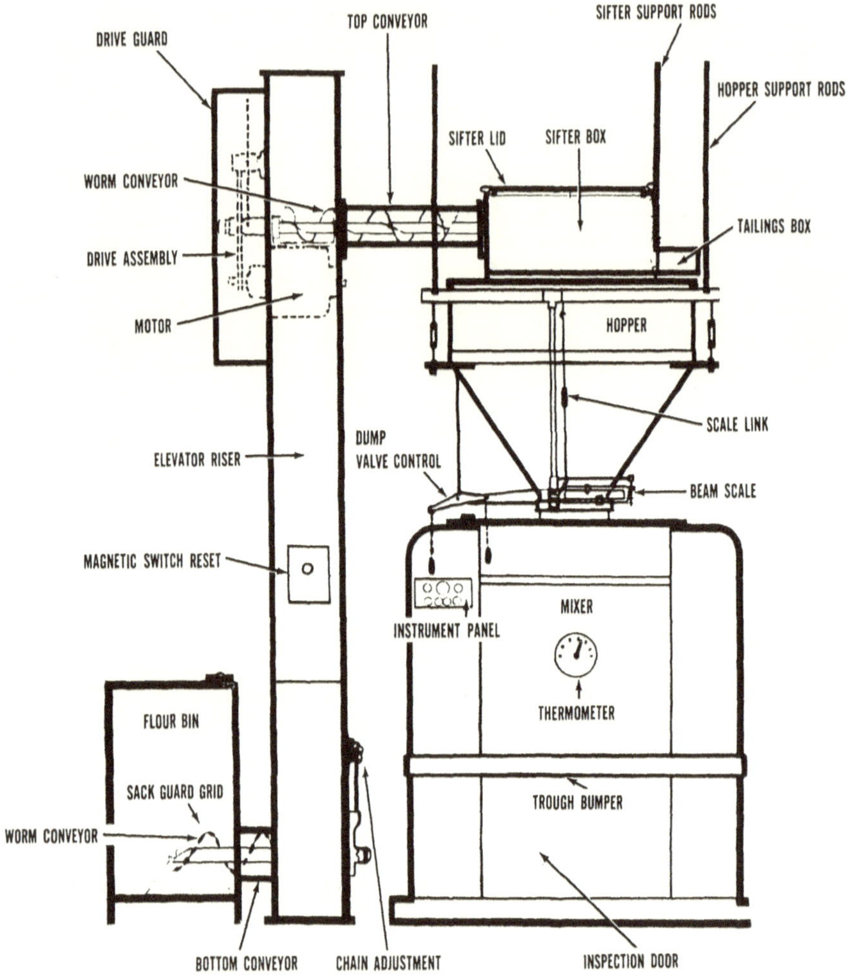

FIG. 3.3. HORIZONTAL DOUGH MIXER WITH WEIGH HOPPER, ALSO SHOWING FLOUR ELEVATOR LEADING FROM BAG DUMPING STATION.

dumping station is combined with conveying means leading to a sifter and a weigh hopper over a horizontal dough mixer. There are many hundreds of these machines in daily operation mixing all types of yeast-raised doughs, including bread, roll, pizza, pretzel, and sweet goods doughs. High speed horizontal dough mixers are popular because they have been specifically designed for mixing extensible doughs, and they do this job very well. Furthermore, these mixers are sturdy, some units having been in almost constant use for decades. They have the additional merit of being fairly versatile, since they are adaptable to producing quite a few different kinds of doughs—in fact, they are often pressed into duty for mixing materials for which they are really not very well suited.

The usual high speed horizontal mixer configuration of agitators, which consists of round bars running parallel to the bowl surface, is ideal for stretching, pressing, and folding the dough without tearing it—just the type of action that is best for developing doughs. It is a configuration which is not, however, very efficient in mixing powders, batters, and other materials needing an intense, randomly directed, dispersing action, or doughs which should not be developed, like pie doughs.

Several manufacturers in the U. S. and foreign countries offer mixers of this type to the baking industry. Whether they are used to mix sponges, doughs, straight doughs or liquid ferment doughs, the mixer construction is basically the same. A horizontal mixing bowl, roughly U-shaped in cross section, is mounted in a rigid frame over a compartment holding the drive motor and transmission. A single horizontal agitator shaft passes through the bowl axially and is turned by a sprocket and chain drive leading from the transmission. Mixer arms are affixed to two spiders mounted on the agitator shaft. The various mixer models have 2, 3 or 4 rods attached to these spiders—some of the rods may be free to rotate but usually they are fixed. One or more of the mixer bars may be angled to push dough from one end of the bowl to the other. Some mixers have stationary braker bars or baffles mounted inside the bowl parallel to the agitator shaft to prevent the dough from hanging on to the mixer bars through several rotations and to assist the mixing action in other ways. The axle can be mounted off center so that the free space on one side of the mixer is greater than on the other side; as a result, there is an increased squeezing and pressing action on one side.

Two-speed motors are almost always included to permit high and low speed mixing. Timers, or more sophisticated controls, are always included in the circuits and, additionally, there is a provision for jogging, or giving the agitators a partial turn, to assist in throwing the dough out of the chamber during the discharge operation. Bowls and bowl covers can be fabricated of mild steel or stainless steel. The latter material seems to be much more common nowadays.

See Figure 3.4 for a front view of construction details of tilt bowl horizontal dough mixers.

The standard agitator configuration which has been described above is specifically intended for high speed mixers used mostly for develop-

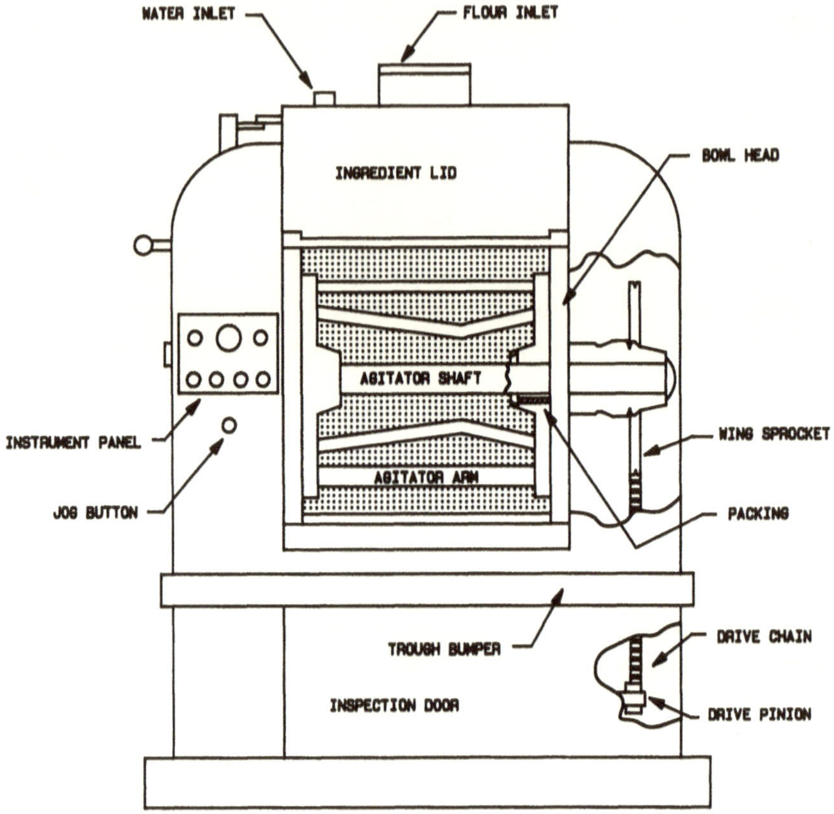

FIG. 3.4. IMPORTANT FEATURES OF A TILT-BOWL HORIZONTAL DOUGH MIXER.

ing doughs. Slow speed horizontal mixers can be equipped with very different kinds of agitator systems; they may even be equipped with two sets of axles. Double-arm mixers were formerly called creamers because they were used specifically for batters. They are now sometimes used for stiffer doughs, though heavier drives and motors must be supplied in these cases. Air is incorporated by agitator arms of this conformation, as distinct from the action of figure-eight type arms. The latter type reduces gluten development and heat buildup. The range of speeds in commercial mixers of this type will be from about 14 to 60 rpm, with an average of about 25 rpm.

The bowl is totally inclosed during the mixing process for sanitary and safety reasons as well as to prevent the dough from being thrown out. There are two basic types of discharge methods. In some mixers, the bowl is tilted by a separate motor until the open top is in a position to allow the dough to be dumped into a trough or other container or into a chute leading to make-up equipment on the floor below. Another type has a tight fitting front panel which can be moved to expose the interior of the mixing chamber. Loading of ingredients can be performed by dropping them through ports in the top cover or by manually placing them in the tilted bowl.

Horizontal tilt bowl mixers of traditional design discharge their contents by turning the bowl through 90° so that the opening faces the side instead of the top. "Jogging" the agitators (i.e., moving them through a partial turn) throws the dough out into a trough or other container. Sometimes, the dough must be cut out or pulled out of the bowl by the operator. A more recent development is the 140° tilt feature which almost turns the bowl upside down so the dough falls out with minimal operator intervention. To assist in discharging the contents, the agitators can be jogged while the bowl is tilting. The lower front center part of mixers employing the 140° tilt had to be redesigned so that troughs could be moved directly beneath the bowl or chutes could be installed for gravity feed to a lower floor. Furthermore, the "jogging while tilting" action required the installation of more powerful motors to turn the bowl. Mixers are also being made with discharge action to either side, that is the bowl tilts 140° either backward or forward.

Obviously, hand feeding ingredients into a mixer bowl that is facing downward is going to be a little difficult. These mixers are best adapted to automatic feeding of ingredients through ports in the top cover. The best way to feed in the sponge is by tilting the bowl backward (relative to the discharging position) about 45° to open a space between the back of the top and the mixer bowl so a trough elevator can dump the trough's contents through the opening (Booth 1987). Another method is to have a special sponge door at the top front of the mixer, as illustrated in Figure 3.5. The pictured mixer is not, however, a 140° mixer.

A wide range of mixer sizes is offered. Size is usually specified by the amount of bread dough which can be processed in one batch. The most common sizes are probably 8, 10, 13, and 16 hundred pounds, normally referred to as No. 8, 10, 13, and 16. Motor size, in horsepower, varies with the capacity of the bowl, but can be tailored to the requirements of special doughs such as the generally stiffer pizza, pretzel, and bagel doughs.

Quite a bit of engineering effort has been expended in designing improvements for horizontal mixers. A considerable amount of it has been directed toward improvement of sanitary features. The packing gland seals between bowl and agitator shaft have been recognized as problem areas and have received considerable attention. Although many mechanical improvements have been made over the last few decades, the basic mixing principles have remained much the same.

The action of the agitator bar on dough ingredients varies through the mixing cycle. During the first 2 or 3 minutes of a typical dough mix, the principal action is wetting of the flour. This is the least efficient part of the cycle, mechanically speaking. The powdered ingredients, fat, and liquids are flung throughout the enclosed space in a more or less random fashion. When enough of the flour has been wetted, a sticky lump of dough forms. This mass tends to be rotated around the agitator shaft by the action of the mixer arms. The ball grows as it cleans up the ingredients which are sticking to the sides until finally the agitator picks up the whole mass and throws it around the walls of the bowl. From this point on, the agitator rolls, folds, and kneads the dough, stretching and shearing it so that dry lumps are broken down and moistened, gas bubbles are ruptured and flattened, and gluten is developed. Although this phase is more mechanically efficient than the wetting stage, there is still a waste of energy inherent in the system resulting from the lifting, throwing, and slapping around of the dough mass—actions which use power but contribute little to mixing the dough. The cost of power is not a very large part of the expense of the mixing operation, however.

Mixing times for bread doughs might typically run 10, 11, or 12 minutes, with a frequent goal being the completion of four batches in an hour. The dough is considered to be ready for discharge when the rate of electric power usage by the motor reaches a peak. Experienced operators can estimate the status of the dough by listening to the sounds made by the dough mass as it is flung about the bowl, and, to some extent, by listening to the transmission noise. Mixing beyond the point of maximum power requirement usually results in breakdown of the gluten and weakening of the dough structure as shown by a progressively lower rate of power consumption.

One result of the friction and shearing action which take place in dough mixing is the generation of heat. Most of the work done by the mixer motor is translated into sensible heat, and, if this heat is not removed during mixing, the dough could reach a temperature in excess of 100°F. Hot doughs do not process well in conventional make-up equipment. They tend to be sticky, soft, and weak. They also generate gas at a higher rate than usual. The difference in performance of the first part and the last part of a hot dough passing through the divider, rounder, and molder would be pronounced. Hence, it is necessary to cool the dough during mixing, either by using chilled ingredient water (or ice) or by refrigerating the mixer jacket.

FIG. 3.5. DOUGH ELEVATOR WITH TROUGH PREPARING TO DROP SPONGE THROUGH SPECIAL DOOR AT TOP OF MIXER.

Auxiliary equipment is available for these tasks. Ingredient water chillers using vapor compression refrigeration equipment are available from mixer manufacturers and others. Mixer bowls are normally provided with jackets, for chilled water in some designs and for direct expansion of refrigerant gas in others. The direct expansion units seem to be more popular, and are available in sizes up to 25 horsepower for the largest mixers. In more recent types, the flat ends of the bowl are also jacketed to achieve even greater cooling capacity. Older jacket designs moved the refrigerant back and forth through channels of uniform cross-section. The newer style of direct expansion jacket has cooling channels that begin with a small cross sectional area which increases as coolant flows downstream and continues to evaporate. The abrupt right angle turns that the coolant had to make have been replaced with gently contoured radii, allowing the coolant to flow at a maximum rate with minimum pressure drop (Broaddus 1978).

Mixers can be obtained with two thermistor temperature probes; one probe extends into the dough to sense dough temperature while the other makes contact with the inside of the bowl. During the early part of mixing, it is possible to freeze ingredients to the bowl surface if the temperature is reduced too rapidly. The bowl probe senses the surface temperature and will, if necessary, reduce the coolant flow to keep the dough from freezing to the bowl. A temperature can be set on the control panel to indicate the desired final dough temperature, so that all doughs can be delivered to the divider in the same condition.

Bread produced from dough processed in horizontal dough mixers is the conventional product of relatively coarse, non-uniform grain which has been accepted as "normal" by the industry. That this normal grain is as much a function of the mixing action as of any other factor is a point that was generally overlooked until continuous developers became available and permitted the production of fine grained bread with very uniform cell structure.

Other types of batch dough mixers—It is possible and practical to mix small batches of yeast-raised dough on vertical planetary action mixers. These machines are discussed in more detail in the section describing mixers for batters. Some retail bakers who need a single utility mixer for all products mix sweet goods, pie doughs, and even bread doughs in such equipment. In a relatively few cases, wholesale bakers mix sweet yeast-raised doughs in batches of up to 200 lbs total weight on the largest size planetary mixers.

The agitator used for extensible doughs such as bread doughs is called a dough hook. It is a single curved arm of bronze, aluminum, or stainless steel. The power of the agitator drive, which is as much as ten horsepower on the largest units, would not be sufficient to drive an agitator with two or more arms at sufficient speed to mix a bread dough. The type of transmission used in the largest cake mixers does not deliver the full horsepower of the motor at low speed. The mixing action is efficient, however, and removable bowls add convenience.

Spiral kneaders have recently made their appearance in the U. S. after a somewhat longer history of development in Europe. The kneading element is a strong stainless steel spiral which rotates rapidly in a rather shallow and wide bowl. A slow speed is available for blending the liquid and dry ingredients. The bowl itself rotates and it has a large protuberance, about a third the height of the bowl, rising from the bottom center. Another version has a heavy rod that is fixed in the mixer head and goes to the center bottom of the bowl as the head is lowered. Advantages are said to be very short kneading times (3 to 8 minutes), low temperature rise, and excellent dough development. An example of a spiral kneader is shown in Figure 3.6.

FIG. 3.6. SPIRAL KNEADER.
SOURCE: HOBART CORP.

Some of the mixers which have been used for high speed dough conditioning are vertical dough mixers with planetary action. The following description applies to one of the earlier models of U. S. mixers used for high intensity developing of bread doughs. In construction it is reminiscent of the early arch type large planetary cake mixers. A stainless steel cylindrical bowl is mounted on casters so it can be moved in and out of the mixer frame and through the fermentation area. This bowl holds a maximum batch of 600 lbs of dough. A stationary center post and a double arm planetary agitator extend cantilever from a two speed, 50 hp, overhead planetary drive. The entire drive system with its agitator can be hydraulically raised and lowered to permit placing the bowl in position. At high speed, the agitator rotates at 140 rpm and the planetary head rotates at 40 rpm for a precession ratio of 3.67 (very close to that of the single-ratio planetary cake mixers). Low speed is one-half the high speed. Speed, mixing time, and hydraulic lift equipment are controlled from a central panel.

Several bowls are used in the mixing cycle. At the ingredient scaling station a bowl is first charged with all the ingredients except sugar and salt, using conventional weigh hoppers and meters. The bowl is then rolled into the mixer and the dough is put together with a short mixing cycle of 15 seconds at slow speed and 18 to 30 seconds at high speed. The bowl is removed from the mixer, allowed to stand 2.5 hr and then returned to the mixer. Sugar and salt are added and the mixer operated 15 seconds on low speed and three minutes on high speed. These times are applicable to white bread doughs, less time is required for whole wheat and rye. After the remix, the bowl is allowed to stand 12 to 15 min before it is finally discharged to the divider hopper. Maximum production rate with this equipment is approximately 6000 lb per hr of dough.

In the process just described, a special conveyor carries dough from the divider directly to the molder, eliminating the conventional rounder and overhead proofer. There are a number of other important differences between this equipment and horizontal mixer lines. Many of the differences can be regarded as advantages, while others are disadvantages, depending to some extent on the goals and demands of the particular bakery operation. Because the batch is smaller, more scaling is necessary than for a larger straight dough. All of the ingredients are scaled at once, however, and, in the process in which this equipment is used, there is actually less scaling than in the conventional sponge and dough process using large horizontal mixers. Because the dough ingredients remain in the mixer bowls throughout the process, no troughs are used.

One factor that at first thought would appear to be a disadvantage is the impracticality of applying a coolant (other than chilled ingredients) to the portable bowl. So much energy is put into the dough by the intense mixing action of this high speed equipment that the heat generated could not be removed through the available bowl area during the short mixing time. Therefore, the dough is discharged at a relatively high temperature (e.g., 92°F for white bread dough). The relatively high temperature is not the disadvantage that it would be in a 1600 lb. batch. Typically a mixer load of dough goes through the divider in 6 min instead of 15 min, and though it develops gas at a higher rate, it is divided before it can cause more than the ordinary scaling errors.

Elimination of the intermediate proofer cycle of, typically, 12 to 15 min results in a younger dough at the molder. The higher dough temperature leads to a much shorter proofing time, since much of the time bread doughs spend in conventional proofers is needed just to bring the dough up to proofing temperature. With this dough process, dough make-up is less critical and higher dough temperatures can be tolerated, it is said. Among other advantages reported the manufacturers and users of the equipment are: (1) The equipment is easier to clean and can be changed over faster, (2) Equipment and installation costs are significantly lower for the mixer and make-up line, (3) floor space requirements are reduced, and (4) Labor costs are reduced.

Much European dough mixer research has been concentrated on the Tweedy mixer, and its variants, and the Chorleywood bread process, which is one version of the "no-time" or mechanical dough conditioning procedures using these mixers. Such systems depend on high speed, short time development of a straight dough which is "conditioned" with oxidizers, reductants, enzymes, and emulsifiers so as to eliminate the need for some of the fermentation and rest periods. The chemical aspects of these processes have been discussed in considerable detail in "Formulas and Processes for Bakers," the preceding volume in this series, and will not be repeated here.

Fish (1982) gave a review of one of the systems using the Tweedy mixer, and the following discussion is based, in part, on his paper. One form of the high speed mixing complex consists of the mixing chamber, a top frame unit, a microprocessor to control the system, and an auxiliary fermentation and ingredient feed system. The mixer itself consists of a cylindrical bowl having a high speed rotating impact plate at the bottom and a series of baffles on the bowl's interior surface to force the dough onto the agitators. The system mixes to a predetermined energy input as measured by a watt hour meter. Desired input depends upon the formula, weight of dough, etc., but 6 to 7.5 watt hours per lb was required in a typical plant. Mixing times of less than five minutes are obtained. During mixing, a vacuum of ten inches or more is drawn on the chamber by a liquid ring vacuum pump.

A typical cycle is described as follows. The operator notifies the computer of the product to be made, after which the formula is withdrawn from the memory and displayed on the monitor. The operator then types in the required flour weight, the proportion of water required, the dough temperature required, and the watt hour input desired. Total weight of ingredients to be manually added and the type of flour is also inserted.

The flour quantity and type required is automatically delivered to the mixing chamber, after which the computer determines the weight of water from the percentage figure already entered. Water is taken from a holding tank where chilled and tap water has been blended to arrive at a temperature 2°F below the final desired temperature. During delivery of the flour, its temperature has been checked and the calculated water temperature achieved by adding either hot or cold water to the holding tank. The temperature rise due to intensive mixing is counteracted not only by pre-chilling the liquid ingredients but also by use of a jacketed bowl using a circulating coolant or a direct expansion refrigerant.

Crumb structure is largely a function of the vacuum drawn on the chamber. Increasing the vacuum apparently increases the fineness of the grain up to about 25 inches of vacuum, after which deterioration of the grain is observed. Less vacuum leads to a more open grain. After mixing actually starts, it is very difficult to draw the air out of the dough; therefore, a vacuum is drawn on the chamber about 15 seconds before the rotary impeller is started. This withdraws the gases before a cohesive dough exists, after which little additional gas is removed.

Post-mixer Development

The desirability of making the dough's physical characteristics more uniform as it starts into the make-up equipment has led to the introduction of post-mixer devices which knead the dough just before it passes into the divider (Campbell 1979). Although one function of these machines is the completion of development which has been begun in a horizontal dough mixer or the like, they also reduce the variability resulting from changes in the bulk dough as it waits for processing. In conventional systems, the first part of the batch is denser, firmer, less gassy, and differs in other respects from the last part of the batch to be processed through the divider. These problems were made worse by the everpresent tendency to introduce larger batch sizes and by the trend away from sponge doughs and to flourless pre-ferment systems. By kneading and compressing the dough up to the time it enters the divider, the post-mixer developer substantially eliminates the changes resulting from dough aging. It appears post-mixer developers have been applied mostly to bun lines, probably because of the longer time required to process a batch of dough through these lines.

Current designs are either self-contained devices that mount over bun divider hoppers and include both pumping and developing units or work within the divider hopper. In the latter case a separate pump is not needed. Developer heads adapted from continuous mixing systems have also been used as post-mixer kneading equipment. Post-mixer developers which fit on the divider in-feed consist of a hopper which is filled with dough from a trough dump or other means, a stainless steel helix which pushes the dough through a cylinder, and vane-shaped rotors that turn within a chamber having protrusions on the wall to resist rotational dough movement. Dwell time, and therefore throughput, is controlled by adjusting the rotor speed.

Devices which fit in the divider hopper consist essentially of two sets of paddles affixed to shafts mounted directly over the hopper. This method is best suited to de-gassing trough-fed doughs.

There is also available a free-standing unit that will handle 15,000 lb per hr of dough. It has a cooling jacket but must be fed by a dough pump since it has no internal pumping capability. This has been used on both bun and bread lines as well as in production of doughnuts, hard rolls, and French bread. The separate pump required by such units is generally a screw-fed, ball-type unit, which de-gases and meters the dough.

The devices constituting the post-mix developer, or kneader, unit are often inclosed in one frame and mounted on casters. They are available in both horizontal and vertical configurations, and with motors sized to fit the needs of the system. In a kneading chamber holding about 120 lb of dough, two stainless steel rotors turning in opposite directions mesh with one another to stretch and press the dough in a manner somewhat similar to the action in a horizontal mixer. At a typical rate of about 7,200 lb per hr, each pound of dough gets about one minute of kneading action. Throughput is controlled by varying the rotor speed. A cooling chamber is provided to counteract the heat resulting from the work performed on the dough in the kneader.

Since the dough can be removed from the horizontal mixer before it is fully developed if further development is applied in the kneading equipment described above, mixing time can usually be decreased and the number of mixer cycles increased.

TEMPERATURE RISE DUE TO MIXING

In mixing bread doughs in small mixers, it is easy to observe that the dough becomes warmer as mixing progresses. Heat rise also occurs in batters, pie doughs, etc., but seldom to the extent it occurs in bread doughs and the like. There are three significant causes of the temperature rise observed as dough is being mixed: (1) Frictional heat directly related to the amount of work performed on the dough, (2)

Heat of hydration of the flour, and (3) Heat of solution of ingredients such as sugar or salt. Heat of solution can, in fact, be negative, i.e., result in an absorption of heat (cooling) as the substance dissolves. In any case, heat of solution is a relatively minor factor in temperature change and is often ignored when calculations of the cooling requirements are being made. The factor causing most of the heat rise in bread dough is the work done on the dough by the mixer agitator.

Heat input during mixing must be offset by some method of cooling if doughs of uniform handling qualities and fermentation rates are to be obtained. It is very common practice to use chilled ingredients, especially refrigerated water, for this purpose. Adding part of the water as ice is also common. Ice is very effective in reducing dough temperature because of the relatively large latent heat of fusion of water (heat absorbed as ice at 32°F changes into water at 32°F), but development is interfered with as a result of the late entry into the hydration process of the part of the water which has been added as ice. The undesirable effects of delayed hydration can be somewhat reduced by using the smallest possible particle size of ice, so it is common to add crushed, flaked, or shaved ice. Under no circumstances should large chunks be used. The initial temperature of the ice is also a factor which must be considered, in that ice at 30°F can be assumed to melt quicker than ice at 0°F. There will be some loss of cooling effect if the warmer ice is used, of course. If the sequence of addition is subject to control, ice should be added to the dry ingredients and the mixture given a few turns before the chilled water is added.

Nearly all high speed dough mixers and many other dough and batter mixers are equipped with jackets containing a circulating refrigerated liquid or which can be cooled by direct expansion refrigerants.

The baker using ice in his mix will find it necessary to calculate how much of this material should be added or how much other cooling means should be applied to reach the dough temperature which has been selected as the standard. In the literature can be found abbreviated methods which yield rough approximations of the cooling requirement, but the reader will doubtless be helped more by a detailed calculation which takes into consideration all of the important factors. This is best illustrated by using an example such as the one given by Valentyne (1959) for calculating the heat balance for a mix composed of 850 lb sponge, 350 lb flour, 280 lb water, and 120 lb other ingredients. The mixing time will be ten minutes and there will be four mixes per hour. A target temperature of 80°F has been selected. The flour temperature is assumed to be 100°F, the sponge temperature is 90°F, and the ingredient water is at 40°F.

First, it is necessary to calculate the temperature rise which would occur in the absence of cooling. The heat generated as flour takes up water varies according to the original moisture content of the flour—the lower the moisture content, the more heat is generated. If the flour has an original moisture content, of 11 to 12%, an average heat of hydration of 6.5 Btu per lb can be used for practical calculations. The approximate specific heats of some of the materials in Btu per lb per °F are: sponge 0.60, flour 0.42, and water 1.00. The heat generated by the mixing action is 42.5 Btu per min per hp, as determined by the electrical input which is to a large extent converted to heat in the dough mass. We now have enough data to make a heat balance calculation according to the following steps.

Since the sponge is at 90°F and it is necessary to cool it 10°F in order to reach the targeted temperature of 80°F, we multiply 10° by 850 lb of sponge and multiply that result by the 0.60 specific heat of the sponge. The result is 5,100 Btu.

Heat of hydration generated by mixing 350 lb flour with water can be estimated by multiplying this poundage figure with the specific heat of flour (6.5) to get a figure of 2,275 Btu.

The requirement for cooling flour from 100°F to 80°F is determined by multiplying 350 lb of flour times 20°F temperature difference and then multiplying that result (7,000) by the 0.42 Btu specific heat of flour to get 2,940 Btu.

Cooling other ingredients from 100°F to 80°F will require 120 lbs time 20°F difference times 0.4 Btu average specific heat, or 960 Btu.

If the average motor horsepower is 60, then we can multiply 60 hp times 42.5 Btu times 10 min times 0.9 (an arbitrary figure representing 90% efficiency) to get 23,000 Btu.

The total of the heat added to the dough mass during mixing is the sum of 23,000 + 960 + 2,940 + 2,275 + 5,100, or 34,275 Btu.

Now, we can consider the cooling agents which are already present or which must be added. The ingredient water being added has been chilled to 40°F, so we can multiply 280 lb water times 40 degrees times the specific heat of water, which is 1.00, to get 11,200 Btu withdrawn from the dough. This leaves 23,075 Btu of cooling to be obtained from some other source; either ice or refrigeration of the jacket.

For each pound of ice added to the dough, we get a cooling effect of 144 Btu per lb from the change of ice at 32°F to water at the same temperature. This is the so-called latent heat of fusion. Ice will normally enter the mixer at some temperature below 32°F, and the cooling effect of moving from that temperature, say 0°F for the purpose of this illustration, to the melting point of 32°F will be 32 degrees times the specific heat of ice (0.5—note that this is different from the specific heat of water) or 16 Btu per lb of ice. To this must be added the difference between 32°F water and the 40°F that we have assumed for ingredient water in the previous calculations (which the ice replaces); the 8 degrees difference is multiplied by the specific heat of water (1.00) to get 8 Btu. Under these conditions, the cooling effect of substituting one pound of ice for one pound of water is 144 + 16 + 8 or 168 Btu per lb. Since we need 23,075 Btu of additional cooling in the batch (see preceding paragraph) the simple calculation to determine the pounds of ice needed is to divide 23,075 by 168. The result is 137 lbs. Since an equal amount of 40°F water is replaced, the final mix will be 143 lb of water and 137 lb of ice added to the 850 lb sponge, 350 lb flour, and 120 lb of small ingredients.

Mixer jackets can be refrigerated by circulating through them brine, propylene glycol, or other liquid of low freezing point which has been cooled by passing it over coils or plates which are themselves chilled (generally to a temperature considerably below freezing) by a direct expansion refrigeration system. Although this method has some negative features, it has the substantial advantage that the storage capacity of the jacket and other parts of the system allow the central refrigeration unit to be somewhat smaller. The chilled liquid can be recirculated through the jacket, with of course some loss of efficiency, while this is not possible with direct expansion units.

Direct expansion cooling systems are composed of a condensing unit having a motor driven compressor and a water cooled condenser, an evaporator which is in the jacket of the mixer bowl, and a refrigeration control system with a thermostat, solenoid valves, etc. The compressor turns the gaseous refrigerant returned from the jacket into a liquid. Heat gained by the liquid during this compression is transferred by the condenser to water which is circulated around its coils. The cooled liquid refrigerant is then transferred to the mixer jacket and there expands, taking up heat in the process. If it is not carefully controlled, this process can easily bring the inner surface of the bowl to a temperature so low that the dough (or any free liquid in the bowl) freezes onto the metal. Obviously, this situation is to be avoided at all costs. An average jacket temperature of 25°F is considered acceptable.

Refrigeration equipment must be chosen in a size sufficient to meet the maximum cooling demand which will be imposed upon it. For plants mixing a single formula of dough in a known amount of batches per hour, the calculation is fairly simple. Using the example shown earlier in this section, in which 23,075 Btu of cooling is required per mix (in addition to the cooling derived from 40°F ingredient water), we can simply multiply the figure 23,075 by the number of batches per hour, say 4 batches, to get an hourly requirement of 92,300 Btu. Some heat is added by the pump and pipes leading from the condenser to the evaporator, this can be calculated or determined experimentally, but a 10% rate of loss is within the expected range for normal operations and will be assumed to apply here, giving an initial requirement of slightly less than 102,600 Btu per hr. From this requirement, the size of the refrigeration unit can be quickly calculated by any supplier of such equipment. Of course, this figure is a minimum and must be increased to take into account fluctuations in demand throughout the mixing cycle, variations in external temperature, etc.

Where refrigeration is to be used, which would be the case in nearly every installation of horizontal dough mixers, a decision must be made as to whether or not chilled ingredient water is to be made available or tap water only is to be used. In a bakery where a variety of goods—cakes, cookies, pastry, etc.—are to be made in addition to bread, it is very likely that a source of chilled ingredient water will be at hand. Even so, it is probably the safest alternative to size the refrigeration unit on the assumption that it will be responsible for meeting the total dough cooling demand. It is also probable that doughs will mix faster and better if the mixer and ingredients (including water) are started at room temperature, or a few degrees lower, and the temperature of the dough kept below the predetermined level solely by jacket refrigeration.

The temperature indicating devices on mixer jackets mostly indicate the jacket temperature, or just the refrigerant temperature at some point, ideally at the exit pipe. This is true even if there is a sensor on the inside surface of the bowl. Dough temperature is not necessarily indicated by the meter—in fact, it is almost certainly higher than the indicated temperature. Heat transfer is fairly good, however, if the mixing action is vigorous and the dough makes good contact with the jacket. There is no closed loop on most of these mixers; the sensor has no direct effect on the supply of refrigerant to the jacket and therefore does not modulate the refrigeration unit. Changes in cooling must be made by the mixer operator or are programmed as some function of the mixer cycle. Recent developments have changed this situation and temperature control is much more nearly automated. The following discussion is based in large part on an article by Booth (1987).

In mixers having automated temperature control, a dual output temperature controller and indicator work in conjunction with a timer on the refrigerating system. One set point on the controller is preset to the desired final dough temperature, while the other set point is preset to an upper temperature limit beyond which the dough is considered unacceptable. The refrigeration system is controlled by the temperature controller as well as by the programmable controller. The latter device controls the refrigeration until about half way through the mixing cycle, after which the temperature controller takes over. This insures that refrigeration is operating throughout the complete cycle.

If, at the completion of the normal mix cycle, the dough temperature is higher than the top limit, the mixer, instead of tilting and discharging automatically, will resume running in slow speed with full refrigeration. When the dough temperature finally falls below the top acceptable limit, the machine will stop mixing, tilt, and discharge.

Mixer contents can also be chilled by injecting compressed carbon dioxide into the ingredients (Baron 1983), although this is not a common practice. Mixers can be fitted with equipment which allows injection of just enough carbon dioxide to keep the dough temperature within a prescribed range. Heat is taken up as the gas expands, the same phenomenon that is used in preparing dry ice from gaseous carbon dioxide. It has been said that the action of reducing agents is facilitated by the anaerobic conditions created by the high concentration of carbon dioxide (or low concentration of oxygen) in the mixing chamber.

CONTINUOUS DOUGH MIXERS

Discussions of continuous bread dough mixers in the United States generally refer to the complete systems developed by American equipment manufacturers beginning in the 1950's. These systems, some of which are still operating, include equipment for scaling, premixing, fermenting, mixing, depositing, and panning. They replaced batch system weigh hoppers, ingredient pre-mixers, batch mixers, fermentation troughs and rooms, dividers, rounders, intermediate proofers, and molders.

Although the mixer was just one of the novel features of the continuous breadmaking systems, it is the aspect which will be dealt with in greatest detail in the present discussion. The mixing system in continuous lines has two components, the premixer (or incorporator) and the final mixer (or developer). The premixer is, essentially, a stainless steel screw in a long stainless steel cylindrical housing. Ingredients are delivered to the feed end of this screw in three or more streams, then roughly incorporated into a dough form and discharged under pressure by the screw.

The three major streams fed to the screw are (1) liquid ferment (broth, brew, or liquid sponge), (2) liquid shortening, probably containing some emulsifiers, and (3) flour and certain other dry ingredients which have not been incorporated in the liquid ferment. After these ingredient streams are blended together, they are forced through large stainless steel tubing to the inlet side of a positive displacement gear pump. Since there is some slip in the screw, this force-fed pump is the basic rate-measuring device for the dough production system and it must be kept hydrostatically full.

It is not practical to synchronize all feed rates perfectly with this metering pump, so the feed streams to the screw are alternated in rate through a high-low cycle, with the high-rate time interval kept constant, and the switch from high to low triggered by build up of material in the screw feeder.

The metering pump forces the roughly combined dough to the final mixer (developer) through stainless steel tubing. This final mixer has an internal cavity, oval-shaped in cross-section, with two counter-rotating double-arc paddle arms extending axially through the cavity. The dough mass enters the cavity near one end and exits near the other. A swing door opens in the front of this mixing head for easy access to the interior.

The dough exits from the final mixer through an intermittently operated extruder slot that extrudes a dough piece broadside and drops it into the pan below. Extrusion is synchronized with the pan feed conveyor. Since the developer head will hold only a few gallons of dough, the average residence time in the chamber will be less than one minute. The mixing action for this short period is so intense, however, that the dough emerges in a homogeneous, developed condition.

One important advantage of systems using continuous mixers is that more accurate scaling of dough pieces is obtained, as compared to batch systems. This is understandable when it is realized that all of the dough is precisely the same age as it passes through the extruder cut-off slot. There is no cyclic change from old, gassy dough to new dough, as in a divider operating on a succession of dough batches. Another advantage is a labor saving of at least two operators per shift.

Differences which early in the introduction of continuous mixing were regarded as important advantages but later came to be regarded as possible disadvantages were the more uniform grain and finer texture, and a whiter appearing crumb. These effects can be understood as among the results of the mixing action in the continuous systems. The extremely intense, efficient stirring produces a more uniform mix of all components, including air and gas. Since the air is distributed in finer bubbles and there are more bubbles, the total area of the cell walls will be greater and the cell walls themselves will be thinner (i. e., the same amount of dough will be spread over a larger area). The bread crumb will thus appear smoother, silkier, and softer. The finer grain also accounts for the whiter appearing crumb. Finer grained products always appear whiter, other conditions being equal, because the cells on the cut surface are relatively small and shallow, and reflect (and refract) light more efficiently.

Aside from efficient mixing, an additional cause of grain uniformity in continuously mixed bread dough is that the continuous mixer operates on only one physical phase whereas the batch mixer operates throughout the mixing cycle on two phases—the plastic dough phase and the gaseous phase. The continuous mixer functions at 50 or 60 lb pressure and is kept hydrostatically full. Since the chamber contains no gas-filled space, no new gas is incorporated into the dough. All existing bubbles receive the same dividing and dispersing action in the mixing cycle and are more uniformly dispersed at the finish. Much more of the carbon dioxide is dissolved in the dough's liquid phase at 60 psi than in the batch process dough which is at atmospheric pressure. In the batch mixer, the agitator constantly draws new, and often large, gas bubbles into the dough during the mixing period, and the bubbles enfolded near the end of the process do not have time to be completely dispersed. One result is that bread from batch mixed dough is generally coarser and the grain is less uniform than it is in bread made from continuously mixed dough.

It has been shown experimentally that a small batch mixer operated hydrostatically (i.e., completely full of dough) will produce uniformly mixed dough, and the dough will produce bread having the characteristically fine, uniform grain of continuously mixed bread. This principle is used in some of the mechanically conditioned dough systems described elsewhere.

The continuous mixing systems discussed above were designed to produce commercial white American bread which is relatively high in specific volume due to efficient incorporation of gas in the dough and retention of the gas in the baking process. For proper gas retention, the gluten must be well developed, and this is best accomplished by the kneading action characteristic of the dough-developing type mixers used in the American equipment. In the European market, there is less demand for light, fluffy white bread and, consequently, less need for highly refined gluten development equipment. Continuous bread dough mixers of relatively simple design were developed in Europe to fit their particular needs. The Strahmann mixer is one of these. It incorporates premixer and mixer-developer in one unit. The shaft has a plurality of screw-type paddle-bladed units spaced axially along it. They rotate with the shaft to drive dough ingredients from the inlet port at one end to the discharge port at the other end. Near the discharge end of the mixer, the paddle blades are separated along the shaft by perforated baffle plates, mounted concentrically with the shaft but rotating in a direction opposite to the shaft rotation.

The effect given by this mixer is an intense shearing, cutting, mixing action. The baffle plates may be provided with any of a variety of shapes of perforations to give the desired action. It is not believed that the Strahmann mixer achieved any significant market penetration in Europe and it probably is not being used at all in the United States at this time.

Another continuous mixer developed in Europe, the Ivarsson machine, uses an ingeniously modified horizontal screw agitator to mix, knead, and propel the dough through the cylindrical casing. After discharge from the mixer, the dough is proofed and sent through conventional dough make-up equipment.

The ration development technicians of the U. S. Defense Department have, from time to time, tried to develop a portable continuous baking plant which can be used to feed troops in the field. When the author was Chief of the Cereal and General Products Branch of the Quartermaster Food and Container Institute, a continuous mixer was designed for use with a chemically leavened white bread mix. The mixer was light, compact, and simple so as to better fit it for field use. The dry mix metering feeder was volumetric, and had sufficient accuracy to keep the finished dough within acceptable limits. The premixer was a screw type machine which fed a Waukesha type metering pump. The pump forced the roughly mixed dough through the developer, a device which had many similarities to the developer used in commercial continuous bread plants and operated under 40 to 60 psig pressure. Dough was extruded from the discharge end through a rubber hose directly into pans.

Because there was no conventional dough make-up equipment in the military process, variations in the absorption rate were not as critical as they would have otherwise have been, so the volumetric dry feeder gave adequate accuracy. There were no provisions for cooling the developer chamber, and the dough came out much warmer than in commercial plant. This shortens the baking time, however, and was not a source of major difficulties in the absence of the usual dough make-up procedure.

MIXERS FOR BATTERS

General Considerations

Mixers discussed up to this point are particularly suitable for mixing yeast raised doughs made from relatively strong, high gluten flours. The efficient mixing of cake batters and the like requires an altogether different kind of agitator action. Less energy is required to mix batters, and their viscosity is much less than that of the usual dough. Batters do not require development, in the same sense that a bread dough must be developed, but they do require that air be enfolded in the fluid and then subdivided into small bubbles. Horizontal dough mixers do not function at all well in mixing fluids of this type. Pie doughs present a still different set of problems, needing to be kept cool and to be blended without any development action.

From the standpoint of their response to mixing action, cake batters can be divided into two types: (1) "Shortening" type cakes such as layer cakes, pound cakes, cup cakes, etc., and (2) Sponge type cakes including conventional sponges, chiffons, and angel food cakes. Generally, the same mixers can be used for both types, the different mixing actions being achieved by using specialized agitators and different agitator speeds. Injection of compressed air, which is especially helpful in mixing such products as angel food cake batter, meringues, and marshmallow, requires a hermetically sealed mixing chamber. Both batch mixers and continuous mixers can be used for nearly all types of batter.

Batch Type Batter Mixers

Vertical dough mixers, a class identified by the perpendicular orientation of the mixer shaft, are widely used in the food industry. They are relatively inexpensive, fairly versatile, and can be constructed in a wide range of sizes. A feature common to most vertical mixers is the use of removable bowls. Their other characteristics may be quite diverse—there may be one or more beater shafts, the beater shafts may move in a planetary pattern or remain stationary, and the designs of the agitators can be varied over a wide range. The types of vertical mixers of chief interest to bakers are the planetary mixers capable of preparing most batters and some doughs and often used for adjuncts such as icings, and the spindle mixers commonly found in cookie and cracker bakeries.

Vertical planetary mixers with removable bowls are used universally for the batch mixing of all types of cake batters. The agitator or beater action is described as "planetary" because the beater has two motions: it revolves on its own vertical axis at a relatively high speed, and its axis is moved around the inside of the bowl at a relatively slow speed. The latter, planetary action, is in the opposite direction of the agitator rotation. The combination of motions is very effective in stirring all parts of the bowl's contents and reducing dead spots. With a rubber-edged beater, the entire inside surface of the bowl is wiped progressively as the planetary cycle is made. A small area at the bottom of the bowl may remain out of contact with the beater blades, but material in this area is generally picked up as the batter increases in viscosity.

This general type of mixer is produced by several United States companies and by manufacturers in several foreign countries. It may be obtained in bowl sizes of 20, 40, 80, 120, 140, 160, and 340 quarts. Even smaller sizes are available for home kitchens, bakery laboratories, and the smallest retail bakeries. In wholesale bakeries, the 140 and 340 quart sizes are most often found. A typical example of a 340 qt size is shown in Figure 3.7.

On the larger units the bowls and mixing heads are raised and lowered automatically with an auxiliary motor. Bowl hoisting and dumping equipment is available for the largest mixers.

Some of these mixers have four geared speeds from low to high, while others have infinitely variable speeds. On a 340 qt mixer of this type, the beater speed is continuously variable from 45 rpm to 325 rpm. The planetary action is only 25% as fast as the beater rotation (4 to 1 precession ratio) in the standard single ratio machine. In the dual ratio machine, a selection can be made between 4:1 and 2:1, or, the planetary action can be stopped completely to create a vortex facilitating the addition of dry materials.

A convenient feature of most planetary mixers is the easy removal and replacement of the agitator element. A large selection of agitator types is available, making this machine the most versatile of all mixers in terms of versatility. The most commonly used mixer elements are probably the dough hook, the wirewhip, and the batter beater. The dough hook is a single complexly curved arm which provides the stretching, kneading action necessary for developing gluten with a minimum of tearing. The wirewhip is an assembly of wires, wide at the top and coming to a rounded point at the bottom, a design intended to give maximum air incorporation and bubble dividing action. Batter beaters are generally of cast aluminum, though the larger ones may be formed of bars attached to a frame. They have a somewhat shield-shaped outline; they may have either 2 or 4 wings shaped to fit the inside of the bowl.

Batter beaters used for shortening type cakes are of either the two-wing or the less common four-wing design. The outer edges of the wings are shaped to match the curvature of the bowl side-wall. They are heavy enough in construction to withstand high starting torque. Frequently these stainless steel or cast aluminum agitators have a white rubber insert along their edge, for wiping the inside of the bowl wall.

There are two principal methods for mixing shortening type batters. The older method requires slowly creaming the shortening, flour, and sugar together to thoroughly coat the flour and entrap some air before mixing in the eggs. Then the water is slowly cut in to obtain the

FIG. 3.7. A 340 QUART VERTICAL PLANETARY MIXER. RAISED POSITION SHOWING BATTER BEATER.

SOURCE: AMF

necessary "emulsion." If aerated plasticized shortenings with emulsifiers are being used, as is most often the case, the procedure can be simplified. Except in the case of white cakes containing no egg yolks, the whole mix can be put together in two simple stages. The first stage, containing all the ingredients except the water (in some instances, about a third of the water is added in the first stage) is mixed at low speed just long enough to achieve a consistency that will not splash out of the bowl. Then the speed is shifted to high and mixing is continued until all the lumps are dispersed and a smooth homogeneous batter is attained. Water is then added, generally through a meter, and mixing is resumed—first at low speed to prevent splashing and then at high speed to get the desired aeration.

Using these methods, the large 340 qt mixer will complete four batches of cake batter per hour, 400 to 550 lb per batch depending on specific volume reached. Most commercial cakes, even some nominally "white" cakes contain some whole eggs and can be mixed in this way. However, the true white cakes are frequently mixed in a different pattern, with the water being cut in more slowly.

Whipped type cakes (sponge, chiffon, and angel food) usually contain no plasticized shortening, though some varieties may contain butter or oil. A major requirement is to whip enough air into the batter so that a foam of specific gravity as low as 0.30 is produced. The beater used is a wire ship shaped to conform to the inside of the bowl; some models have center posts and others do not. The wires are 0.125 to 0.1875 inches in diameter and are set in a pattern designed to give the maximum turbulence.

Whipped type cake batters are batch mixed by first adding all the ingredients except the flour and whipping at high speed until a light foam is obtained. Flour is then charged as quickly as possible and "folded in" at slow beater speed, breaking the foam as little as possible. If added first, the flour would inhibit the foaming action, making it difficult or impossible to obtain the necessary specific volume. It is considered impractical to obtain an angel food cake of the best possible texture by batch mixing all of the ingredients in one stage. Continuous mixers with pressurized air accessories can satisfactorily whip angel food cake batters in one stage.

Bowls may be fitted with baffles to restrict the swirling motion of batters. This problem is greatest in the large wholesale type vertical mixers where very thin batters tend to swirl badly unless baffles are placed in the bowl to resist the rotation of the liquid. Standard tinned baffles are available for the largest mixers, along with smaller diameter whips which are required when the baffles are installed. Using a lower planetary ratio, say 2 to 1, is another option to prevent swirling, if the mixer is geared for this ratio. Bowl covers can be used to cut down on splashing of liquids or dispersion of dusty ingredients.

Horizontal cake mixers, with kneader blades for shortening type cakes and a squirrel cage whip for foam batters have been all but completely replaced by the more efficient and versatile vertical planetary mixers. Another batch mixer that gained some popularity for whipping cake batters, particularly angel cake batters, is the air pressure whisk. These mixers are comprised essentially of a vertical tank with pressure-tight lid, and three rows of rod-shaped beaters—two of them powered and one stationary. Mixing action is intense; the drive unit for a 200 qt unit is a 10 hp motor. In making angel food cake batter in this device, flour is charged to the mixer at the proper time from an auxiliary tank. Operating pressure is 5 to 22 psig. A batch of angel food cake batter can reportedly be mixed in 2.5 min, and a batch of pound cake batter in 1.5 min. Superior aeration and better uniformity are claimed. The batter is delivered under slight air pressure through a plastic hose from the mixer to the depositer.

Wilkinson (1987) points out there are still some types of cakes best made by batch mixing, in spite of the advances in continuous mixing technology. He enumerated the previously known problems with use of planetary mixers: (1) Scaling of ingredients was not consistently accurate, (2) Mix speeds could not be controlled accurately, (3) A stagnant film was always left on the sides of the bowl, requiring the operator to scrape down the bowl regularly, (4) Design of the mixing bowl left an unmixed or poorly mixed residue in the bottom, resulting in variable top-to-bottom quality, (5) Batch sizes were too small for large-scale production, (6) Sanitation and cleanup were constant problems, (7) Operator safety is a concern, and (8) The entire mixing process was, and in many bakeries still is, operator dependent.

Wilkinson described a new type of mixer based on traditional planetary mixer design but with features which give greater control over: (1) Variations in quality of the ingredients, (2) Accuracy in scaling the formula, (3) Time in the mix cycle, (4) Completeness of mixing the batch, (5) Temperature of the final mix, and (6) Attention and skill of the operator. This includes "complex planetary mixing" together with electronic control systems. Complex planetary mixing refers to (among other changes) the addition of a second tool in the planetary head to give versatility that can't be achieved in simple planetary mixers. For example, a wire whip for foaming can be combined with a cross-beater for development. A tool for constantly scraping the sides is added. Figure 3.8 illustrates the type of action occurring in double planetary mixers.

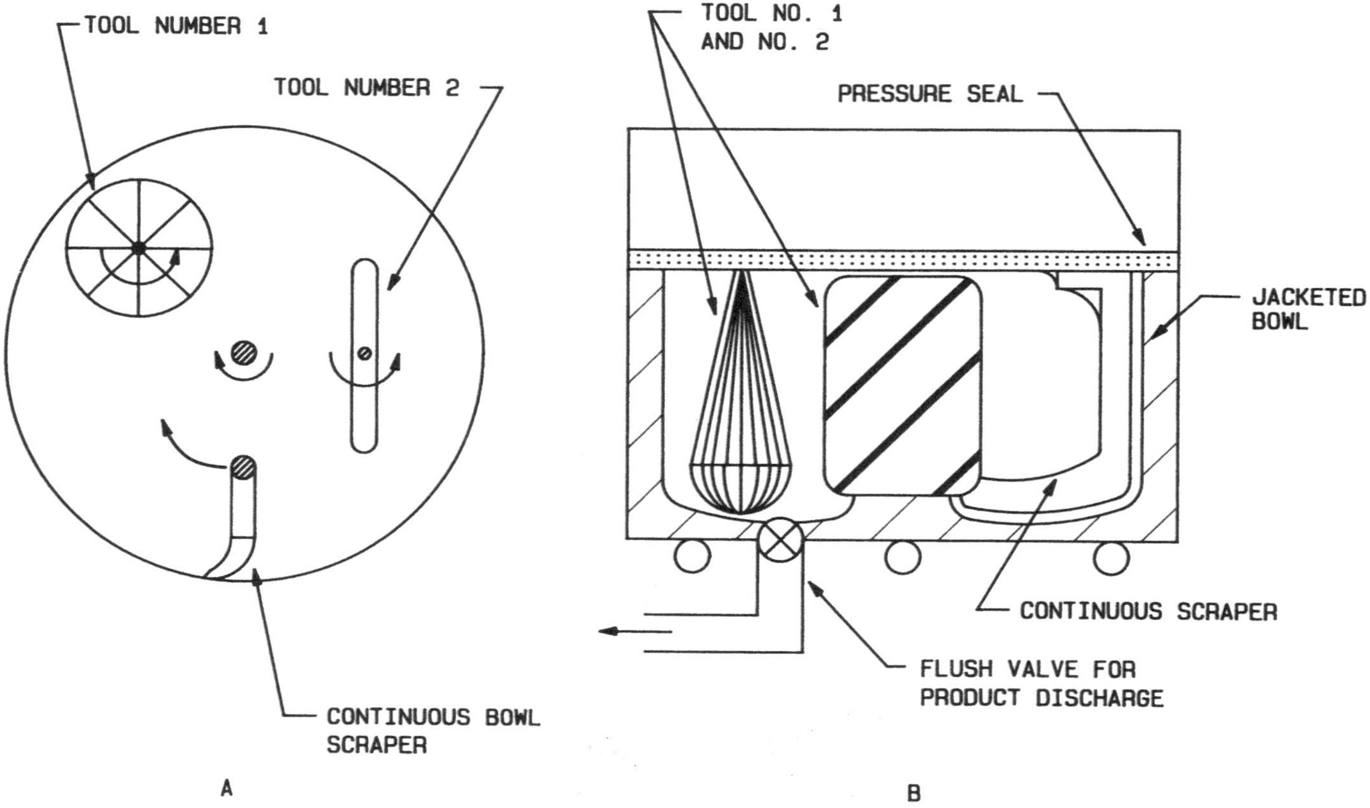

FIG. 3.8. MIXING PRINCIPLES OF A DOUBLE PLANETARY MIXER.

SOURCE: WESTERN BAKERY IMPORTS, INC.

Traditional open-bowl mixing depends on the incorporation of air at atmospheric pressure, so the time and speed of mixing and the type of agitator are the main factors determining the specific gravity of the finished batter. Much greater control can be obtained if the bowl is pressurized. This requires some fairly elaborate modifications but is being done routinely in many plants. The usual practice is to raise the pressure in the bowl to about two atmospheres, or, say 15 psig, during part of the mix cycle, for the time required to get the predetermined specific volume. Consideration must be given to the expansion which occurs when pressure is released. In these systems, the batter would normally be conducted through tubes from the bottom of the pressurized bowl to the depositer. As the batter emerges into an atmospheric pressure environment, whether at the depositer or elsewhere, the gas in the bubbles (which is at a higher pressure than the atmosphere) will cause the bubbles to expand very significantly. Some of the bubbles will coalesce or burst, and it is conceivable that specific volume could decrease because of these phenomena. These complex planetary agitator, sealed pressurized bowl, automatically fed systems are said to be applicable to sponge cakes, variety cakes of any type, cremes and filling, wire-cut cookie doughs, rotary molded cookie doughs, and light pastry doughs. Typical mix time is about one-third to one-half that required in a regular planetary mixer. From 5 to 8 batches per hour can be produced with most formulas.

Systems such as described in the preceding paragraph can be automated for the following operations: (1) Delivery of ingredients according to a programmed formula and sequence, (2) Mixing for a predetermined time and at a predetermined speed, (3) Adding pressurized air at a given time, and (4) Discharge to the depositer when the mixing cycle is completed.

Continuous Mixers for Batters

Semi-continuous mixing of cake batters was introduced in the U. S. in the mid 1940's. One of the first commercial continuous cake mixing systems was designed around a Votator scraped-surface mixer based on an ice cream freezer design. Another early system used a modified Oakes marshmallow mixer. In both systems, the mixer operated continuously, but was fed from a premix prepared batchwise in standard batch cake mixing equipment. Later, about 1950, a truly continuous cake mixing plant was designed and installed using the Oakes mixer. Since then, other manufacturers have introduced mixers of similar design.

In the Votator scraped surface equipment, the mixer consists of one or more horizontal, stainless steel cylinders containing axially mounted beater bars which can be arranged to scrape the inner wall of the cylinder. An outer cylinder is placed concentrically with the mixer tube so that a heat transfer fluid can be circulated through the annular space to heat or cool the batter as it is being mixed. Mixing action is intense, and this machine also has the advantage of being operable under relatively high pressures—a feature which is beneficial in the production of highly aerated batters.

The Oakes mixer has a stainless steel mixing chamber, consisting of front and back stator halves and a rotor which is mounted between them. Concentric teeth on both faces of the rotor mesh with teeth on the stators. Premix batter is pumped in at the center of the back stator, and is sheared repeatedly as it is forced out around the rotor and from back to the center of the front stator, where it exits. Diameter of the rotor in the largest machine is 14 inches. Rotor speed is continuously variable over a fairly wide range. Only a small amount of batter (3 qt in the largest model) is present in the mixing chamber at any one time so that dwell time is very short and power requirements are low.

Air, or other gas, may be injected into the mixer with the other ingredients and distributed uniformly throughout the mass. The stators are jacketed, allowing coolants to be circulated for maintenance of a preselected temperature. Capacity of the available sizes, in lb per hr, are 1,000, 2,000, 4,000, and 6,000. There is also a laboratory model with a nominal production rate of 30 to 200 lb per hr. Production rates are, of course dependent on the composition of the batter. High rates are achieved with cheese cake, layer cake, and vanilla wafer batters, and with pumpkin and similar pie fillings. The lowest rates are for meringues, marshmallow icings, angel food batters.

The AMF mixer is similar in external appearance to the Oakes, but the mixing head is quite different. Instead of splitting axially for dismantling, the stator splits radially. The rotor has several rows of teeth on its periphery. They mesh with rows of teeth on the inside of the stator wall. Rotor speed is continuously variable, between limits. These mixers are usually equipped with batter metering pump, air meter, and speed controls.

The intense shear applied to batter in mixing heads of this design is sufficient to emulsify air and shortening in a complete formula blend, whereas conventional batch mixers require that some of the water be withheld until the shortening has been creamed with the dry ingredients and eggs. Their principal advantage, however, lies in the continuous mixer's unique ability to produce highly aerated batters—angel food, sponge, etc.—from a rough premix containing all of the ingredients, including flour. Air or inert gas is metered into the mixing head at a calculated rate, along with a proportionate amount of premix. If it is desired to produce a foamed batter with a final specific gravity of 0.25, the proportion would be approximately three standard volumes of air to one of batter.

When preparing angel food batter, the mixing head operates at up to 90 psig, so that the air in the chamber occupies a volume about one-sixth that at atmospheric pressure. As far as the mixing head is concerned, the desired mixture is 1 part premix to one-third part air, and this mixture is easily accomplished, even when flour is present. The dense foam so formed is delivered to the cake depositer through plastic or rubber tubing of a length chosen to allow a gradual reduction of pressure until the batter is at about atmospheric pressure as it reaches the depositer. As the pressure slowly decreases, air bubbles in the foam grow in volume and the bubble walls get thinner. The end result is a very tender cake of unusually large volume. If a throttling valve is used to relieve pressure in the mixing head, the foam will break down. Sometimes, an adjustable sanitary stainless steel device designed to present a long convoluted path to the exiting foam is used to gradually reduce the pressure. This adds the advantages of compactness and adjustability for different rates and batters.

Continuous batter mixers can be incorporated into completely continuous systems in cases where production schedules warrant (i.e., in plants making a few cake varieties in long unbroken runs). In these plants, liquid ingredients (including plasticized shortening) are premixed in vertical tanks equipped with high speed propeller agitators. The resultant emulsion is metered into an open vortex-type premixer to which the flour is fed from a gravimetric feeder. The premixer forms a rough premix which is metered to the continuous mixer along with a stream of air. Operation is automatic and generally requires only one operator.

In general, cake batter is much easier than bread dough to handle as a fluid through pumps, pipes, valves, and meters. Engineering of a continuous production line is considerably simpler for a cake plant than for a bread bakery. The economics of cake production is such that expenditure of a large amount of money for continuous mixing equipment may not be warranted, particularly for the small to medium size plant handling a large number of different formulas.

An example of a continuous mixing plant for cake batter was described by Shannon (1971). A slurry is prepared in a planetary mixer equipped with a valved mixing bowl of 340 qt capacity. When it is completely mixed, the valve is opened so the slurry can be pumped into a holding tank without moving the mixer bowl. From the holding tank, the slurry is fed to the continuous mixer at the rate desired. The continuous mixer (in this case, a rotor-stator unit) is provided with a valve which admits compressed air at the rate necessary to achieve a predetermined specific volume. A back pressure valve is adjusted to retain the batter in the continuous mixer chamber until it has been fully processed, as measured by specific volume or some other property. Typically, specific gravity and temperature of the slurry, air pressure, pump speed, rotor speed, flowrate setting, back pressure, batter temperature, and batter specific volume would be read or measured (and recorded) at frequent intervals. From the continuous mixer, the finished batter is fed to the depositor.

Particulate ingredients such as nut pieces, chocolate chips, and raisins cause difficulties in continuous mix systems. The usual solution is to by-pass the continuous mixer and feed particles into the tube leading to the depositor at a rate consistent with the formula.

SPECIALIZED MIXERS

Pie Dough Mixers

Mixing of pie doughs introduces a special problem in that a very slight degree of overmixing results in a tough pie crust. Furthermore, the incorporation of air is distinctly inadvisable. For these doughs, special mixers have been developed to simulate the pushing, folding, and pressing action of human hands. Generally two agitator arms are used and they perform a fairly complex maneuver in a shallow bowl which is constantly rotated. The object is merely to make the dough into a cohesive and fairly uniform mass which can be handled effectively in subsequent forming steps. Kneading and rolling to develop gluten are carefully avoided. This special mixer has been found to give the desired results in commercial pie dough production. The Artofex mixer was one of the earliest examples of this type, but many other companies are offering very similar designs, with various modifications to the mixing arms, bowl configuration, motion, etc..

Mixers for Cookie Doughs

Cookie and cracker doughs represent a very wide range of properties. Some of these doughs can be mixed more or less well in vertical planetary mixers, horizontal dough mixers, or continuous mixers of various sorts. There are also special mixers which have been designed to mix certain types of cookie or cracker dough more efficiently. The spindle mixer shown in Figure 3.9 is one of these. This machine is particularly suitable for cracker sponges, but can be used for some heavy cookie doughs. Mixers for cookie and cracker doughs will be described in more detail in a later chapter devoted to specialized plants for such products.

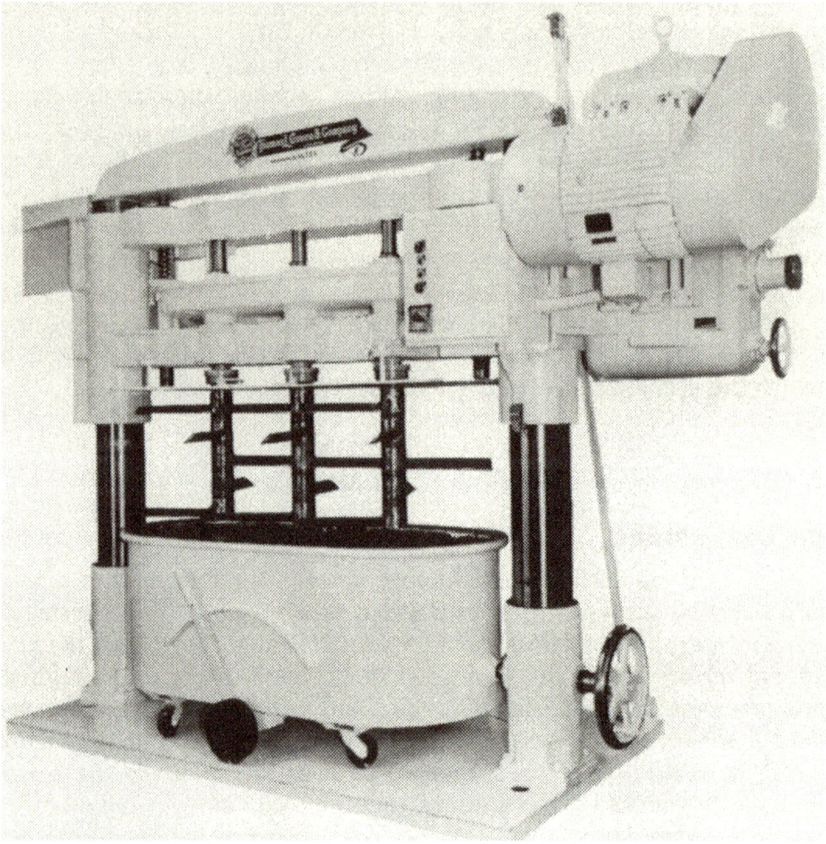

FIG. 3.9. THREE SPINDLE MIXER. VARIABLE SPEED. FRONT VIEW SHOWING PADDLES RAISED.

SOURCE: THOMAS L. GREEN CO.

Other Mixers

There is often a requirement in the bakery for distributing and suspending poorly dispersible ingredients in liquids. Various methods exist for doing this. One piece of equipment which has had considerable success in solving dispersibility problems is the "Likwifier" illustrated in Figure 3.10. The mixing element is a disc shaped rotor with vortex-creating cavities. This impeller is said to give intense agitation with minimal foaming. An interesting application is the reclamation of pan bread "cripples." A quantity of these misshapen or broken loaves are put in the mixing chamber with a double quantity of water and made

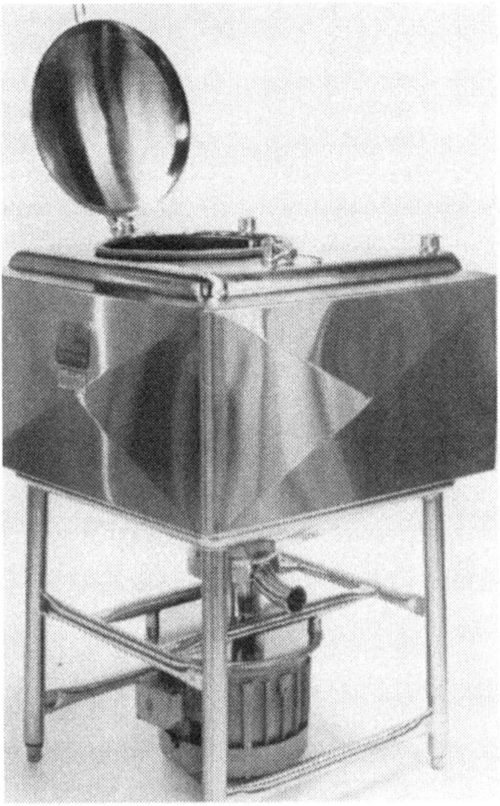

FIG. 3.10. A MIXER FOR HIGH SPEED BLENDING AND DISINTEGRATING.
SOURCE: BREDDO CORP.

into a slurry. It is recommended that the slurry be added to bread doughs at the rate of 9% of the flour weight. It is said that no quality deterioration results.

WHAT ABOUT THE FUTURE?

Mechanization of the scaling, feeding, and mixing operations in wholesale bakeries has reduced the cost of direct labor in that part of the factory until it has become a minor part of the products' total value. Although the equipment required for achieving this efficiency is expensive, its productivity and useful life are so high that the annual costs are almost negligible in cents per pound of bread or cake.

Innovations in mixing equipment design that started perhaps 30 or

40 years ago have continued, but in recent years equipment manufacturers have been mostly concerned with relatively minor refinements of earlier inventions. The one area where substantial advances can be foreseen is in computer assisted manufacture, where automated testing of materials in process can provide the basis for adjustment of manufacturing conditions to correct defects.

There is an expanding market for more efficient, more versatile, better controlled mixing equipment for small shops. At one time, the local retail baker seemed headed for extinction and it made no sense to spend R&D funds on that vanishing business, but recent marketing developments have resulted in an extreme proliferation of in-store bakeries, doughnut shops, muffin shops, cookie shops, biscuit bakeries in fast food restaurants, and pizzerias. All of these operations require small-scale equipment. It is true some of these outlets use pre-mixed doughs or other partially prepared products with some degree of success, but their current practices might be re-evaluated if there were available small scale equipment tailored for their needs and priced so as to be affordable.

Except for the very smallest shops, ready-to-bake (or ready-to-shape) doughs sold in refrigerated or frozen form are not really economical substitutes for on-premises preparation starting with dry pre-mixes. Delivery and storage of frozen and refrigerated doughs also create problems. These disadvantages lead to the conclusion that there are opportunities for selling specially designed mixing equipment to these shops. I do not think, however, that equipment based on cosmetic alterations of original designs of 80 or 100 years ago or on new conformations of mixing paddles will meet the needs of the new breed of bakers. What is needed is a series of mixers especially scaled to the level of production found in certain types of outlets and designed for one type of product—bread or bagels or cookies or biscuits or pizzas. Versatility would not be a prime selling point. The equipment should also be safe enough and user-friendly enough so it can be operated by unskilled help.

BIBLIOGRAPHY

BARON, D. 1983. Dough temperature control systems. Proc. Am. Soc. Bakery Engineers *1983*, 100-105.

BONAVIA, W. 1967. Techniques, operation, and engineering aspects of continuous production of cake batter. Proc. Am. Soc. Bakery Engineers *1967*, 301-307.

BOOTH, M. D. 1987. Automated high volume dough mixing. Proc. Am. Soc. Bakery Engineers *1987*, 132-144.

BROADDUS, M. R., JR. 1978. Soft roll equipment. Proc. Am. Soc. Bakery Engineers *1978*, 122-132.

CAMPBELL, S. P. 1979. Mechanical dough conditioning—new systems. Proc. Am. Soc. Bakery Engineers *1979*, 175-185.

FISH, A. R. 1982. High speed dough development. Proc. Am. Soc. Bakery Engineers *1982*, 130-142.

MATZ, S. A. 1960. Bakery Technology and Engineering. AVI Publishing Co., Westport, CT

OLDSHUE, J. Y. 1983. Fluid Mixing Technology. McGraw-Hill Publications, Inc., New York, NY

PIERCE, W. L. 1974. Pumping of bread and bun doughs. Proc. Am. Soc. Bakery Engineers *1974*, 92-99.

SHANNON, T. A. 1971. A look at cake production by the continuous mixing process—equipment, procedures, and formulas. Proc. Am. Soc. Bakery Engineers *1971*, 138-144.

VALENTYNE, P. H. 1959. Practical aspects of heat balance in dough mixing. Bakers Digest *33*, No. 1, 40-44.

WILKINSON, G. 1987. Cake mixing technology. Proc. Am. Soc. Bakery Engineers *1987*, 100-106.

CHAPTER FOUR

DIVIDING, ROUNDING, AND SHEETING EQUIPMENT

INTRODUCTION

In this chapter, we will consider the equipment applied to doughs after they leave the mixer and before they begin the forming process which defines the final product. Emphasis will be on extensible doughs such as those used for bread, plain rolls, sweet rolls, Danish pastry, puff pastry, and a few chemically leavened products such as saltine crackers and soda bread. Many of the devices have been used primarily in the conventional processing method for bread and rolls, and their basic operating principles are several decades old. This equipment performs functions which precede the final forming operation which is the major determinant of the shape of the final piece as it is seen by the consumer. The final shaping of the dough is done by devices which will be described in subsequent chapters.

DIVIDERS

Purpose

The function of the divider is to separate a dough mass of variable or indeterminate size into many smaller pieces of identical weight. In most, but not all, cases, a dough piece emerging from the divider will be carried through all subsequent processing steps as a single unit until it finally comes out of the oven as one finished product. Of course, the weight of the finished product will not be the same as the weight of the raw dough because volatile materials are lost through evolution of gas formed in fermentation reactions and evaporation of liquids in the oven and elsewhere. This weight loss is offset to a very minor extent by the accumulation of dusting flour picked up by the dough pieces as they travel through the make-up equipment.

Separating dough into pieces of uniform weight is an operation which presents unusually difficult problems. Because of the elastic, sticky, cohesive nature of dough, it is unsuitable for metering by gravimetric procedures, and the constantly varying density that results from fermentation interferes with accurate volumetric measurements. Of the two approaches, volumetric measuring has proven to be the the more practical and economical. Campbell (1981) describes the four commonly used techniques of dough dividing as hand scaling, press or die methods, ram/shear/block dividers, and vacuum assisted rotating drum and piston type dividers. An additional method, extrusion dividing, has been used in continuous systems and in recently developed units for batch systems.

How Dividers Work

At the present time, all commercial dough dividers use some form of volumetric scaling. The principle is not much different from that used by the master baker working at the bench when he rolls a dough into a long cylinder of uniform diameter and then cuts pieces of uniform length to get a dough piece of the desired weight. Experienced bakers can achieve a remarkable degree of accuracy with this type of scaling. The same volumetric principle is relied upon when the baker presses a dough, by a rolling pin or other device, into a sheet of uniform thickness (perhaps using rails to hold the rolling pin at a predetermined distance above the table) and then cuts the sheet with a device having cavities of known (or at least uniform) dimensions. Using a doughnut cutter on sheets of dough is an example of the latter process.

The simplest type of automatic divider is the bench-top roll divider which separates a large dough piece into, e.g., 18 or 36 portions of more or less equal weight, by forcing through the dough a cutting head actuated by a spring and lever assembly. In brief, the process is to place a weighed amount of dough, more or less flattened out, onto the divider platen, lower a slotted pressure plate which forces the dough lump into a uniform thickness, and then release the knives which by spring action cut through the dough sheet. Advanced models are powered by a hydraulic system receiving its energy from a motorized pump. These bun dividers have the advantage that they quickly process all the dough in one small batch so that variations in density due to fermentation changes are not much of a problem. If large batches are mixed, so that they have to be cut into pieces of the size suitable for dividing, this advantage no longer exists.

The pockets are not adjustable in size, which considerably reduces the machine's versatility, although some modification of scaling weight can be achieved by varying the weight of the chunk of dough placed in the divider. Conversely, it is necessary to keep the batch weights within narrow limits, because the piece size will vary proportionately to this factor. In some models, the cutter assembly can be removed and replaced with another one having a different number of cavities. Unless a full mixer batch is always placed in the divider, the dough chunk itself has to be cut to an exact weight if accurate piece sizes are to be obtained. Although these dividers have been successfully used by small retail bakeries for several decades, they are of no importance in large commercial bakeries.

Some models of the bun dividers described in the preceding paragraph have a rounding mechanism built into the equipment. After the pieces are cut, the molding table is driven by eccentric wheels which give it a combined spiral and circular motion. Each dough piece remains in the chamber formed by the cutter blades. As the table rotates, the upper or dividing disk is simultaneously lifted to the limit setting so that enough space is allowed for the height increase as the dough pieces become rounded.

In the case of traditional large-scale methods of bread manufacture, a mixer load of dough is placed in a trough which is transported to a fermentation room. After bulk fermentation is completed, the filled trough is moved on its casters, usually by man-power, to the divider area or to the floor above the divider. If a trough elevator is present, the trough is rolled into its brackets, then the elevator motor is started, first raising the trough above the divider then turning it through an angle of more than 90° to dump the dough into a capacious hopper above the divider. When trough elevators are not available, the dough may be cut into chunks of about 50 lb weight which are manually transferred to the divider hopper. If the troughs are on the floor above the divider room, a chute in the floor carries the dough to the hopper. Doughs can also be pumped from the trough or from a post-mixer developer to the hopper.

Bread dough dividers currently available from equipment suppliers are based on one of the following systems: (1) Dough is forced into a pocket of known dimensions and the excess dough is sheared off, or (2) A cylinder of dough is extruded from an orifice and pieces of uniform length are cut off. The first method is the basis for the design of nearly all the bread dough dividers in use today, while the extrusion method is used in continuous dough processing systems and is also the principle of some recently developed machines.

Figure 4.1 is a schematic diagram one of the most common types of dividers. In these devices, the influence of atmospheric pressure and the weight of the overlying dough mass force dough from a hopper into a compression chamber. At the start of the cycle, a knife moves horizontally to cut off a piece of dough near the hopper bottom. Next the ram or piston moves forward, pressing the severed dough piece into one or more chambers contained in a rotatable cylinder. At the end of the ram stroke, the cylinder turns, cutting off the exess dough. Each cavity will then be filled with a predetermined volume of dough which should weigh within the limits desired for the piece. Finally, the discharge lever ejects the measured dough piece on to a conveyor leading to the

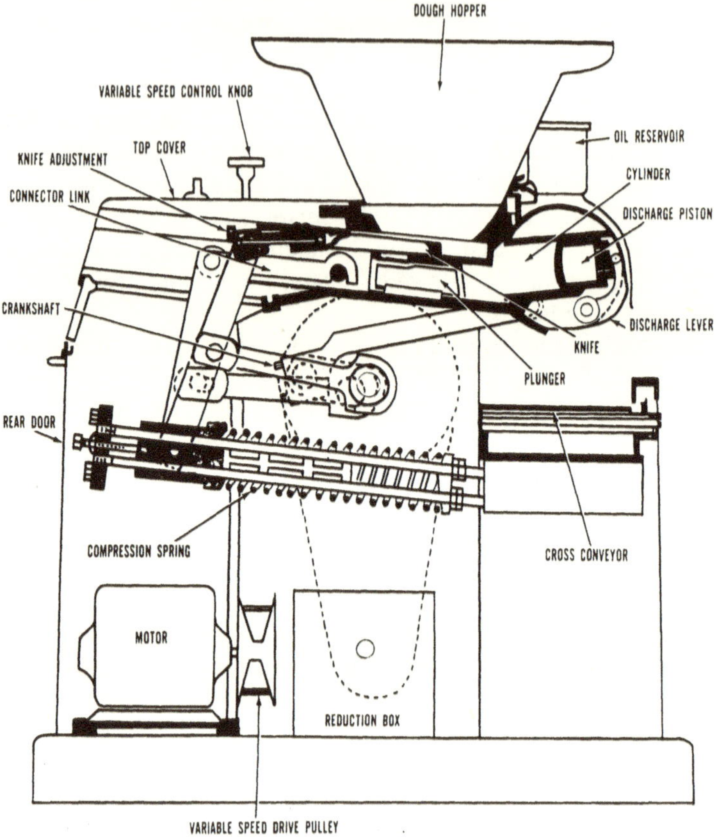

FIG. 4.1. THE WORKING PARTS OF A COMMON TYPE OF DOUGH DIVIDER.

rounder. In the return cycle, the emptied cylinder is turned back to its original position so that the cavities again face the compression chamber. The knife and compression piston withdraw, allowing more dough to be drawn into the cavity by gravity and suction. Then, the cycle repeats.

Commercial models are available which have from 2 to 8 pockets in the cylinder and operate at speeds up to 25 strokes per minute. The scaling range is from 6 to 36 oz, and motors up to 7.5 hp are used. An eight-pocket heavy duty divider is shown in Figure 4.2. This machine will scale dough pieces at a speed of 2,400 to 9,600 loaves per hour. It is 5 ft 3.5 in high, 5 ft 6.5 in long, and 3 ft 5 in wide. A special feature of this model is the variable speed drive on the discharge conveyor.

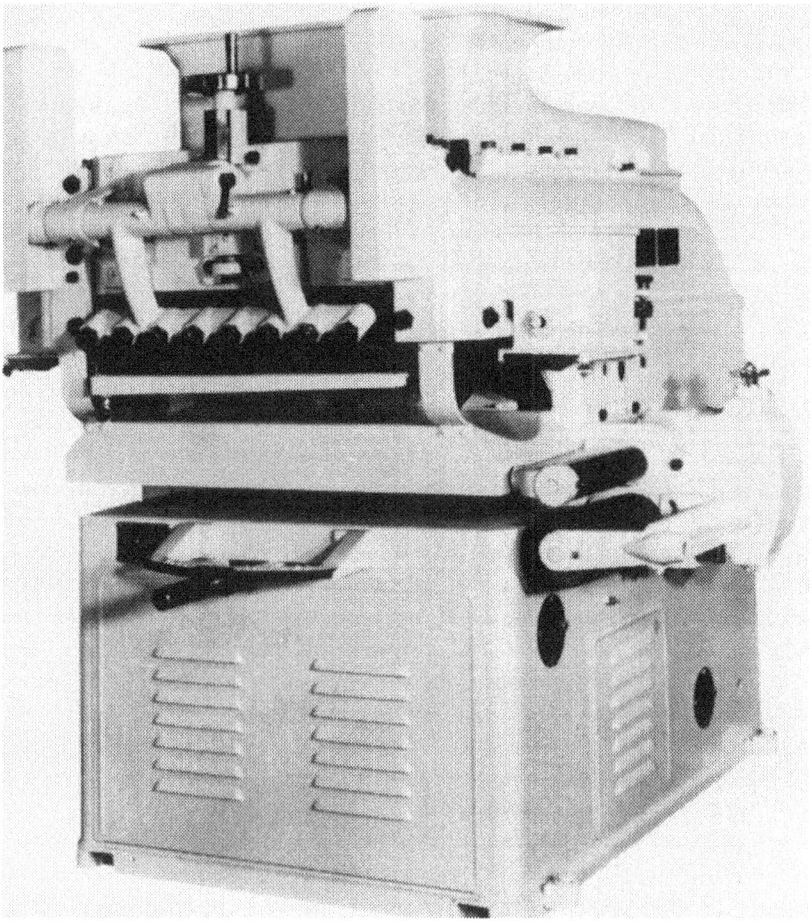

FIG. 4.2. EIGHT POCKET DOUGH DIVIDER.
SOURCE: APV (BAKER PERKINS)

Some dividers substitute a reciprocating division box instead of a rotating cylinder to measure and cut off the dough pieces. In these machines, the box containing the pockets is forced downward after they are filled, shearing off the excess dough as a result of the movement of the pockets past the chamber edge.

The dough pieces leaving the divider tend to be quite irregular in shape and to have damp, tacky surfaces. This condition is very conducive to the sticking together of pieces which contact each other. The formation of doubles in this way causes severe problems at the rounder and must be avoided by careful positioning of the discharge conveyor and by adjusting its speed so that the pieces never contact each other. A warm air current is sometimes provided on the conveyor leading from the divider to the rounder in order to partially dry the cut surfaces. Sometimes the air is applied at the rounder itself to allow reduced flour usage (or no dusting) at this point. The suggested application is 1500 cfm of 122°F air at 80 ft per min. In one system, the blower is located on top of the rounder cone so that the airstream covers all the working surface except for the discharge chute. The main airstream is directed toward the higher rounder tracks. Apparently, a filter is not considered necessary (Anon. 1985).

Controlling and Adjusting the Divider

The volume of the pocket which establishes the size of the dough piece is adjustable to permit varying the weight of the finished loaf and to allow the operator to compensate for changes in dough density. Volume adjustment is made by trial and error changes of the piston depth in each individual pocket. Pistons are adjusted by different mechanisms in different models. Lock nuts or shims are the usual means. Any change in the density of the dough will be reflected in an alteration of the weight of pieces formed at a specific piston setting. Therefore, it is essential to set the machine using a sample having the density of the dough which will be processed during a normal run. Allowing the dough to stand around for a considerable period while making adjustments is a sure way to create scaling errors.

A variable speed drive is provided on all dividers for the adjustment of the scaling cycle. Changing the speed is simply a matter of adjusting a wheel or lever controlling the variable speed drive. The speed of the divider must be coordinated with the speed of other equipment in the forming machine complexus. However, top speed of the divider is limited as frequently by dough characteristics as by machine limitations.

Dough dividers have a special lubrication system for dough contacting areas which provides an important variable in machine adjustment. Oil used in this system has two functions: It lubricates the mechanical parts of the divider which come in contact with the dough, and it forms a seal between the dough box and the divider head. Lubrication systems are either pressure-fed or gravity-fed. Adjustment of the rate of application of either type is simple. The correct setting is that which uses the least amount of oil to secure proper functioning of the divider. Divider oil should be as colorless, odorless, and tasteless as possible. It is usually a special grade of mineral oil, since it has been found that vegetable oils can leave a gummy film which may cause excessive friction or even bind the close tolerance moving parts of the machine. Specifications commonly call for a technical white grade mineral oil with a viscosity of 85 to 120 at 212°F and a flash point of 662°F. Machine bearings and other moving parts which do not contact the dough are lubricated with the usual petroleum oils and greases.

Since divider oil is taken up by the dough and appears in small quantities in the finished loaves, it falls under certain provisions of Federal regulations governing food additives. White mineral oil has been cleared for use as a release agent and lubricant for bakery products at a level not to exceed 0.15%. It must meet the following requirements: (1) It is a mixture of liquid hydrocarbons, essentially paraffinic and naphthenic in nature, obtained from petroleum; (2) It meets the test requirements of USP XX for readily carbonizable substances; (3) It meets the test requirements of USP XX for sulfur compounds; (4) It meets the specifications prescribed in Vol. 45, p. 66 of the Journal of the Association of Official Agricultural Chemists, after correction of the ultraviolet absorption for an absorbance due to added antioxidants; and (5) It may contain any antioxidant permitted in foods by regulations issued in accordance with section 409 of the Act, in an amount not greater than that required to produce its intended effect. The limitation of 0.15% applies to the total of all petroleum hydrocarbons used in the food.

The ram and the knife control the amount of dough that enters the dough box, and they are adjustable. If an insufficient amount of dough is taken in, the weights of the pieces will usually be erratic, while an excessive amount of dough will receive unnecessary punishment, possibly causing poor performance of the dough in later stages of processing. The former condition is frequently accompanied by a knocking sound in the divider as the partially empty pockets allow the gas to be compressed in them. An up and down movement of the dough in the hopper is considered a symptom of poor adjustment, or of dough which is not pliable enough. The adjustments which will produce the best results for any particular dough must be determined empirically.

Maintenance

Maintaining the proper clearances between the knife and the ram and between the dough box and the divider head is very important in assuring proper operation of the machine. Excessive clearance will cause the machine to overheat with consequent bad effects on the dough and extra demands on the motor. Manufacturers will provide recommendations for maintaining the proper clearance between these parts.

Cleaning the divider is a rather complicated job. It must be performed whenever the machine is shut down for more than an hour in order to prevent caking of the retained dough. Proper cleaning requires removal of the ram, knife, dough hopper, and pistons.

Following is a recommended sequence for cleaning one type of machine; the procedure can be adapted for other types.

(1) Stop the divider with the knife and ram all the way back.

(2) Remove the housing covering the knife connecting links and release the knife. Draw the knife out of its guide slots and, if possible, remove it from the machine.

(3) Disengage the ram from its connecting shaft, and draw it out of the dough chamber.

(4) Remove the nut which holds the plunger in its cylinder and then remove the plunger. Since pistons are not usually interchangeable, they should be plainly marked as they are removed, unless the manufacturer has provided them with an identification symbol.

(5) Remove excess dough from all parts with a hardwood or plastic scraper, then finish cleaning with water and soda or detergent. After rinsing and drying, cover parts with a thin film of divider oil.

(6) Place the oil catch pan under the divider head to protect the conveyor belt. Vacuum or blow out all flour from the divider proper, the conveyor belts, motor, drives, and switch boxes.

(7) Clean the inside and the outside of the divider housing.

(8) Replace the plungers, ram, and knife.

So far as sanitary problems are concerned, divider oil will not support insect, rodent, or microbial life. Dusting flour is subject to the usual infestations if it is not carefully monitored. Dirty or hardened dough pieces, if allowed to accumulate, can end up in the finished loaves as readily detectable contaminants.

Serious accidents can be caused if the hazardous nature of the divider is not fully recognized. Consult the safety code features of the American Standard Safety Code for Bakery Equipment for recommendations.

Some Problem Areas

The static pressure of the dough in the hopper has an effect on the weight of the pieces emerging from the divider. It is important to keep this pressure within a narrow range by regulating the rate at which dough is fed to the hopper. A constant level can be maintained by using an intermediate hopper fitted with an exit orifice which is adjusted by signals emitted from a sensor in the divider hopper.

It is important that the dough chute leading to the dough hopper have a proper pitch so the dough will exert an even pressure at all times. The chute should be inclined at an angle of at least 45°, if possible, and under no circumstances should the angle be less than 30°.

The surface finish on the dough chute and divider dough hopper is quite important and should be about a 125 disc grind finish. If there are problems in getting proper dough feed to the the compression chamber, they can sometimes be solved by applying Teflon to the high friction surface.

The amount of flour dusted on the conveyor belt should be kept at the minimum needed to promote adequate performance of the divider.

Breaking of the dough between the divider chamber and the dough hopper is an indication of improperly conditioned dough. Excessively bucky doughs may not fill the divider pockets completely, with resultant erratic scaling. Wide variations in piece weight may also be observed if the dough is full of large gas bubbles. Some dividers have compression pistons which have been equipped with rods projecting from their faces so that large gas pockets in the dough will be punctured and collapse.

Interaction of Doughs and Dividers

Anything that affects the dough density will change the weight of the scaled pieces. Since the dough continues to ferment in the hopper, with the gas produced by the yeast contributing to a lower density, a continuous slight decrease in scale weight can be expected during the processing of a batch of dough. Then, a rise in weight will be observed as a new batch starts to flow into the compression chamber. If the batches are small enough, and the divider operation is rapid enough, the changes in piece weight will probably be within limits that can be tolerated. When a divider with dough in the hopper is shut down for even a few minutes, a considerable error in weight must be expected upon re-starting. Frequent tests of the weight of dough pieces from each pocket should be made by the operator using an "over-and-under" scale placed adjacent to the conveyor or by automatic check weighers.

Dividers have a pronounced effect on dough properties. The compression and cutting result in a considerable loss of gas, not only initially, but continuing from the cut and torn surfaces of the dough piece as it is transported to the rounder. Furthermore, the compression and cutting actions disorient the gluten fibrils, changing the dough's rheology. A temperature increase is observed as the dough passes through the divider, and this affects the fermentation speed, among other things. The temperature change is due to contact with hot divider surfaces and to the mechanical work performed on the dough. The dough also picks up flour and divider oil during its passage through the machine; if adjustments are properly made in the rate of application of these materials, the amounts transferred to the dough should not be great enough to cause a significant change in its properties.

Improving Weight Control

The inherent accuracy with which each individual divider meters dough is determined by (among other factors) the dough pressure within the device, the precision with which the critical elements of the dough measuring parts were machined, the amount of air leakage which occurs, the piece size, and the capacity of the equipment. Accuracy declines with normal wear. In addition, there are at least four major processing factors which affect accuracy. If the dough's absorption varies, its stiffness and specific gravity will also vary. Since the divider measures volume, a change in specific gravity will lead to a change in weight of the dough piece. Stiffer doughs make it more difficult to create the desired amount of suction in the divider, and this causes changes in flow of dough into the measuring chamber. The time required to process each batch of dough must be kept as short as possible, because the specific gravity of the dough constantly decreases due to the generation of gas by fermentative processes. Although the divider is designed to expel some of the gas as it meters the dough, this process is not exact, and the larger the amount of gas in the dough, the less exact it is.

Improved weight control can be achieved by degassing the dough immediately before it enters the divider. This allows the divider to operate on dough that is always of about the same density so that one source of error is practically eliminated. The degassers are usually some type of pump or conveyor which compresses and works the dough, forcing out most of the gas. Post-mixer dough conditioning systems, as described in the mixer chapter, have similar effects but are also intended to add to the development of the dough (Campbell 1979). Strictly speaking, conditioning the dough is a totally separate phenomenon from achieving uniform density at the divider even though both functions can be performed by the same piece of equipment.

It is necessary to check on the accuracy of the divider by occasionally weighing the dough pieces. Of course, this can be done manually by using an over-and-under scale, but there are more efficient ways of handling the problem. Nearly all large scale operations now use automatic check weighers. These devices can be installed either immediately following the divider or after the rounder. The advantage of placing the checkweigher after the divider is that there is quicker feedback to the controller which adjusts the piece size. A disadvantage is that the pieces at this location are longer than rounded dough pieces, and this may affect registration on the weighing belt as detected by the photocell. To correct the latter problem, the distance between dough pieces may have to be increased either by speeding up the conveyor or by putting fewer pieces on the conveyor. Either alternative has its disadvantages. Weight loss in the rounder, where pieces tend to lose some small bits of dough, will be taken into account when the checkweigher is placed after this unit.

When automatic checkweighers are being used, the dough ball is weighed on a short conveyor belt which rests on a load cell or other mass sensing mechanism. The infeed conveyor, which has no direct connection with the weigh conveyor, carries the dough piece past a photoelectric sensor immediately before it reaches the checkweigher belt. At this precise moment, the checkweigher is tared so that any contamination (dust, small pieces of dough, etc.) on the weighing belt at that time is taken into consideration. The dough piece continues its progress, passing on to the weigher belt. When it arrives at the weighing zone, just over the load cell, another photoelectric sensor directs the mass sensitive device to take a final reading. If the dough piece is within the preselected range, it is transferred to a conveyor leading to the overhead proofer. Out-of-standard pieces are diverted to a reject conveyor from which they may be returned to the divider or used in other ways. The weights are recorded in a computer memory and averages periodically compared with the target weight. The operator receives a signal when the average is above or below the unsatisfactory limit. In the most sophisticated systems, a feedback loop allows the signal to adjust the divider so as to correct the weight (Benier 1983).

Recent Advances

During the past five or ten years, there have been some innovations in divider design which appear to have improved several of the less desirable features of these devices. The following discussion was taken in part from Campbell's (1983) paper on this subject.

Although the common ram/shear/block dividers are reliable and sturdy, they have several well-recognized shortcomings including scaling variance between multiple pistons, limited speed range, metal to metal contact in the product zone, incorporation of divider oil into the product, and a limited ability to compensate adequately for changes in dough density. Because of constant fluctuations in product density and the tendency to form doubles because of the multi-pocket design, an operator must continually monitor the divider and rounder. One improvement has been the replacement of the mechanical drive and linkage with a hydraulic power unit, reducing the number of wearing parts and making the divider more compact, less expensive, and easier to repair. The hydraulic drive allows the ram to make a double stroke, providing improved degassing of the dough in the hopper and in the metering pocket. Another manufacturer is producing a divider for bread doughs which does not use piercing rods to release gas from the dough. Instead, this divider utilizes a combination of multiple stages of dough compression, with a variable compensator to alter the amount of ram pressure depending upon dough characteristics. There are 16 gas relief valves in the division box to expedite gas escape. The manufacturer has also opted for very heavy construction and close tolerances, and included an arrangement to make removal of the ram and shear mechanism a one-man task so as to facilitate cleaning.

Campbell also describes a novel single-pocket high speed divider which operates on principles entirely different from those previously discussed. In this new system, dough is automatically pumped to a stainless steel hopper having at its bottom two interfering screws made from a copolymer. A vacuum degassing system assists in loading the dough uniformly into the auger mechanism. By using electronics to control functions which were previously controlled mechanically, a degassed dough of consistent density is fed by a metering device to an orifice where a knife severs a portion. In other words, the basic principle is the familiar combination of extrusion at a fixed rate and cut-off at fixed intervals which is so widely used in the plastics and snack food industries (see Fig. 4.3). The scaling accuracy is said to be in the range of plus or minus 3 gm. The device is electronically controlled so that it is only necessary for the operator to enter the scaling weight and production rate before starting the machine. Dough piece sizes from 6 to 52 oz are feasible. This machine does add a significant amount of development to the dough, so it is commonly necessary to reduce the amount of development in the horizontal mixer. The reduced mixing time allows the dough to come out of the mixer 5 to 7°F cooler than normal. Heating due to the action of the transfer pump degasser and the new divider will usually be just about enough to bring the dough ball temperature up to point it would be in exiting from a conventional divider. The "one-pocket" divider uses neither dusting flour or divider oil. The new divider incorporates a clean-in-place system which can be activated with a minimal amount of disassembly, but startups after short term shutdowns need be preceded only by extrusion of any remaining dough. About 50 to 100 lb of scrap dough is generated in cleanup.

A new design of positive pressure dough divider has been patented by Jones, *et al.* (1985). This machine consists of a dough hopper, a pair of sheeting rolls beneath the hopper, and a transfer chute through which

FIG. 4.3. EXTRUSION DIVIDER. DOUGH IS FORCED THROUGH THE ORIFICE AT UPPER LEFT AND CUT OFF BY THE ROTATING KNIVES ON EACH SIDE.
SOURCE: AMF

the dough passes to a measuring chamber in a rotating metal wheel. The metal wheel which contains the cylindrical chamber rotates slowly while the dough is entering the chamber, then rapidly moves to the discharge position. The measuring chamber is provided with a piston, which, in the usual fashion for dividers, moves downward to create suction when it is brought in contact with the dough, then moves in the opposite direction to discharge the dough piece as the wheel turns about 180°. The powered sheeting or pressurizing rollers are of relatively large diameter to provide adequate flow of dough through the passage to the measuring chamber while at the same time maintaining good dough quality.

Degassing Equipment

Several times in the preceding discussion, the subject of degassing doughs has been mentioned. The subject was also treated in Chapter 3, as part of the discussion of post-mixer development. Dough continues to generate gas as it stands in the trough or hopper waiting to be processed by the divider. This means that the last part of a dough batch will be lighter, less dense, than the first part. Since dividers operate on a volumetric measuring principles, the last part of the dough will yield pieces a few grams lighter than the first part, even though dividers have some provisions for expelling some of the gas from the dough. If a large part of its content of fermentation gases can be removed from the dough mass immediately before the dough is acted upon by the divider, the variation in density caused by the accumulation of carbon dioxide will be reduced, and the divider will deliver dough pieces of more uniform weight. Gas can be released from dough by working or mixing it under conditions such that the gas bubbles or vesicles are burst and collapsed. It is necessary that the gas be able to diffuse into the atmosphere; if the dough is mixed in a tight container (as in a completely filled pump chamber), some or all of the expelled gas will be re-incorporated into the dough. If a partial vacuum can be applied to the mixing chamber, gas removal will be facilitated. Merely pumping the dough from the trough to the divider hopper is not necessarily a particularly effective way to degas it. Nearly all dividers include some sort of mechanism to reduce the gas content of the dough before it is volumetrically measured. This often takes the form of rods or pins which perforate the dough as it is pressed into a chamber. The rods create pathways through which the gas being pressed out of the dough can escape.

It is also possible to degas the dough in a separate piece of equipment. This alternative has been discussed in considerable detail in a paper by Harris (1985), from which much of the following has been adapted. He describes lists the following advantages of degassing dough: (1) Improves scaling accuracy, (2) Extends run time on doughs, (3) Minimizes production interruptions, and (4) Improves overall quality. The two types of degasser pumps in use at the present time are the lobe-or rotary-type pump and the auger-type conveyor. Rotary pumps generally include two interlocking rotors turning within a closely fitting chamber. The lobed rotors entrap pieces of dough and squeeze and knead them, pressing out the gas which then travels through grooves machined in the cover plate to the low pressure center of the chamber. From the center, gas flows from the pumping head through small tubes or pipes (Fig. 4.4). Degassed dough is pushed through the exit port. Degassing is a function of the pump's speed and no significant texturizing effect is achieved.

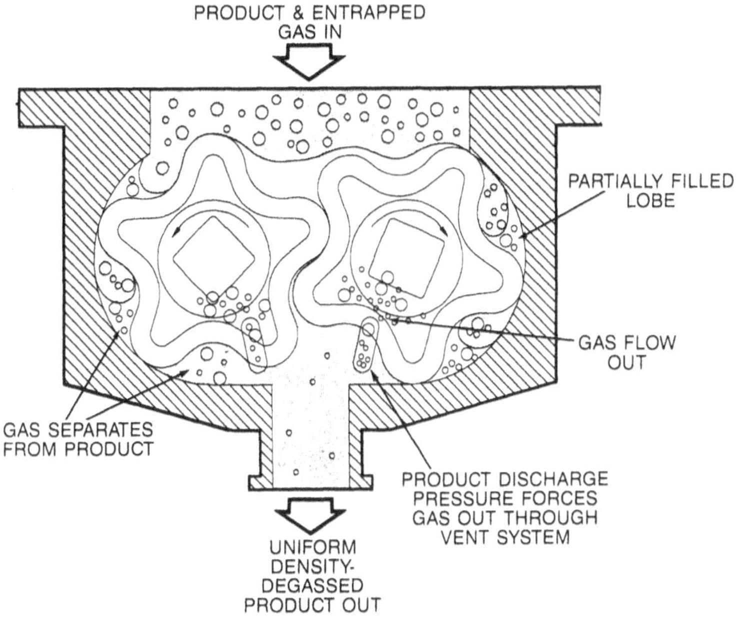

FIG. 4.4. DOUGH DEGASSING PUMP.
SOURCE: APV (CREPACO)

In the auger pump, dough is drawn into the moving flights and is forced toward the exit end of the pump housing. During this process, gas is squeezed from the dough and accumulates in large bubbles before it is released out of a vent hole in the top of the chamber. This type of auger has a second, smaller auger or stator that follows the degassing screw. Some development and texturizing of the dough occurs at this point. Another type of auger degasser has twin screws that handle the dough relatively gently, so as to cause less damage during the degassing and pumping process. Twin-screw pumps degas the dough at two points. While the mass of dough is in the hopper, the rotation of the augers opens up a channel under the center of the dough which allows the free gas to migrate to the rear of the pump and then escape by way of a vent tube. Then, as the dough reaches the exit end of the chamber, gas accumulates in bubbles which are released through vent holes in the top side of the pump housing.

Avenues to Explore

In the author's opinion, inadequate attention has been given to possibility of replacing current divider practice by cutting of pieces of uniform weight from continuous sheets of dough. This is currently being done for yeast raised doughnuts, pizza circles, and crackers. The need for dividers, rounders, degassers, and their associate conveyors would be eliminated. Several advantages, other than the obvious economic one, would accrue. Degassing leading to uniform density is a natural concomitant of the sheeting process and, in most cases, some dough conditioning also occurs. Equipment for sheeting and cutting is relatively inexpensive and is very flexible, since piece weight can be varied over a wide range merely by changing the adjustment on sets of sheeting rollers. The severe punishment the dough undergoes in most dividers would be eliminated. Cleanup should be simpler. Divider oil would not be a contaminant. In the case of bread molding, the loaf molder would be operating on a sheet having relatively uniform characteristics throughout, as opposed to the flattened lump with variable physical characteristics and dimensions which is currently presented to it. Intermediate proofing could take place while the sheet is in continuous form. It is also possible that a separate step for molding could be eliminated.

The most obvious disadvantage of a sheeting and cutting method of dough dividing is that of floor space requirement. It seems very likely that the floor space needed for sheeters and proofing tunnels would be much greater than that required for dividers, rounders, and overhead proofers. On the other hand, the sheeting and cutting line would be suitable for areas with low ceilings.

A bread plant using the suggested system might operate as follows. The bread dough (probably, a sponge method dough would be best for this kind of line) would be dumped into a trough fitted with a pump. After some floor time, the dough would be pumped into the hopper of a set of sheeting rolls. The thick sheet emerging from these rolls would be carried through additional roll stands for a gradual thinning. When the proper weight-to-area relationship is reached, it might be desirable to pass the sheet through a humidified proofing tunnel to allow the dough to relax and to condition the surface. Before or after this intermediate proofing step, the dough would be cut into rectangular pieces, as by disc knives and a guillotine or by cutting dies. The dimensions of each piece will be such as to yield the correct weight for each loaf. If more than one piece is cut across the belt, a slight amount of scrap may be generated at the edges of the web in this operation, and it will be returned to the dough trough by the usual type of continuous scrap conveyor. If the entire width of the web can be used for one piece (in which case only a guillotine, no slitters, will be needed), there is no reason why any scrap should be generated in the cutting operation. Also, the edges forming the ends of the loaf would be sealed. The individual pieces would go into the molder immediately. No sheeting rolls would be needed in these molders. The grain of the baked loaf should be much more uniform when the molder can operate on such a uniform piece. A molder might have to be designed specifically for these dough pieces.

Some complications would be introduced when certain kinds of rolls are being made, but it is at least possible that hamburger buns and hot dog rolls could be made without any additional molding step after the piece is cut. Round hearth breads of the cottage loaf type might not be adaptable to this process, but long thin loaves, as some French breads, might be produceable with much less effort and more uniformity. Pita bread, pizza circles, and the like are already being produced by similar methods.

ROUNDERS

Function of Rounders

When the dough piece leaves the divider, it is irregular in shape with cut surfaces which are sticky and from which gas can readily diffuse. The gluten structure is somewhat disoriented and so is not in suitable conditon for molding. It is the function of the rounder to close these cut surfaces, giving the dough piece a smooth and dry exterior and making a relatively thick and continuous skin around the dough piece. The rounder also re-orients the gluten structure and forms the dough into a ball for easier handling in subsequent steps. It performs these functions by rolling well-floured dough pieces along the surface of a drum or cone, controlling their upward or downward movement by a spiral track. As a result of these actions, the surface is dried by an even distribution of dusting flour as well as by the dehydration occurring from exposure to the atmosphere, the gas cells near the surface of the ball are collapsed forming a condensed layer which inhibits the diffusion of gases from the dough, and the dough piece assumes an approximately spherical shape. After this processing step, the dough ball passes to the intermediate proofer.

Rounders seem to be indispensable parts of traditional loaf and roll forming lines. They are dispensed with in continuous forming lines of various types where the dough is extruded from a homogeneous bulk in the form of a long piece which is the unbaked form of the loaf.

Types of Rounders

There are several kinds of rounders being offered for retail bakeries. Some of these are combined with simple, small dividers. One example of these is described in the divider section of this chapter. Another is the Eberhardt rounding system, which is said to have an output up to

1,000 per hour of pieces from 4 oz to 5 lbs in weight. The machine consists of a rotating center portion, circular in shape but having sides slanted inward, and an outer portion slightly slanted outward. The two parts create a circular channel which is about twice as wide at the top as at the bottom. The inner drum rotates off-center so that a dough piece placed in the channel is rolled and kneaded to form it into a ball.

Belt-type rounders are used in some of the most successful designs of roll and bun forming equipment. These rounders consist essentially of a fabric conveyor belt which moves the small dough pieces under long metal guides having a rounded under surface. These guides are slanted with respect to the direction of belt travel. The combination of the forward motion imparted by the belt and the sidewise force exerted by the stationary guide, causes the dough piece to roll. The slight adherence of the dough to the belt causes the surface of the dough to be pulled into a relatively smooth outer layer. As a result of these combined actions, a more-or-less ball-shaped piece is formed, very suitable for later molding into some sort of fancy bread roll or for proofing and baking without further forming to yield plain hamburger rolls and the like. These systems are rather sensitive to dough consistency, but when the formula is properly adjusted, they are very efficient. Registering the dough pieces as they come onto the belt is necessary to avoid doubles, which are very disruptive to efficient processing. This type of equipment has generally been regarded as being more suitable for dough pieces smaller than about 9 oz in weight and 4 in in diameter than for loaf sizes, but there are now being offered belt type rounders for pieces from 13 to 29 oz. These loaf rounders use tunnels which are constructed of Teflon and are capable of being changed in curvature in two dimensions.

Other types of rounders more suitable for bun size pieces than for bread are based on cups which move in a circular motion above a fabric conveyor belt. The hand action used by bakers for rounding dough pieces on a bench top is duplicated quite closely by this equipment. The cups are roughly hemispherical. They can be made of Teflon or lined with Teflon to prevent sticking, and they often have ridges radiating from the top center of the inside to the edge to give the cup a better grasp of the dough piece. Several cups are attached to a support extending across the belt, and several supports are attached to a power source which moves the whole assemblage in a circular pattern. These rounders must be coordinated with the divider or with some other registration system so that the dough pieces are deposited on the conveyor belt in the exact positions which will later be covered by the cups. This is the main drawback to the method, since the rounding itself is a gentle, non-destructive action applicable to doughs having a wide range of

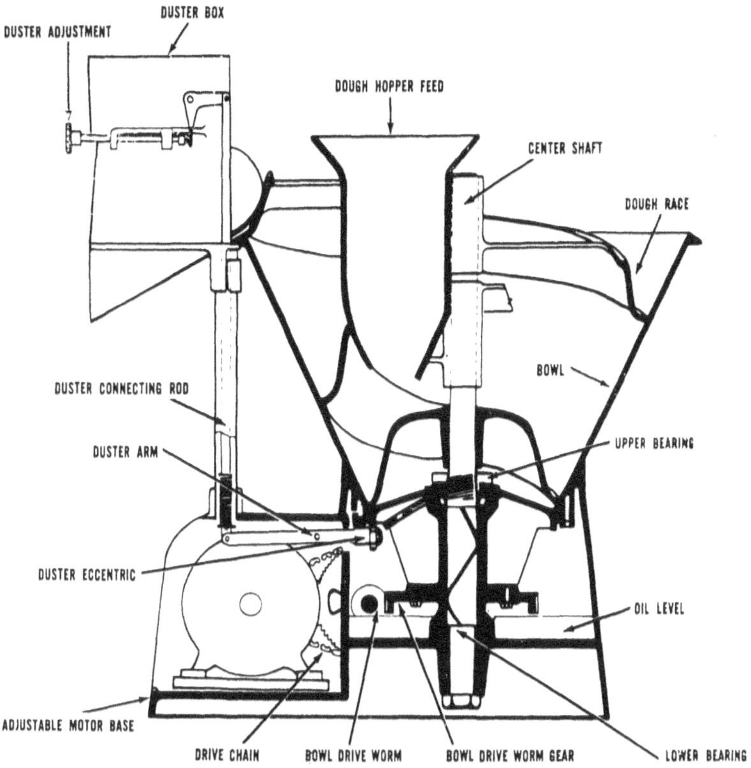

FIG. 4.5. SCHEMATIC DIAGRAM OF A BOWL-TYPE ROUNDER.

rheological properties. If the dough piece is out of registration when the cups come down, jams can result which require the machine to be shut down for cleaning. The good side to this feature is that the dough pieces come away from the rounder in perfect alignment, making them easy to feed to forming equipment which requires registration at the infeed.

Rounders used in medium to large commercial bakeries with traditional loaf bread equipment may conveniently be classified as bowl-, umbrella-, or drum-type. The bowl variety consists of a rotatable cone-shaped bowl around the interior of which is placed a stationary spiral track or "race." Figure 4.5 schematically illustrates such a device. From the conveyor leading from the divider, the dough pieces fall into the feed hopper of the rounder and then drop to the bottom of the rotating bowl. The chunks of dough are tumbled and rolled along the dough race as they attempt to follow the rotation of the bowl. Finally, the rounded pieces emerge from the top of the bowl and are discharged onto the belt leading to the intermediate proofer.

A second type of widely-used rounder is the so-called umbrella or inverted cone variety. These machines differ from the preceding kind in that the dough piece is carried around the outside surface of a cone which has its apex facing upwards. Since the dough first contacts the rounder at the largest diameter of the cone, as shown in the schematic diagram of Figure 4.6, its initial movement is more rapid than it would be in the bowl-type rounder. Opinions vary as to the relative merits of the two designs. It is certain, however, that many examples of both types have been performing satisfactorily for many years.

A third type of rounder is the drum type. These machines differ from the bowl and umbrella varieties in that the cone segment has very little slope to its sides, i.e., the sides are almost vertical. The dough

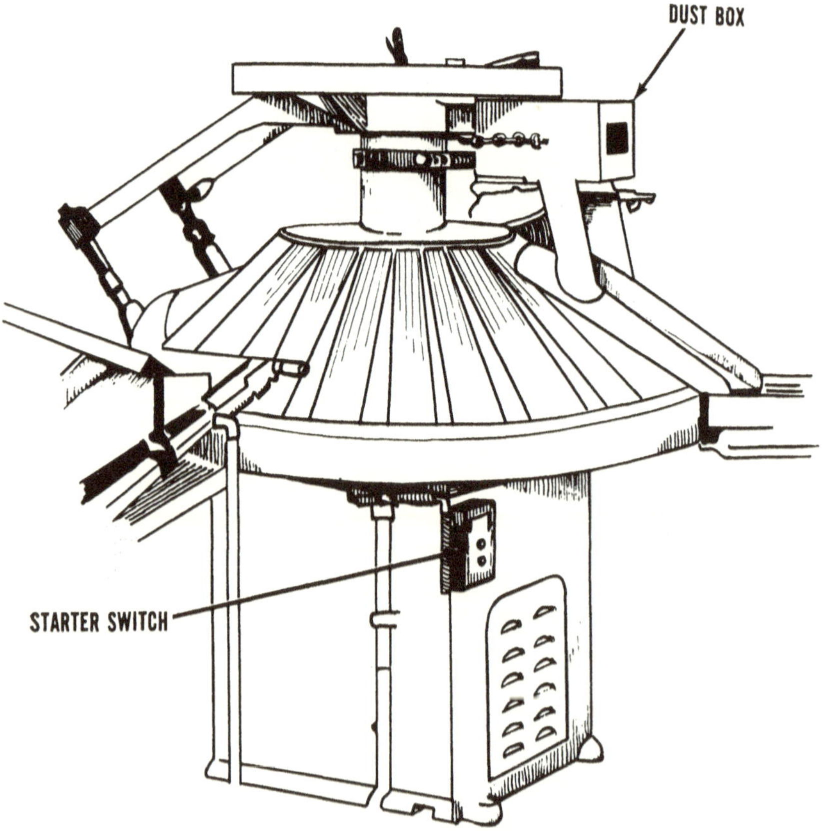

FIG. 4.6. UMBRELLA-TYPE (INVERTED CONE) ROUNDER.

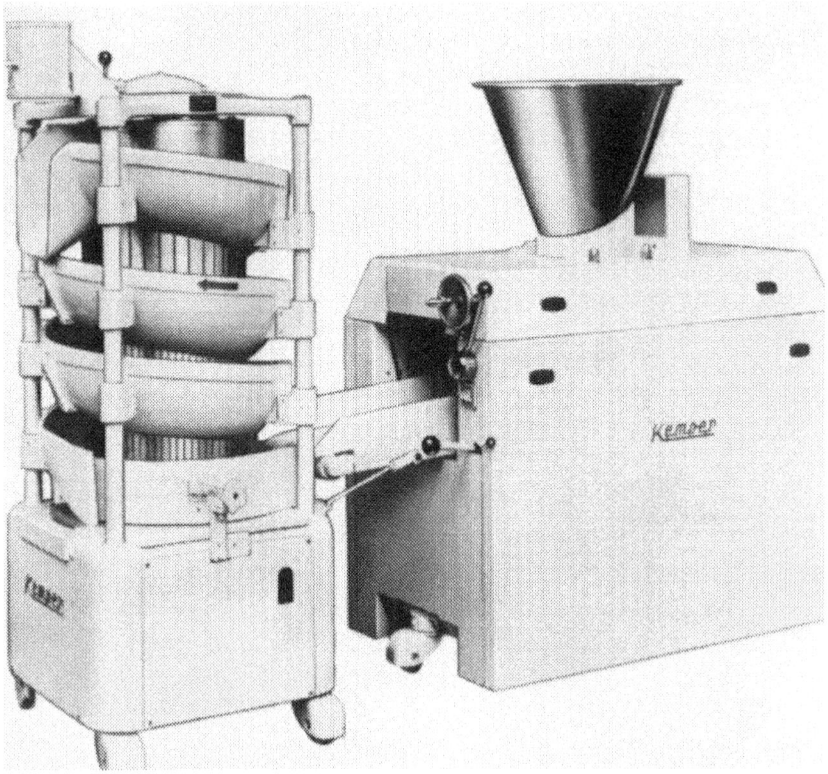

FIG. 4.7. DRUM-TYPE ROUNDER WITH DIVIDER.
SOURCE: EMIL KEMPER

piece enters the curved raceway near the bottom of the drum. It is obvious that the ball travels at a more uniform rate in these machines than in the other two types. Among the advantages claimed for the drum rounders are that they require less floor space, allowing greater latitude in positioning the machine with respect to other makeup equipment. Not as many of these units as of the other two varieties are to be seen in bakeries. Figure 4.7 shows a rounder of this type having a capacity suitable for a medium-sized bakery and accompanied by a divider of similar capacity.

The diagram in Figure 4.8 illustrates the positioning of the three preceding types of rounders relative to the divider.

A fourth style of rounder has concave sides. The round dough race is constructed in two or more sections on different planes. The dough

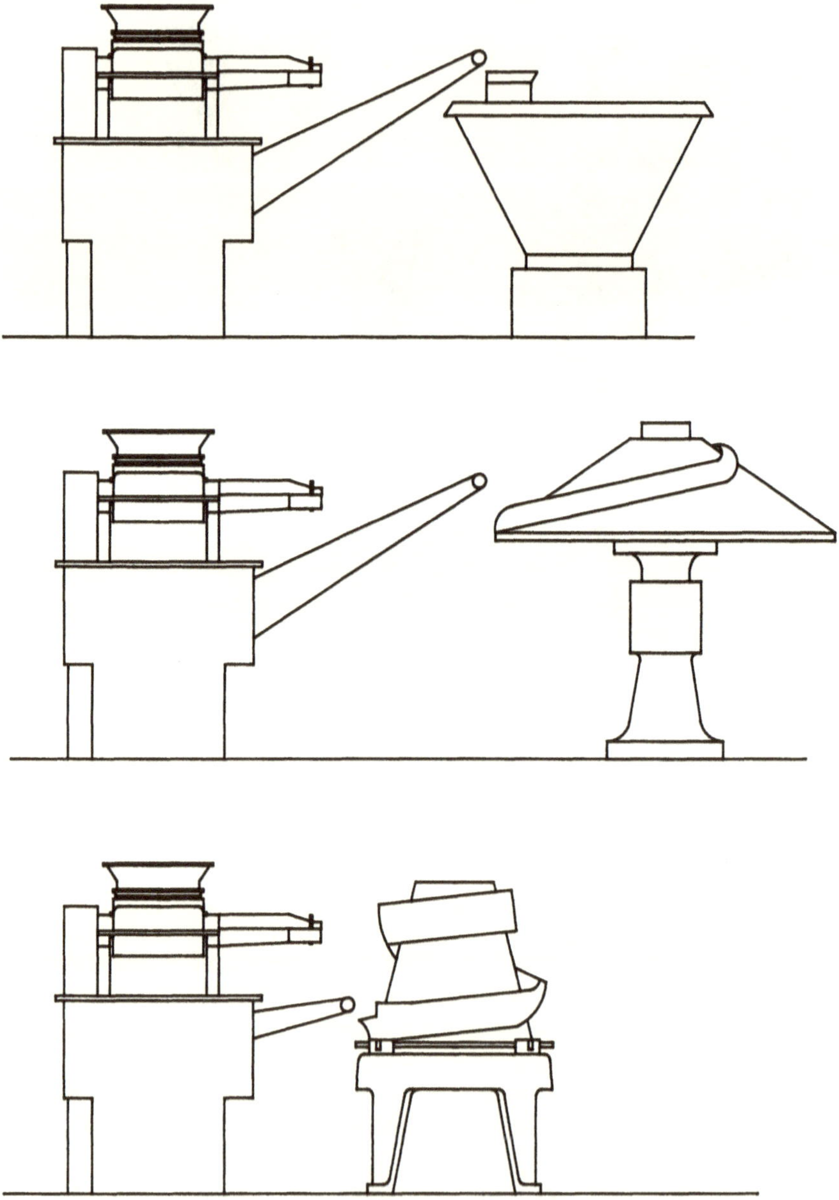

FIG. 4.8. THREE TYPES OF ROUNDER SHOWN IN RELATIONSHIP TO DIVIDER.

pieces are fed in at the top, move through the sections of the dough race, and finally are discharged into the feed drum of the intermediate proofer.

In addition to their form, rounding machines may vary in the texture or composition of the rotating surface, in the means provided for adjusting the relationship of the dough race to the drum or cone, in the method of applying dusting flour, etc. The rotating surface is usually corrugated vertically or horizontally to give it a better grip on the dough piece, but the design and size of the ribs varies considerably from one manufacturer to another. The surface may be waxed or it may be coated with a plastic such as Teflon to reduce sticking. Frequently, a device to shunt aside oversize dough pieces (doubles) is fixed at the exit chute.

Controlling and Adjusting the Rounder

Most rounders provide a means for adjusting the distance of the race from the rotating surface. This furnishes a method for controlling the formation of "pills" or small pieces of dough which are pinched off between the race edge and the drum. As a rule, the race should be just as close to the drum as it can be placed without actually touching it. It should be recognized that a film of dough builds up on the surface of the rounder after it operates for some time, and this reduces the amount of clearance between the race and the drum.

The speed at which the rounder rotates is usually not easily changed, a constant speed sufficing for all applications. The amount of dusting flour which the machine applies is capable of adjustment in most cases. Another important variable is the rate at which pieces are supplied to the rounder. Indications of a need for adjustment are the formation of doubles, tearing of dough pieces (usually due to an accumulation of dough film on the drum surface), cores in the finished loaves (too much dusting flour), and stick-ups.

Maintenance requirements for rounders are relatively simple and consist fundamentally of proper lubrication and adequate cleaning. Dusting flour accumulations should be removed at every convenient opportunity. Accumulations of dried dough on the cone surface are particularly harmful to reliable operation and should be carefully removed during each shutdown. To avoid scoring the dough-contacting surface with cleaning tools, hardwood or plastic scrapers should be used. If special maintenance procedures are necessary for a machine, the manufacturer's service booklet will describe them.

Rounders are inherently less dangerous to operate than are dividers and molders. Nonetheless, an occasional operator will find some ingenious way to get a finger or a hand caught in the mechanism. As in the case of all machines involving moving parts, the operator should be directed to keep his hands away from shear points and pinch points and to assure himself that the power will remain off while he is cleaning the device.

SHEETERS, DOUGH BRAKES, AND LAMINATORS

General Considerations

There are many applications in the bakery which require that dough be shaped into a sheet, web, or strip of relatively uniform thickness either for the purpose of cutting from the sheet many pieces of constant dimensions or for preparing the sheet for further operations, such as interleaving with fat or flour, topping with some sort of material, or stacking with other dough sheets. The equipment used for these operations can be categorized as sheeters and always consists of some combination of rollers designed for pressing the dough into thinner layers without tearing it. Dough brakes are a form of heavy duty, intensive action sheeter. Laminators are devices which automatically stack, lap, or otherwise combine sheets or webs in vertical assemblages.

Dough Brakes

Dough brakes are quite simple in principle, being a set of long horizontal steel cylinders with provisions made for adjusting the clearance between them. These cylinders rotate rapidly compared to regular types of reduction rolls and may be, for example, of 8 inches diameter and 20 or 26 inches width. They are mounted on heavy springs and a set of plates or chutes (and, sometimes, belts) is provided for guiding the dough into, and out of, the rolls. Usually, the brakes include automatic return devices which bring the dough back in front of the operator. In one design, a revolving plate in front of the operator simplifies the tasks of folding the strip and turning it 90° for the next pass (see Figure 4.9). The job of operating a vertical dough brake is both arduous and dangerous. Guard rails in front of the rolls as well as braking or reversing switches and other effective safety devices are essential parts of this equipment (see Figure 4.10).

At one time, dough brakes were widely used for conditioning bread doughs and for laminating puff pastry and certain kinds of cookie and cracker doughs. They are less popular now, but can still be found in some small and medium size bakeries in this country, and in some biscuit factories elsewhere. They have lost out to better controlled sheeting devices which require less manpower. Although they can be regarded as high speed, heavy duty sheeters, they are not particularly suitable for sizing dough sheets.

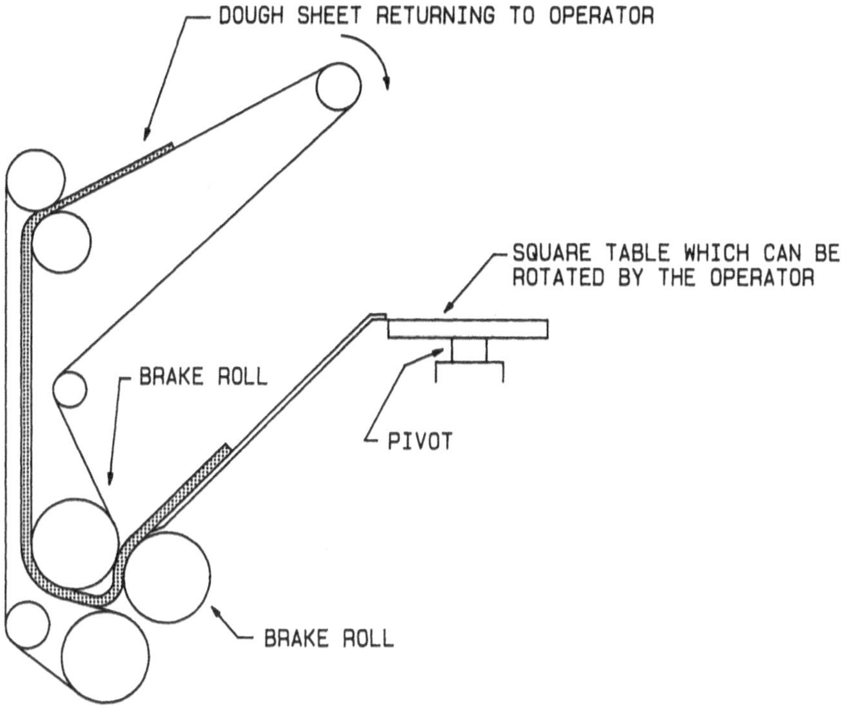

FIG. 4.9. DETAILS OF WORKING PARTS OF ONE TYPE OF DOUGH BRAKE.

In using dough brakes, a given quantity of dough, usually about 40 lb, is formed into a thick strip by an initial pass through the brake. If laminating ingredients such as shortening are to be used, they are applied to the strip by hand, and the dough is folded so as to enclose them. The folded assemblage is sent through the brake again in a direction perpendicular to the initial sheeting. The folding and sheeting is repeated one or more times, with or without cooling and rest periods in between the steps. Unless carefully adjusted for gradual reduction, these machines severely punish the dough, making it difficult to retain recognizable laminations after two or more passes. The rapid reduction which occurs between the high speed rollers does not give the fat time to flow and it tends to break up. Also, the finished shapes are not uniform.

Sheeting Rollers

Sheeting rollers are used to change the dimensions of a dough sheet, making it thinner, wider, and longer. Their action is slower and more

FIG. 4.10. HIGH CAPACITY DOUGH BRAKE.
SOURCE: VICARS

gentle than that of a dough brake. In the simplest form, they consist of a pair of rollers, either horizontally or vertically aligned, with provisions for adjusting the separation between them. Their axles are always parallel. A conveyor belt brings the dough piece into the nip of the rollers at a constant rate, and a take-away conveyor removes the sheeted dough at a faster rate consistent with the length extension resulting from the change in the web's dimension. Dusting flour is often applied just before the dough enters the roll stand. Sheeting is also useful for combining strips of dough end to end, so that an continuous web is presented to die cutters or other processing units.

Sheeting rollers tend to develop the dough. They cause a rearrangement of the protein molecules (Levine 1985 and 1987; Feillet *et al.* 1975 for pasta doughs) which increases with the number of passes and with reduction in roll clearance. This can be desirable or undesirable, depending on the state of the dough as it enters the sheeting system. Dough which has been well developed in the mixer and then subjected to repeated sheeting and folding will break down, tearing and becoming sticky. Stenvert *et al.* (1979) described the use of dough rollers for development. In their article, which reviewed earlier work of other investigators in the field, it is pointed out that dough rollers are very efficient dough developers in terms of energy consumption, requiring only 10 to 15% of the net energy used in a high speed mixer. Dough developed by sheeting can produce bread comparable to that produced by the Chorleywood bread process.

When used as the principal means of dough development, roller processing can yield bread of exceptionally fine texture. In passing dough through rollers, one aim is to remove all large gas bubbles. This requires repeated sheeting and folding steps. Only very stable doughs—those resistant to overdevelopment—are able to tolerate the work input required to remove excess fermentation gases and still produce satisfactory bread.

The major use of sheeting rollers is not for developing inadequately conditioned doughs, but for sizing dough sheets. In large-scale operations, there are many requirements for reducing the thickness of dough sheets and for forming continuous webs of uniform thickness which can be further processed into more complex products. Gauging rolls (sizing rolls) are set in heavy steel frames and have provisions for adjusting at least one of the two cylinders in each set so the separation between them can be changed. Usually, indicators will have been included by the manufacturer to indicate the setting and provide the opportunity to duplicate it in future runs. Efficiency of performance is related to roller speed relative to dough speed, diameter of the rolls, surface texture of the rolls, width of the gap between them, rheology of the particular formula being processed, and amount of dusting flour applied. Other factors may also influence the action of the gauge roll set.

Rolls may be made of stainless steel, chrome plated metal, or other materials, and they may be Teflon-coated for difficult-to-handle doughs. The rolls may be chilled or warmed by circulating fluids internally, although this is not often done in conventional dough handling systems.

The web of dough is usually delivered to the rollers by a conveyor belt and the thinned down web is taken away by another belt. In some lines the speeds of rollers and delivery and receiving belts are synchronized, while in other setups there is no direct gearing or connection between these units. Whether preset or manually adjusted, the absolute and relative speeds of these pieces of equipment are critical to proper performance. Usually, the web falls free a few inches before it contacts the offtake belt and the dough is often allowed to form wrinkles across the belt at this point so it can relax somewhat from the punishment of the reduction. If too much relaxation is permitted, there is some risk the dough will be carried under the roller with messy consequences. In other cases, the offtake conveyor may be speeded up to slightly stretch the dough in order to prevent unwanted contraction with consequent thickening.

As a first approximation, a bakery technologist will frequently aim for a 50% reduction at each roll stand. It is not often that this is the final setting. The important thing to avoid is dough breakdown, which is evidenced by a rough surface and, especially, the formation of cracks or crevices at the sides, giving ragged edges. Often, the surface becomes wet and sticky about this time, more proof of dough deterioration. It is a mistake to try to overcome this sticky condition by increasing the dusting flour or by drying the surface, since these expediencies will either exacerbate the condition or lead to poor product quality (probably both).

Designers of sheeting lines must carefully consider the uncertainties involved when they scale up processes developed in a laboratory or pilot plant facility (Levine 1985). One factor to consider is the thickness of the relaxed dough after it has passed between a set of rollers—this will be a minimum of 15% greater than the roll spacing, but is usually much greater, and it is dependent on the design of the rolls including such factors as roll diameter. For very elastic doughs and small roll spacing (relative to initial dough thickness) the final thickness of the sheet might be four times greater than the spacing. The maximum reduction of thickness which can be achieved in one pass between a pair of rolls without causing unacceptable dough breakdown can be as high as a factor of six, but a factor of two is more commonly used.

Levine (1987) states that larger diameter rolls require smaller gaps than smaller roll sets to achieve the same sheet reduction. Production size rolls also impart much more energy to the dough than do the smaller diameter rolls typically used in the pilot plant or laboratory. Levine points out that problems arise when the assumption is made that dough reacts to sheeting as though it were a solid material. Dough behavior should, instead, be considered as primarily that of a very viscous liquid. The primary mechanism associated with deformation of dough between rolls is shear development. The most important factors governing sheeting roll behavior is not the absolute compression ratio that the dough experiences but the length of dough contact, spacing, and speed of the rolls. Larger diameter rolls reduce the dough more gradually and so abuse the dough less. The designer of sheeting lines should also take into account deflection of the mechanical components (rolls, shafts, and bearings), which can be very significant when dealing with very thin products, such as tortillas.

Reversible sheeters perform functions similar to those of dough brakes, but they are slower. This can be a real advantage in preventing dough breakdown, but it is, of course, more expensive in terms of labor. Reversible sheeters ordinarily consist of two conveyor belts, one on each side of a pair of rollers which can be brought closer together in a series of steps as the dough sheet is gradually reduced in thickness. Machines are available with electronic programming of the reduction steps. The conveyor belts move at different speeds to accommodate the lengthening of the dough as it passes through the rollers.

In using these machines, a sheet of dough is brought through the rollers at a relatively wide setting, then the rolls are brought closer together and the sheet sent through in the opposite direction. This process is repeated until the dough has assumed the proper thickness. Because of the bi-directional nature of the process, reversible sheeters are not suitable for including in continuous lines. They are widely used in retail shops, supermarket bakeries, pilot plants, and medium sized variety bakeries. There are many brands of reversible sheeters. They vary in sturdiness, materials of construction, capacity, degree of automation, dimensions, roll size, and price.

The lengthwise extension and sidewise extension of a strip of dough are interrelated for any given set of sheeting conditions. It often happens that the width of the sheet is unsuitable for the process when the correct thickness is obtained. Manipulation of various factors may yield the proper dimensions without the necessity for adding other equipment, but use of cross-rollers gives added flexibility in designing lines and setting process conditions. Cross-rollers are short cylinders which are rolled back and forth across the width of the belt—the axis of the cylinder being parallel to the direction of travel of the belt, of course. Such equipment tends to extend the dough more in the direction perpendicular to the belt direction than it does along the belt, thereby offsetting to some extent the lengthening of the dough piece which occurs at the gauge rollers. The effect of cross-rollers tends to be somewhat uneven, occasionally leaving striations or other non-uniformities in the dough strand. This can cause variations in weight of pieces cut from the dough and visible discontinuities in the finished product.

Dough stretchers or sheeters of the general type described by Hayashi (1986) have come into use in recent years as a means of obtaining rapid thinning of dough without subjecting the sheet to excessive stress. This device consists essentially of a group of rollers of small diameter which rotate on their axes while simultaneously being carried about an oval path. At the bottom of their circuit, they contact the dough and "massage" it to a thinner dimension by their repeated wiping and rolling action. The manufacturers state that a device of this type only "500 mm long accomplishes the work done by a machine 10 meters or more in length and equipped with pairs of rollers 100 mm long and combined parallel in 70 stages." There is probably a natural excess of parental enthusiasm in this comparison, but one may say that the device does accomplish rapid reduction of certain types of doughs. It has been applied in conjunction with a laminator to form puff pastry doughs.

Laminators

Laminating machines are found on every saltine production line and can be used for several other types of of bakery items, such as croissants, puff pastry, and Danish pastry. For years, the standard equipment for laminating doughs was the hand-fed, high speed dough brake, but it is not presently regarded as the equipment of choice for mass production of laminated doughs.

Combination machines for sheeting bulk dough and then lapping and cross-rolling it to form the structure which is characteristic of soda crackers are widely used. Saltine dough is not interleaved with fat or other material when it is laminated. Some types of cookie or cracker dough (e. g., cream crackers) do receive deposits of interleaving material in the laminating process. Several ingenious methods have been developed to form the laminations. Figure 4.11 shows two ways in which layers can be formed automatically. Further discussion of this class of equipment will be found in the chapter describing cookie and cracker equipment.

Puff pastry dough, Danish pastry dough, and some other types of rich doughs receive heavy applications of shortening at the first laminating stage. Specialized equipment is available to extrude two sheets of dough simultaneously and apply a layering material between them. Such units include a housing supporting two sets of primary in-feed rollers and two sets of secondary rollers. An aperture is provided in the housing above the inner end of the transport rollers for supplying the interspersing device with any kind of filling (fat, margarine, cream, etc.). The initial dough sheet will be between 2 and 3 mm in thickness. A reciprocating sheeting device is located between the secondary rollers and above the transport conveyor. The oscillating device grasps the dough sheet between two sets of endless belts mechanized so as to move the emerging sheet back and forth across the width of the transport conveyor, approximately at right angles to the conveyor belt motion. By adjusting the relative speeds of the sheeting device and the take-off conveyor, piles consisting of any number of dough sheets can be made as long as they are not thicker than 1.5 in. The folded sheets are carried through gauge rollers and then to a rotary cutter or stamping machine.

The vertical dough sheeter-laminator described above was originally

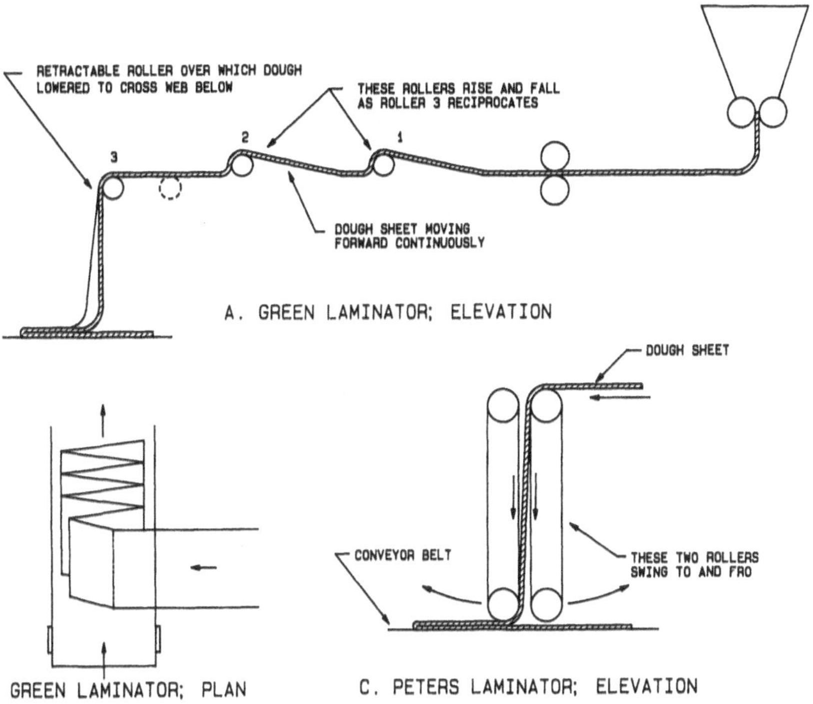

FIG. 4.11. How the Peters laminator and the Green laminator work.

designed for hard biscuits and crackers, but is now available in a specialized version for the production of cream crackers. The model for cream crackers forms two bands of dough, rolls them to the required thickness and then deposits the filling between them. The layers are then brought together and the edges are sealed. The multi-layer complex of dough and filling is then folded transversely onto the conveyor.

Hydropneumatic laminators are available. These devices take sheeted dough, cut it into rectangles of appropriate dimensions, and deposit them in an overlapping stack on a belt moving transversely to the depositing conveyor. These devices utilize a hydropneumatic drive, thus dispensing with complicated cam-and-gear arrangements and enabling them to achieve high operating speeds. Up to 15 laminations per minute can be made and a full range of adjustments is available to modify speed, length, and position of the dough sheet. Straight-out or cross laminating can be performed. Other laminators or dough lappers receive a continuous sheet from sets of gauge rolls and deposit partially overlapping layers on a belt traveling at a 90° angle.

Shortening may be applied as preformed sheets or as a strip of plasticized material extruded by appropriate means. A mixture of flour and shortening can be used, this being the laminating component for cream crackers. Beaded or flaked shortening and liquid oil have been suggested, but the author knows of no cases in which these ingredients have been successfully applied.

Danish products can be made with a continuous flow system which starts with extrusion of dough as a strip onto a conveyor belt. This strip is thinned and widened by one or more pairs of reduction rollers. Subsequently, several narrow bands of shortening—covering perhaps half of the total area—are extruded onto the surface of the dough. The roll-in fat should equal at least 25% of the dough weight. A curling roller forms the dough and fat into a cylinder which is then flattened out by other sets of rollers. This is similar to a three-fold operation. A cutting wheel cuts the strip of dough in lengths to fit a bun pan which is then placed in a retarder. Some operators give the dough a little floor time (room temperature rest) before sending it to the retarder. After 1 to 4 hr rest in the refrigerated room (34° to 38°F), the dough pieces are folded in four layers, sheeted out, then folded and sheeted again. The dough can now be retarded until it is scheduled for makeup, or it can be cut and shaped immediately. The pieces are filled, formed, and topped as required for the particular item being made, and then proofed. Room temperature Danish should be proofed for about 50 min to 1 hr at approximately 95°F dry bulb and 85°F wet bulb temperatures.

A review of four patents on dough laminating equipment and processes may be of interest. Steel (1972) described a laminator and additive depositor useful, for example, in making cream cracker doughs. In this device, a continuously moving peeler conveyor band is caused to reciprocate with a reciprocatory laminating conveyor belt. The depositor is active during movement of the laminating device in one direction only. Albrecht and Stanley (1974) described equipment consisting essentially of a vertically oriented oscillating conveyor which delivers a strip of dough onto a horizontally traveling belt. The belt is concave, forming a trough along the length of the upper surface. Because the radius of the concavity is designed to equal the arc described by the bottom of the vertical conveyor, the dough strip travels an equal distance regardless of whether it is deposited on the center or the edge of the belt. This avoids the uneven stretching that would occur if the dough were deposited on a flat belt. Morgenthaler and Seewer (1976) patented equipment consisting of a set of reversible conveyors adjustable in angle so that a piece of dough could be folded and lapped automatically in a predetermined sequence of steps. The three-layer dough piece formed in this operation is said to be less subject to tension breakage at the edges and to exhibit less squeezing out of the fat in subsequent sheeting steps. Gugler (1975) described a method for forming folded dough sections for Danish pastry and the like. A relatively thin sheet of dough is severed into sections on a conveyor belt. Periodically, a portion of the conveyor belt system is swung upward and to the rear, then forward and down, to fold a forward portion of each dough section back over the following portion.

BIBLIOGRAPHY

ALWES, M. 1985. A Quick Guide to Pumping Viscous Products. APV Crepaco, Inc., Chicago, IL

ANON. 1985. Hot air on rounder. Am. Soc. Bakery Engineers EIS Report *72*.

BENIER, J. 1983. Automated checkweighing. Proc. Am. Soc. Bakery Engineers *1983*, 113-118.

CAMPBELL, G. P. 1983. Dough dividing. Proc. Am. Soc. Bakery Engineers *1983*, 106-112.

CAMPBELL, S. P. 1979. Mechanical dough conditioning—new systems. Proc. Am. Soc. Bakery Engineers *1979*, 175-186.

FEILLET,P., FEVRE, E., and KOBREHEL, K. 1977. Modification in durum wheat properties during pasta dough sheeting. Cereal Chem. *54*, 580-587.

HARRIS, R. W. 1985. Degassing and texturizing doughs. Proc. Am. Soc. Bakery Engineers *1985*, 57-64.

HAYASHI, T. 1986. Apparatus for producing strip of dough having constant dimensions and flow rate. U. S. Pat. 4,583,930.

JONES, D. A., WAWRA, G. F., and MOSS, W. R. 1986. Positive pressure dough divider. U. S. Pat. 4,573,898.

LEVINE, L. 1985. Throughput and power consumption of dough sheeting rolls. J. Food Process Engineering *7*, 223-238.

LEVINE, L. 1987. Sheeting of doughs. Cereal Foods World *32*, No. 5, 397.

STEELS, G. 1972. Dough laminating. U. S. Pat. 3,698,309.

STENVERT, N. L., MOSS, R., POINTING, G., WORTHINGTON, G., and BOND, E. E. 1979. Bread production by dough rollers. Bakers Digest *53*, No. 2, 22-27.

CHAPTER FIVE

FERMENTATION AND PROOFING EQUIPMENT

INTRODUCTION

General Considerations

In the traditional sponge and dough system of breadmaking, there are three stages which are specifically intended to allow the combined ingredients opportunities for undergoing a more-or-less quiescent fermentation. These stages are: (1) Sponge fermentation, (2) Intermediate proofing, and (3) Pan proofing. If there is no sponge, the straight dough may be allowed a fermentation period, instead. Of course, the dough also continues to ferment and undergo other reactions in the intervals between the formal fermentation and proofing periods, and these changes can have significant effects on the dough's interactions with processing equipment and on finished product quality.

Other yeast leavened doughs, such as those for saltines, sweet doughs, and Danish pastry, also rely on fermentation and rest periods for some of their characteristic features.

In order to maintain some control over the many reactions which occur during sponge fermentation, intermediate proofing, and pan proofing so that predictable results can be obtained, the dough is usually kept in enclosures where the temperature and humidity can be regulated. Temperature must be controlled so that the yeast activity will proceed at a predetermined rate, and humidity is preferably controlled so that the outer surface of the dough will not be too sticky or too dry. The equipment used to process, enclose, and convey the dough in these three maturing or conditioning phases will be discussed below.

Fermentation of dough either as a sponge or straight dough, has as its principal objects: (1) Saturating the liquid phase with carbon dioxide, (2) Forming the gas bubbles which will be, in the later processing, subdivided to form the basic cellular structure of the product, (3) Developing the fermentation products such as lactic acid, which mellow and soften the gluten, making it more extensible, (4) Providing flavoring substances and flavor precursors which contribute much of the characteristic aroma and taste of bread, and (5) Adapting the yeast to the fermentation of maltose. Doughs are proofed for many of the same reasons and, additionally, to form a shape-retaining skin, to allow gas to expand the existing vesicles, and to permit relaxation of the structural protein network after the violent stretching and pressing it undergoes during molding.

There are many factors to consider in designing systems for conducting or replacing fermentation, intermediate proofing, and pan proofing. Among the most important factors are temperature and humidity, although the mechanical shock imparted to the dough by conveying mechanisms is also significant and some other effects may enter the picture in certain cases. When considering the influence of temperature and humidity on the dough, it is necessary to understand that the speed with which the product approaches equilibrium with its environment is as important as the actual conditions existing in the enclosure. Although every reader will understand that the temperature and the humidity of the atmosphere in the enclosure are fluctuating constantly, it easy to lose sight of this uncertainty when discussing theoretical conditions. It should always be kept in mind that only by chance do the temperature and water activity of any individual dough piece match the temperature and relative humidity of the enclosure atmosphere at any given time.

Variations

In a traditional sponge procedure for breadmaking, the sponge in its trough, the rounded dough piece in its tray, and the raw loaf form in its pan will be held for specified periods of time at predetermined temperatures and humidities. There are many variations which can be made in the system while still producing satisfactory finished products.

Not all breadmaking processes involve sponge fermentation, intermediate proofing, and pan proofing. Sponge fermentation is omitted in all straight dough processes as well as in the so-called mechanical development methods. A formal intermediate proofing step is dispensed with in the no-time systems, although a certain amount of floor time always occurs and performs some of the same functions. It is difficult to imagine a system in which pan proofing would not occur, but the time can be short and the process can take place on conveyors leading to the oven rather than in proofing rooms or cabinets. It is possible for all of these steps to take place in uncontrolled environments, in which case reliance must be placed on the starting dough temperature and on operator judgement of the completion of dough maturation in order to insure proper dough qualities at the time the material is taken to the next step. For example, it is not uncommon for the sponge to be placed in a dough trough which is then covered with a piece of plywood or a plastic sheet before it is moved to a warm area of the bakery. This procedure leads to some fluctuation in sponge properties, but a skilled operator can usually compensate for these by adjustments at the doughing up stage with, perhaps, additional adjustments at the intermediate proof stage.

The sponge, which normally appears as a stiff dough when first mixed, is replaced by pumpable liquids in those processes which rely on liquid sponges, broths, brews, etc., as the first fermentation step. Continuous or semi-continuous breadmaking systems make use of such intermediates, but modified batch processes can also use them. Temperature control is an essential part of these systems. Humidity control, of some sort, is also a factor, though it may not be maintained within a narrow range by equipment specially operated for that purpose. In fully continuous processing, only the final dough piece is proofed, while the brew tanks take the place of the sponge room.

THE ROLE OF HUMIDITY

Significance of Humidity

Failure to control temperature and humidity at the fermentation and proof stages results in increased variability of finished products and in processing difficulties. It is in the best interest of management to keep the dough under close control at all times, and this cannot be done if fermentation is placed at the mercy of widely fluctuating temperatures and humidities.

The effectiveness of fermentation rooms, intermediate proof cabinets, and proof boxes for their intended purposes depends upon the accuracy with which they control humidity and temperature. To be sure, in practice the controls may be rather crude and poorly responsive to adjustment, but the value of maintaining the humidity and temperature within fairly narrow ranges is well recognized. In proofing enclosures, an atmosphere of high humidity is required to prevent drying out or crusting of the surfaces of dough pieces during the time they are in the enclosure. On the other hand, if the humidity is allowed to rise to near the dew point, there will exist the danger of water condensing on the dough with the resultant development of a sticky or spotted surface. The solution lies in a careful control of the humidity within a narrow, high range. Control is made more difficult by the fact that, in some of these installations, only a limited choice of dry-bulb temperatures is available to the operator because artifical cooling of the air is not provided. The ideal situation would be to have both cooling and heating means available, but this is seldom found in practice. There will always be substantial differences in the relative humidities at different points in the room; one measure of the effectiveness of the temperature and humidity control mechanism is the range of these differences.

Measurement and Control of Humidity

Water vapor—The amount of water vapor which can be contained in a given space is affected by the temperature and pressure of the gases filling the space. At ordinary pressures, the maximum content of water vapor is independent of the kinds and amounts of other gases which are present, for all practical purposes. The terms "absolute humidity," "specific humidity", "relative humidity", "dew point", and "water activity" are used to describe moisture vapor concentrations. Absolute humidity is the mass of water vapor present per unit of space. Specific humidity is the weight of water per unit weight of gas. Absolute humidity and specific humidity, though fundamental measurements of importance to the scientist, are probably less useful and certainly less often used by the bakery technologist than is relative humidity. The latter datum is the partial pressure of water divided by the saturated vapor pressure of water at the same temperature; it relates the amount of water vapor which is present to the theoretical amount of water which could exist as vapor under the conditions which are in effect. Relative humidity is always expressed as a percentage. Dew point is the temperature at which the partial pressure of water in a gas is equal to the saturated vapor pressure of water. In other words, a gas at its dew point would have a relative humidity of 100%—it would be saturated. A further release of moisture into the space, or a decrease of temperature, would cause condensation, or "dew", to form. The dew point of a gas is thus a measure of the moisture in the gas in terms of the partial pressure of the water vapor. It is independent of the type of gas, in almost all cases.

If two or more solids or liquids containing water are present together in a closed system, a transfer of moisture in vapor form will occur until the water contents of all the system's components are in equilibrium with the same relative humidity. It is this effect which permits the adjustment of relative humidity within an enclosure as well as the measurement of relative humidity. The rate at which equilibrium is approached in such systems depends on many factors, among which are temperature, surface area of the solids, gas currents inside the container, spatial separation of the components, and spread between the initial equilibrium relative humidities. This leads us to the concept of water activity.

Water activity equals equilibrium relative humidity divided by 100, and is denoted by a_w. An aqueous solution (salt in water, sugar in water, etc.) or any material containing water, if held in a closed container at a constant temperature will tend to establish a stable relative humidity, when the rate at which water vapor is leaving the material exactly equals the rate at which the material is absorbing water vapor from the air. This presumes there are no chemical changes or biological reactions going on in the material. If two or more such materials are held in separate compartments which prevent them from mixing but allow free interchange of their vapors, they will tend to gain and lose moisture until the head space reaches an equilibrium relative humidity. Under those conditions, the water activity of each material will be that relative humidity divided by 100. Moisture contents of the two materials will very likely be different, possibly very different, at the same water activity. The same situation will exist whether there are two or a hundred different materials in separate compartments connected by the same head space.

The importance of this concept to proofing technology is that doughs of different composition will have different tendencies to absorb or lose water vapor to atmospheres of the same relative humidity. It would generally be expected that doughs having large quantities of hygroscopic ingredients, such as sugar or corn syrup, would tend to gain water from atmospheres of lower relative humidity than would lean doughs. One of the many factors affecting the fermentation rate of yeast is the water activity of the solution in which the cell finds itself.

Measuring water vapor parameters—The most reliable means for determining the quantity of water vapor in a known volume of air is to collect it and weigh it directly. Although it is possible to collect and weigh water as ice on surfaces held at temperatures so low that the vapor pressure of the water is negligible, it is much simpler to remove the moisture by passing the gas mixture through efficient desiccants such as phosphorus pentoxide. If sufficient contact between the gases and the desiccant is obtained, all water (within the usual limits of accuracy of weight measurement) will be removed. When temperature and volume of the gas sample are known (and these are easily measured), the difference in weights of the desiccant before and after the experiment will provide sufficient information to permit the calculation of absolute humidity. Practical difficulties of instrumentation and procedure are encountered in this technique, and it is seldom used in food laboratories.

For most purposes, there are methods of sufficient accuracy which are based on less sensitive and less elaborate instruments than gravimetric hygrometers. One of these, the psychrometer, has been much used in food studies and, especially, in measuring the relative humidity of bakery proofing rooms. A psychrometer consists essentially of two thermometers, one of which (the dry bulb) is operated in the usual manner while the other (the wet bulb) is mounted so that its bulb is constantly in contact with a film of evaporating water. The rate at which the water evaporates is related to the relative humidity of the atmosphere, and the cooling effect which is transmitted from the wet sleeve to the thermometer bulb is related to the rate of water evaporation. Due to the chilling effect, the wet bulb thermometer will tend to display a slightly lower temperature than the dry bulb thermometer at all relative humidities below 100%. By consulting tables which include all possible combinations of dry bulb temperatures and wet bulb temperature depressions, the relative humidity corresponding to the readings on the two thermometers can be determined. At higher humidities, the temperature depression is slight, making the precision of the determination less than one would like. When based on mercury thermometers, the psychrometer is not readily adaptable to automatic feedback control, but versions using electronic temperature sensors (thermistors, thermocouples) can be so employed.

Usually, the required contact of the wet bulb with water is obtained by inserting the lower part of the thermometer into a cloth tube which extends into a container of distilled water. This tube, or wick, is kept wet by capillary action. The water reservoir must be maintained at the temperature of the dry bulb thermometer, i.e., at the temperature of the gas being measured. Use of colder or warmer water, even though the difference is only a fraction of a degree, can introduce appreciable error into the determination. The wick should be changed frequently, since accumulation of soil will affect its performance. Use of tap water instead of distilled water leads to the deposition of dissolved minerals on the wick slowing the response of the wet bulb thermometer and leading to possible errors. The length of the wick, as measured from the top of the water in the reservoir to the top of the mercury in the bulb, should be as short as possible since the movement of water by capillary action is not instantaneous. On the other hand, no part of the thermometer bulb should rest in the water in the reservoir.

Means for ensuring a rapid flow of air over the sensing region of the psychrometer must also be provided. A sling psychrometer consists of the two thermometers attached to a metal base which is connected to a handle by a swivel fitment. By whirling the base in the conditioned area, sufficient air flow is created to make the wet bulb reading fairly accurate. It has been found that a stream of air having a velocity of about 600 fpm causes the water held by the wick to evaporate at a rate sufficient to yield an accurate and stable wet bulb reading. In fixed permanent installations, fans are used to move a stream of air over the thermometer bulbs. If the fan is mounted on a motor shaft, it should be placed so that the air is pulled over the wick rather than blown over it, since heat radiated from the motor will otherwise affect the temperature of the air and affect the reading. If the psychrometer is located in an area of the proof box where there is a constant circulation of air, a fan may not be needed. Of course, a single psychrometer reading tells nothing about the variations in humidity in different parts of the enclosure.

Although the psychrometer is usually sufficiently accurate for measuring humidity in the bakery, it is subject to several sources of error. The two thermometers double the chance of misreading and, at low temperatures, misreading by a few tenths of a degree can lead to large errors in the relative humidity calculation. Insufficient air flow, dirty wicks, wicks that are too thick, and impure water lead to inaccurate readings.

When approximate indications of relative humidity are acceptable, hygrometers based on measurement of the contraction or relaxation of hair or of certain other animal tissues can be used. These instruments are relatively sturdy and inexpensive, function over a wide range, and give direct readings. The principle is simple. Strands of hair are brought from a fixed point around a spring-loaded cylinder or lever which has an indicator attached to it. As the humidity changes, the hair contracts or lengthens, exerting greater or less torque on the cylinder so that the latter turns the indicator to a point on a scale which has been calibrated for the system. These devices are subject to appreciable amounts of error. For example, the hair expands with increasing temperature, and its response to changes in humidity is not strictly linear. Its response is also rather slow, the lag time increasing with decreasing temperature. Such devices are commonly found in the desk and wall hygrometer-thermometer-barometer combinations offered for home display.

Another approach to hygrometer construction utilizes a diaphragm of hygroscopic animal membrane clamped between two aluminum retaining rings so that it forms a flat, truncated cone surface. Changes in the humidity of the air surrounding the assembly cause approximately proportional changes in dimensions of the diaphragm. Movements at its apex are communicated to the indicating or recording mechanism by a multiplying linkage connected to a small ring affixed to the center of the diaphragm. These devices are about as accurate as the hair hygrometers.

There are also hygrometers which depend for their action on the changes in electrical parameters occurring in certain substances as the relative humidity of their environment varies. Most of these devices are readily adaptable to use in automatic control of humidity generation equipment. The basic electrical hygrometer consists of an insulator supporting two metal electrodes joined by a film of hygroscopic salt which absorbs moisture in amounts related to the temperature and relative humidity of the air. The quantity of water absorbed by the film governs the electrical resistance of the system.

A typical electrical sensor for humidity has a hygroscopic film of lithium chloride coating a bifilar-wound noble metal (e.g., gold) resistance element. Resistance of the hygroscopic film responds to small changes in the water vapor pressure. In another form, the wire electrodes may be wound on a textile sleeve impregnated with the hygroscopic substance. Another sensor is constructed of two intermeshing gold grids stamped on plastic. It is coated with a complex film containing hygroscopic salts and plastic. As the relative humidity of the surrounding air becomes greater, the film becomes more conductive and electrical resistance between the grids is lowered. Variations in electrical resistance are interpreted by recording or controlling units. This sensor is said to be capable of detecting a change of a small fraction of 1% in relative humidity.

Dew point or frost point hygrometers measure the temperature at which dew or frost is deposited on a surface cooled relative to the ambient air. The surface is usually a highly polished mirror, and moisture deposition can be sensed photoelectrically since the dew or frost substantially reduces reflection of light by the mirror. Measurements can be entirely automatic and can be used to feed back signals to a humidity controller, but these devices are necessarily more elaborate than both the common psychrometer and the electric devices based on hygroscopic films. They are also susceptible to fouling by atmospheric contaminants such as smoke.

Humidity adjustment—When a stream of air of controlled humidity is required, as in environmental cabinets to be used for testing storage stability or in proof cabinets, accurate results can be obtained by mixing appropriate quantities of two streams of air, one of which has been thoroughly dried (essentially zero relative humidity) and the second of which has been saturated with water vapor, as by bubbling it through a reservoir of water. By varying the relative amounts of the two streams of air, it is possible to obtain any desired relative humidity. Humidity control according to this principle requires fairly elaborate and costly installations. In proof cabinets and fermentation rooms, it is usually sufficient to maintain the relative humidity at a relatively high level;

reducing the humidity is not often needed in most bakeries. In these enclosures, humidity is maintained at the desired level by the intermittent injection of low pressure steam or of atomized water.

In the simplest type of humidity modification, as found in the small cabinet proofing boxes used in many retail shops, the air is conditioned by a manually adjusted heating coil and a valve or cock permitting steam to be injected for raising the humidity. In larger fermentation and proofing installations, both the temperature and relative humidity are automatically controlled. Water vapor can also be dispensed by intermittent injection of low pressure steam or by vibratory or centrifugal dispersion of liquid water. Forming water vapor by boiling water in the cabinet is seldom satisfactory as the amount of heat is usually excessive and the amount of vapor created is usually inadequate.

CONSTRUCTION DETAILS

Some Design Principles

The principal goals of the designers of proof boxes and fermentation rooms are: (1) To provide for the efficient delivery and removal of dough containers, (2) To maintain the relative humidity at a uniform level, (3) To maintain a uniform temperature, (4) To provide sanitary conditions, and (5) To minimize physical shocks to dough pieces.

A major design problem is selecting proper capacities for the air conditioning unit which will control humidity and temperature. The demands on this unit can be expected to be heavy and cyclic, since it will be called upon to compensate for the sudden changes occurring as dough troughs are moved into and out of the enclosure.

The major unwanted causes of temperature changes in these rooms are: (1) Infiltration of air as doors are opened or through other openings, (2) Entry into the room of troughs and sponges, and (3) heat flow through the insulated walls, ceiling, and floor. Cold air infiltrating into the conditioned room or contacting the walls subtracts from the sensible and latent heat and can result in condensation, loss of heat, and reduction of humidity. Flow of heat through walls, roofs, floors, doors, etc., depends on the area, the difference in temperatures, and the heat conducting properties of the wall. The troughs and their contents, when moved into the enclosure, will not be at the same temperature as the atmosphere. Heat must then be added or subtracted to bring the dough to the desired temperature. Steel has a specific heat of 0.12, dry air of 0.24, and the average dough about 0.88. Generally speaking, the weight of the trough, or pans in the case of final proofers, will exceed the weight of the dough, but the higher specific heat of the latter makes it an important contributor to the heat exchange. All of these factors must be considered in calculating the capacity of the air conditioning unit which will be required.

The design temperature is the temperature which the system must be engineered to maintain under the most difficult conditions expected to be encountered. It will be affected by the load placed on the system by the dough and pans, the climatic conditions (which influence the temperature of the infiltrating air and the temperature differences between the cabinet and the environment), and the effectiveness of the insulation. Much of the following is based on a contribution by Fred D. Pfening (1972) and the general conclusions are applicable to cabinets for intermediate fermentation and pan proofing enclosures as well.

Air Conditioning Units

Fermentation rooms require a temperature-adjusting system and a humidity-adjusting system. The former cools or heats the recirculating air to maintain, e.g., an 80°F dry bulb setting. The other provides moisture to maintain the relative humidity at, say, 76%. The air conditioning components can be assembled in a relatively small housing, which should be made of stainless steel. The radiator and the cooling coil can be located in the same housing about 12 inches apart, the best distance depending on the relative face of each. Sometimes a damper is installed to divert the air stream over one or the other of the coils, depending on which one is activated. Since the radiator face area is smaller than the face are of the cooling coil and the room requires considerably less heating than cooling, it would appear that dampers have little value.

An air conditioning unit should provide sufficient air circulation and should provide heat, moisture, and refrigeration to the air in the required quantities. In some systems, humidity can be controlled at three levels: low, high, and no addition. Lower relative humidity is preferred for use with silicone-coated pans or glazed pans, and for certain other purposes. On the optional lower humidity, a 98°F dry bulb setting will provide no added humidification. This setting is sometimes used for proofing raised doughnuts, where a dry atmosphere is preferred to obtain a crusty surface.

Some conditioning units do not have air cooling facilities. Provision of this capacity should be considered, since the amount of cooling which will be required is slight, and the added cost is not great. Cooling can be obtained by providing coils to circulate chilled water or Freon-12 cooling medium. All the cooling permissible with as little dehumidificaton as possible is the objective.

Blower size is determined by calculation to obtain the desired velocity at the static pressure of the system. The blower runs continuously, while the heating or cooling functions intermittently subject to the direction of the thermostatic controls. Figure 5.1 is a diagram of one type of air conditioning unit for a bulk fermentation room.

When the wet bulb thermostat causes moisture to be injected, a certain amount of evaporation occurs, dropping the dry bulb temperature of the atmosphere as much as 3°F. When the thermostatic control is sufficiently sensitive, the momentary cooling effect will cause the heating coil to react to restore the dry bulb temperature, if required. When the water supplied for humidification is at the same temperature as the room atmosphere, superior results are obtained.

To prevent condensation from forming on the side walls when the proof room is started up after a shutdown, it is very important to be sure that the walls have heated up to operating temperature before humidity is injected into the chamber. The air heats quicker than the steel and panels; the thermostat reflects the air temperature only until the steel equals it. A "humidity lockout" feature can be provided to prevent the humidification from being called for when a cold proof box is started up after a shutdown period. The humidity is locked out until the previously selected dry bulb temperature is reached. Thus, an undesirable dew point and the accompanying condensation are automatically avoided.

It is neither economically nor atmospherically desirable to dehumidify the air in a fermentation room, but, as it happens, the cooling coil condenses moisture out of the air to an extent dependent on the temperature of the coil. The excessive elimination of moisture from the atmosphere, passing through the coil face area, causes the humidifying system to operate more nearly continuously to restore the cyclically eliminated water. Designers will usually plan to give the fermentation room six air changes per hour; therefore, the amount of air in the room will pass through the cooling coil six times per hour. Of course, the higher the ceiling, the more volume of air there is to circulate, and more air changes per hour are permissible. In the event the cooling load is above normal, this condition will prove to be advantageous since the velocity through the cooling coil can be increased and more Btu are obtained at lower discharge temperature, thus decreasing the elimination of water.

Designing Room Enclosures

Fermentation rooms must be constructed so that they isolate the interior from unwanted sources of temperature change. To accomplish this purpose, the room should be built with steel structural members which support insulated steel-covered panels. This room should be made as airtight as possible to prevent loss of heat from within and to exclude heat gains and infiltration from without. The floor area must be of the proper dimensions to accommodate the number of troughs required to satisfy the production rate of the bakery. Heat accumulates from warm floors, walls, equipment, motors, people, troughs, dough, light bulbs, and infiltration. The designer has control over these heat gains only to the extent he can insulate. The total quantity of chilled air to be supplied to the conditioned room is determined solely by its internal sensible heat gains. The entire basis for the design of the air conditioning systems rests on the calculation of these heat gains. An accurate survey of the premises must be made. Undersized air conditioning systems are never satisfactory; oversized systems are unnecessarily costly.

Location liabilities—The fermentation room should be located between the mixer area and a point where the sponge can get back to the mixer easily and the dough can be transferred to the dough chute leading to the divider. Perhaps only in a newly designed bakery can each piece of equipment be ideally located, however. In any event, the fermentation room exterior should not be exposed to heat from ovens, to sun heat through a roof, or to a large glass area in a west or north building wall. The western sunload at 4:00 PM can be the peak load for the cooling system, necessitating a larger installation than would otherwise be required.

Generally, windows parallel with adjacent fermentation room walls should be bricked in, insulated, and sealed against outside elements. It is costly to heat more than is necessary and equally, to cool an area from a high temperature down to 80°F. In many existing bakeries, the construction engineer has no choice concerning the most favorable location of the fermentation room, if the building walls must be a part of it.

Insulation—A definite key to the required capacity of the air conditioning unit lies in the amount of insulation used in the walls, ceiling, and sometimes the floor. The better the insulation used in the room, the less cooling coil capacity is required. If more and better insulation is used in the construction of the room itself, less evaporator capacity is required, and, therefore, less dehumidification occurs. If there is less dehumidification, there is less requirement for humidification. If more can be spent for insulation, there will be savings on conditioning equipment and it will be easier to control the room.

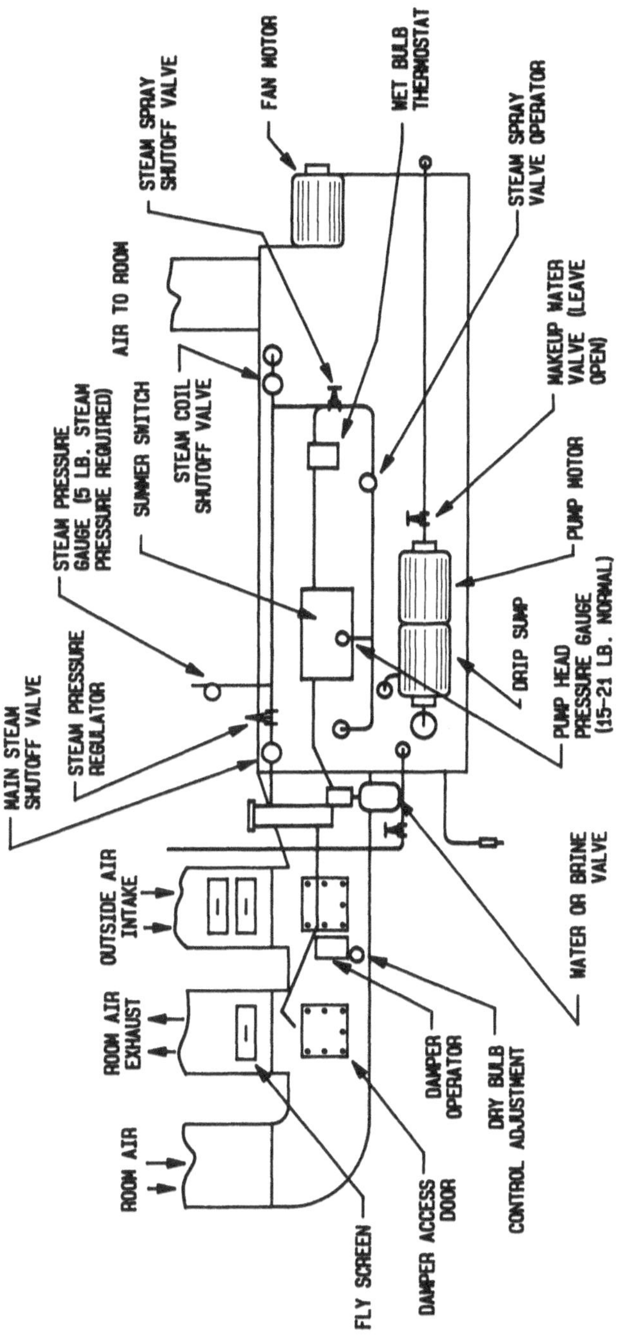

FIG. 5.1. AIR CONDITIONING UNIT FOR A BULK FERMENTATION ROOM.

Height—A prevailing belief among the uninformed is that the lower the fermentation room ceiling height, the easier will be the job of maintaining proper temperature and humidification because the volume of the room will be less. This might be true for other conditioned spaces, but it is a fallacy in fermentation room design. High air velocity has a marked drying effect on dough. Greater height is a distinct advantage because it permits dispersion of the air and permits the additional air changes required to obtain the necessary cooling.

Floors—All floor areas on which the proof room is to rest should be of equal elevation, particularly at the perimeter where the major portion of the load will be carried. The elevation points should be established by a transit level. A solid floor must be present under the proof box. It must not only support the structure, but also have sufficient insulation and moisture-excluding properties to prevent a constantly wet floor from developing during use. Water vapor which condenses under these conditions must be replaced, causing an additional load on the system. Laying new floor, rather than making do with a doubtful existing floor, is advisable when a new proof box is to be installed, or when an old box is moved to a new location.

Floors laid directly on the earth will rapidly condense overlying vapor to water. The remedy is to properly insulate the floor area under, and for 2 to 3 ft beyond, the perimeter of the proof box. The higher the dry bulb temperature at which an air-conditioned enclosure carrying 82% relative humidity is operated, the greater the flow of heat to and through the concrete floor. Condensation potential under such conditions is great. Insulation should be applied as the manufacturer recommends, not only to prevent condensation, but also to prevent the loss of heat. If proofing temperatures are as high as 120°F, it is suggested that heat conductors be installed under the floor surface to maintain a floor temperature of 112°F.

Some construction features which are desirable for a proof box floor are: (1) A tile surface to take the wear of caster mounted racks and to provide the utmost in sanitation, (2) The surface slab of solid concrete or the slab used as a sub-base for tile should be a minimum of 3 inches thick in order for the lag screws to have secure anchorage, (3) Under the slab, there should be an insulating type of concrete not less than 8 inches thick and extending 24 inches beyond the outline dimensions of the proof box on all four sides, and (4) If the proof box is less than 10 ft away from an outside building wall, the wall and floor should be separated with 12 inches width and 8 inches to 12 inches thickness of Zonolite insulating concrete to prevent transfer of cold to the wall panels and floor of the box. If the box is close to a cold wall, the separation strip must be included. Figure 5.2 shows some of these construction details.

When the soil on which the concrete is to be laid has recently been disturbed and there is danger of settling, a reinforcing steel mesh should be placed in the concrete. If the slab is below grade, in a poorly drained location, or where there is a high water table, the soil should be stabilized by raking dry cement into the top inch of soil then sprinkling it with water and rolling it to a hard crust. Over this, mop down two layers of waterproof paper with hot asphalt. This membrane should be turned up against the outside wall to the height of the concrete.

If the slab goes over a well-drained area, the insulation would probably remain drier if no vapor-seal membrane is installed. When the proof box temperature is 110°F dry bulb and 104°F wet bulb, and the earth temperature is approximately 55°F, there is a tendency for the moisture to go toward the earth. It is, therefore, important to ascertain the moisture condition, or probability of what will occur, in order to decide whether or not to use a vapor-seal membrane between the earth and the insulation. This phenomenon is true regardless of what type of insulation is used. A vegetable material insulation will decay in time. A wet insulating material loses its insulating quality.

Air distribution—The reconditioned atmosphere should be distributed evenly in the proof box to prevent drafts or stratification. The number and design of air-diffusing outlets depends on the height of the room. An air stream should never be directed over a dough surface since it would have an evaporating effect and, therefore, would crust the dough. The fewer the air changes the more beneficial the atmosphere remains. A contradictory situation arises between setting up fewer air changes per hour and, at the same time, obtaining sufficient air velocity over the coil to transfer heat at a rate sufficient to maintain the room temperature at 80°F.

Carbon dioxide—Since carbon dioxide is 1.5 times heavier than air, a blanket of carbon dioxide will form over the sponge surface until the sponge rises to the top of the trough. Then, the gas spills over the edge, flowing toward the floor. If sufficient gas accumulates, it may interfere with breathing of personnel who must enter the room. This situation is easily remedied by providing an opening in the room doors of about 10 inches in diameter. The opening should be a little higher than the highest trough. Gas reaching the level of the opening will flow to the outside and disperse into the atmosphere before it can rise to the breathing level.

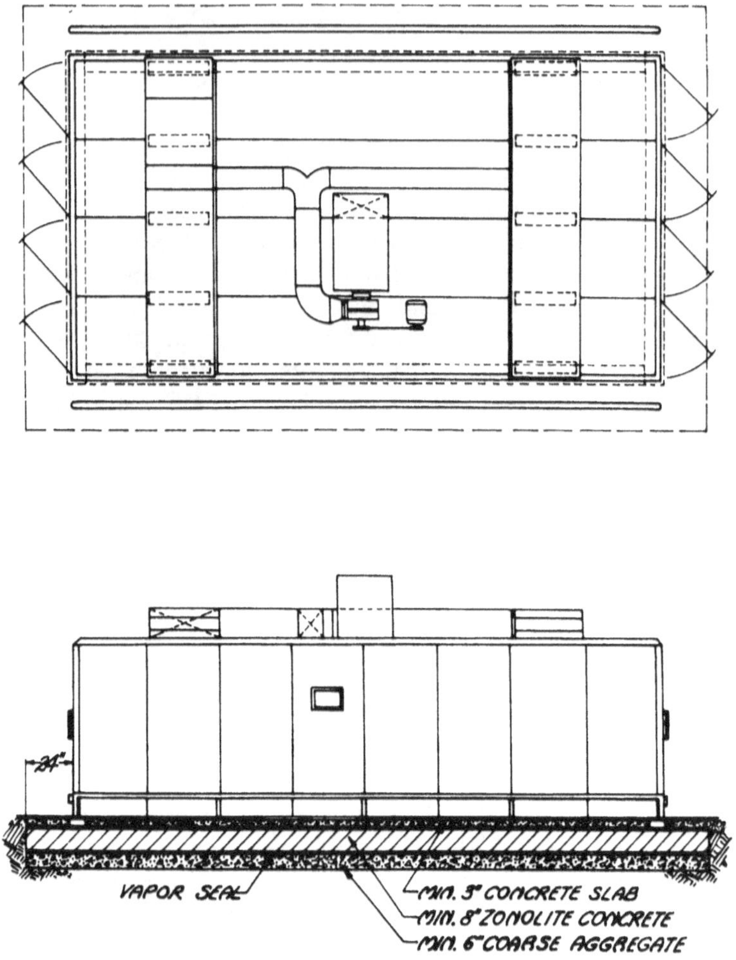

FIG. 5.2. CONSTRUCTION DETAILS OF FERMENTATION ROOM.
SOURCE: FRED D. PFENING, SR.

Dough Troughs

The dough trough is an important piece of equipment, but it is rarely given much consideration by the bakery technologist because it seems to have little direct effect on product quality. The production man gives it some thought, but he is mostly concerned about the sanitary aspects and manpower requirements for trough handling. In sponge fermentations for saltine cracker manufacture, the trough does have an effect on product quality because it serves as a source of inoculation for the sponge, furnishing bacteria of the types necessary to give the characteristic flavor and some of the gassing power to the dough. These factors may function to a much lesser extent in bread sponge fermentations.

Troughs are customarily made from 10 gauge steel, either standard or stainless. They should be welded, with polished seams, and the rims should be rolled and sealed for sanitary reasons. Capacities of stock models range from about 14 to 56 cu. ft. to hold from 180 to 750 lb of sponge or 400 to 1680 lb of straight dough.

Considerable variation is possible in trough design. Most of the modifications are intended to afford better control over the removal of sponges from the trough. Several familiar types are described below.

Slide-end—The whole of one end of the trough can be raised to give an opening through which the sponge can flow. Ordinarily, there will be no mechanical provision for removing the gate or raising it in increments.

Sloping bottom—This design feature can be applied to most of the other variations described here, and is intended to facilitate complete removal of the dough through one end.

Gate-end—The bottom part of one end is formed into a gate hinged at its upper edge. Spring action keeps the gate sealed until it is released to allow the dough to run out.

Chute-end—The lower one-third (approximately) of one end can be opened, leading the sponge to an inclined pan or chute which guides it into the mixer. While the sponge is fermenting, the chute can be turned up, out of the way.

Drop-side—Approximately the top third of one side is hinged and can be dropped down to make it easier for an operator to remove the trough contents.

Controlled flow—These troughs are designed primarily for overhead hoist operation. They have a gated end which can be raised in increments by a rack and pinion arrangement.

Conveyorized—The most advanced models have screw conveyors in a channel at the bottom of the trough to force dough directly from the trough to a processing line hopper.

Accessories—Various accessories can be obtained. Top extensions, or domed covers of aluminum, stainless steel, or plastic are examples. These are helpful in sponge fermentations. Bails can be welded on the sides as points of attachment for hoist hooks.

Trough handling—Several different systems have been devised for conveying dough troughs into, through, and out of the fermentation room. Manually pushing the dough trough on its casters from one point to another is, of course, the original way. Large fork lift trucks can also be used. Automatic handling is the rule nowadays. In one design, automatic handling in the proof room moves troughs in a line down the feed side of the enclosure, laterally across the end of the room, then back to the discharge door. A carriage, powered by a hydraulic ram, moves incoming troughs forward on their own casters and between two horizontal guides along the sides. Upon reaching the end of the room, the cross-carriage powered by a hydraulic motor moves the end trough transversely and positions it in the discharge line. At this point, another carriage identical with the feed unit moves the procession toward the discharge door. Troughs are moved onto and off the carriages by hand. A diagram of this system is shown in Figure 5.3.

PROOF BOXES

From the following discussion, it will be seen that there are many points of similarity between proof boxes and the previously discussed fermentation rooms. Duplication of the preceding topics will be avoided unless repetition is necessary to insure clarity. Proof boxes can be classified as either intermediate proofing cabinets or pan proofing enclosures. The purpose of a proof box is primarily one of heating and humidifying so as to maintain wet-and dry-bulb temperatures which will cause a dough piece to proof satisfactorily within a scheduled time. The process relies heavily on an air-conditioning operation and is subject to all the vagaries of psychrometric measurement and control previously discussed in great detail. Another major problem is designing a mechanism which will satisfactorily control the movement of many small pieces of dough or of pans.

Intermediate Proofers

Function of the intermediate proofer—When the dough piece leaves the rounder, it is rather well degassed as a result of the punishment it received in that machine and in the divider. The dough lacks extensibility and tears easily. It is rubbery and would not mold satisfactorily.

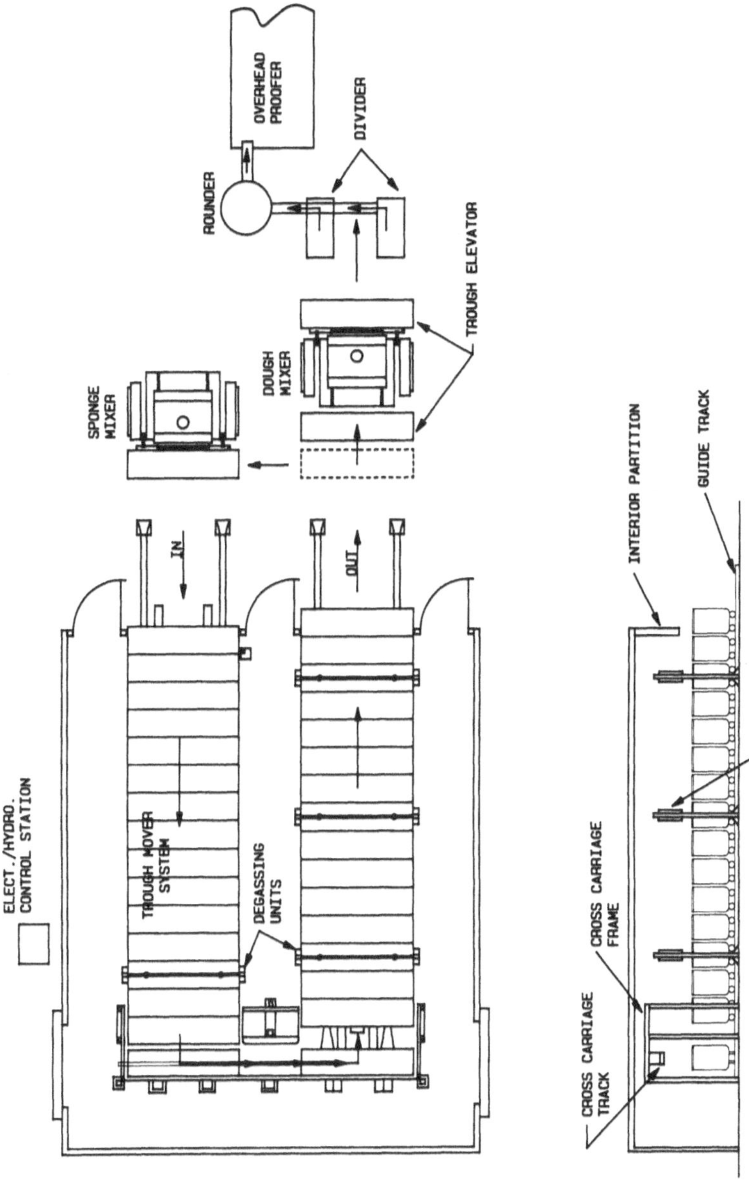

FIG. 5.3. AN AUTOMATIC HANDLING SYSTEM FOR SPONGE TROUGHS.

SOURCE: CLOCK ASSOCIATES

To restore a more flexible, pliable structure which will respond well to the manipulations of the molder, it is necessary to let the dough piece rest while fermentation proceeds and the gluten strands reorient themselves. This is usually accomplished in bread and roll lines by letting the dough ball travel through an inclosed cabinet for several minutes. The physical changes other than gas accumulation which occur during this period are somewhat obscure, but there appear to be some alterations in the submicroscopic structure of the dough which render it more responsive to the subsequent operations. When the dough leaves the intermediate proofer, it is found to be larger in volume due to gas accumulation, the skin is firmer and drier, and the piece is more pliable and extensible.

The essential parts of intermediate proofers are the enclosure or cabinet, the receiving mechanism, the internal conveyor, the power unit, and the controls. In most cases, bread and roll pieces will spend 10 to 15 min in the intermediate proofer. Optimum conditions are thought to be about 80° to 85°F and 70 to 73% relative humidity.

Types of intermediate proofers—Intermediate proofers receive dough pieces from the rounder and deliver them to the loaf or roll molder. A diagram showing the relationship of the intermediate proofer to other makeup equipment will be found as Figure 5.4. Many of the intermediate proofers used at the present time in bread and roll bakeries are the overhead type in which the principal part of the cabinet is raised high enough above the floor to allow space for other makeup machinery beneath it (see the photo in Figure 5.5). When overhead space is not adequate, or when floor space is in plentiful supply, one of the floor level proofing cabinets may be used. Smaller bakeries will use either overhead intermediate proofers or floor level cabinets into which racks of pans containing rounded dough pieces can be rolled.

Intermediate proofers may be conveniently divided into the belt type and the tray type, the latter variety having many subtypes. The former consists of endless belts running in a closed cabinet. The dough pieces are carried forward to the end of the cabinet, then dropped down on the next lower belt, which will be traveling in the opposite direction. This process continues until the dough piece reaches the exit conveyor. The principles of operation of such a machine are illustrated in Figure 5.6. Belts tend to change in length due to various factors and spring loaded idlers may be included to keep the belt relatively taut; otherwise, dough pieces may change position on the tray, creating doubles which jam up subsequent operations.

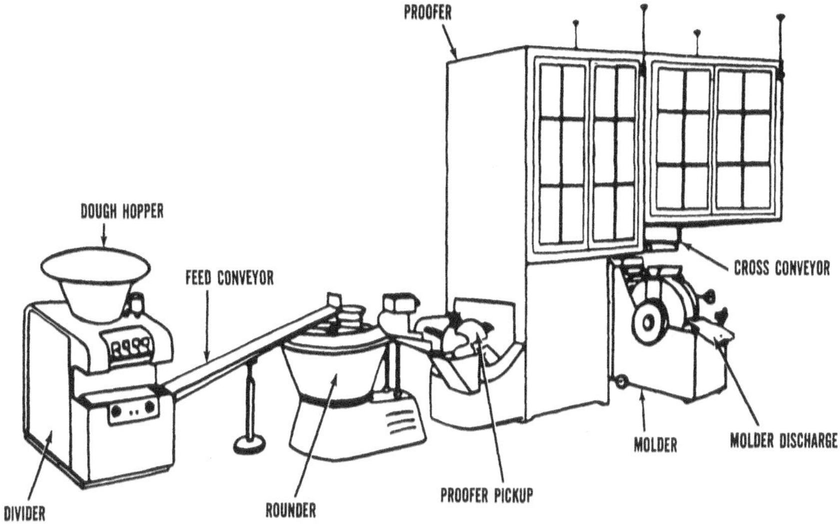

FIG. 5.4. MAKEUP EQUIPMENT SHOWING LOCATION OF FERMENTATION CABINET.

A speed control on the drive mechanism allows adjustment of the time of proofing so that the requirements of different types and sizes of dough pieces can be accommodated. There will normally be an automatic flour applicator to put dust on the trays prior to loading.

In both tray and belt proofers, arrangements can be made to turn the dough pieces over two or three times while they are in the cabinet. The inversions occur at the end of the belt when the dough ball drops to the conveyor below. When trays are used, they are tipped so the dough falls out and into a tray going in the opposite direction. By allowing all surfaces of the dough ball to be exposed to the humid atmosphere there will be less crusting and the skin tends to be thinner. There is also, probably, a more uniform distribution of the gas bubbles.

Tray type equipment is a class of intermediate proofer having conveyors made up of segmented areas for carrying dough pieces. They include equipment which moves the dough in metal pans, troughs or buckets, wooden trays, or canvas loops. For sanitary reasons, most intermediate proofers use solid or perforated aluminum, stainless steel, or solid or perforated plated steel. Paraffined wood trays or plastic-coated wood trays have been used. The trays are usually divided into segments or molded cups to prevent dough pieces from sliding together under the influence of the vibration in the proofer. In some cases, liners of molded plastic are fitted to the trays to minimize or

FIG. 5.5. TYPICAL ARRANGEMENT OF MAKEUP EQUIPMENT. SIX POCKET, THREE PANEL, DOUBLE LAP PROOFER.
SOURCE: APV (BAKER PERKINS)

eliminate the need for dusting flour. Figure 5.7 is a schematic diagram of the intake section of an overhead proofer which has perforated metal pans or trays into which the dough pieces are dropped.

Proofer enclosure panels can be obtained in brushed aluminum, stainless steel, painted steel, or clear plastic. Some or all of the panels should be hinged or on slides for easy cleaning and examination of the interior.

Each of the types has its special virtues and disadvantages and consequently has its proponents and detractors. Among the attributes of the dough carrying unit which are of importance in evaluating the relative worth of an intermediate proofer are: (1) The ease of cleaning and of keeping it clean, (2) The ease of replacement and economy of repair, and (3) The effectiveness of the unit in preventing doubles and retaining the desired form of the dough piece.

Location of overhead proofers of this and other types must be carefully planned. Distance to the rounder should be minimized; ideally only a few feet of conveyor should intervene. Discharged dough pieces should have only a short distance to travel to the molder or sheeter. These arrangements not only save space, but they reduce the adverse influences on the dough pieces as they pass through the uncontrolled

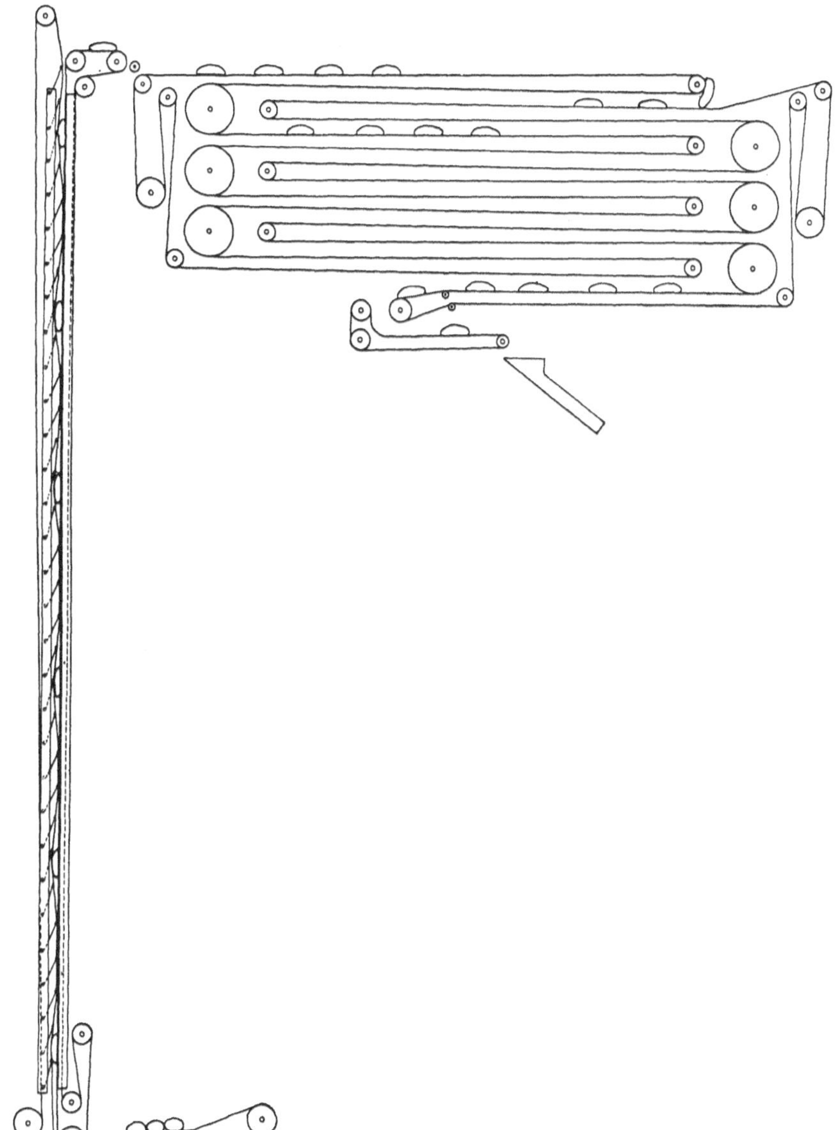

FIG. 5.6. Belt type intermediate proofer.
Source: Dutchess Bakers Machinery Co.

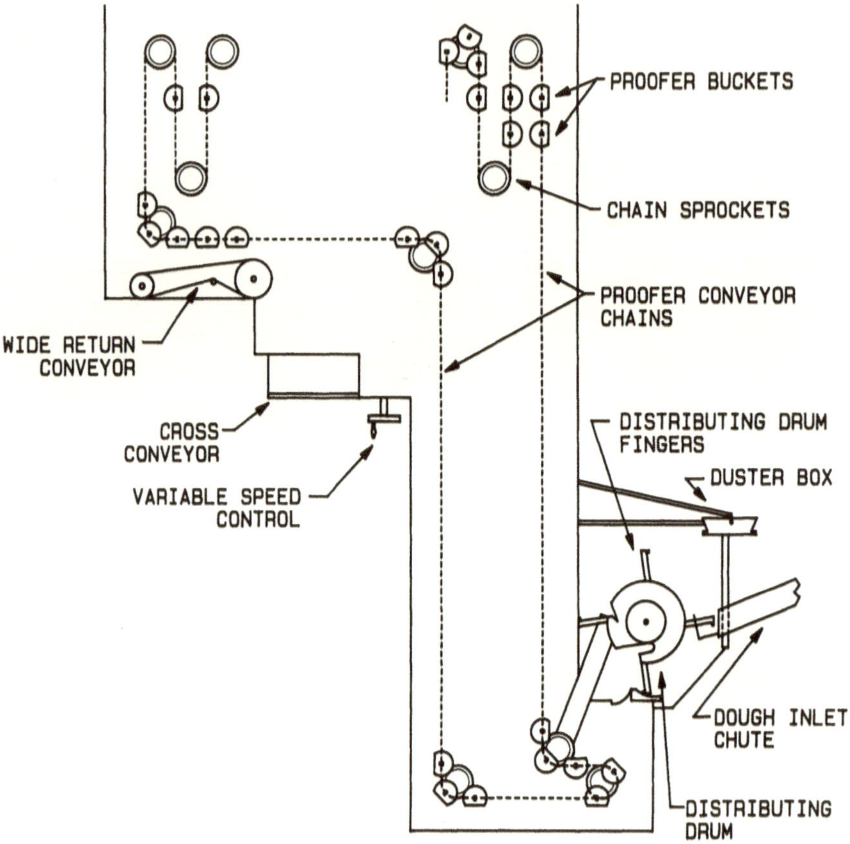

FIG. 5.7. INTAKE SECTION OF TRAY TYPE INTERMEDIATE PROOFER.

bakery atmosphere. Intermediate proofers should not interfere with free movement of personnel beneath or around them, or with placement and function of other makeup equipment. They should be easily accessible from all sides for cleaning and maintenance.

An important part of the intermediate proofer is the mechanism which receives the dough ball from the rounder and puts it on the conveying mechanism. This loading mechanism must take the dough pieces which arrive in single file and arrange them in rows on the tray of the intermediate proofer. Proofer trays may contain spaces for 2 to 8 dough pieces distributed along the width of the tray.

Also important is the discharge mechanism. In a typical arrangement, dough pieces are automatically deposited onto single or double discharge conveyors for either right-hand or left-hand delivery to the molders. This allows flexibility in routing the discharge to equipment which is required for the particular dough being processed.

Controlling and adjusting the intermediate proofer—Intermediate proofers of all types are equipped with variable speed controls which can be used to determine the length of time dough pieces spend within the cabinet. An absolute maximum of dwell time is dictated by the maximum capacity of the proofer and the rate of output of the divider and rounder. If an attempt is made to slow the machine below this point, doubles will accumulate in the dough pockets or the belts. Below this maximum level of capacity, the correct time for the dough piece to stay within the cabinet is determined by the period necessary to condition the dough properly. About 6 to 20 min is the range within which most proof times can be adjusted. This covers nearly all requirements.

It seems that no large capacity (over 3,000 lb per hr) intermediate proofers are available with air conditioning equipment built in. Temperature and relative humidity are thus at the mercy of conditions in the room surrounding the cabinet and of the state of the dough pieces entering the proofer. Fortunately, an absolutely uniform and carefully controlled environment is not essential for the dough at this stage. A moderate range of variation in temperature and humidity can be tolerated without causing significant observable differences in the finished loaf. Since room conditions usually remain fairly constant over extended periods, and since the temperature and equilibrium relative humidity of the dough pieces do not usually vary much, insuperable difficulties in adjusting the intermediate proofer seldom occur. Because of the large mass of dough traveling through the cabinet, the atmosphere in the cabinet tends to approach an equilibrium state which reflects the temperature and water activity of the dough. The temperature is usually related to, but not the same as, the temperature of the dough coming out of the mixer, and the equilibrium relative humidity is established by the composition of the dough, i.e., the batch formula. Most intermediate proofers are provided with panels which can be partially opened to bring internal conditions more in line with those existing in the surrounding space.

The dwell time of the dough pieces in the intermediate proofer is normally short, so environmental conditions do not have as much opportunity to influence the material as they do at the sponge and pan proof stages. Nonetheless, it would seem to be desirable to have means for controlling the temperature and relative humidity of the interior of the cabinet which is not entirely dependent on the dough characteristics. Some modern cabinets of relatively low throughput do have air conditioning units, or provision for ventilating with air from an outside source. Others could be improved by the addition of steam outlets to provide additional moisture. Opening the doors of the cabinet to reduce the humidity or temperature has the disadvantage that it creates drafts with resultant erratic behavior of the dough pieces.

Maintenance of intermediate proofers is largely a matter of keeping the machine cleaned and lubricated. The pockets or belts accumulate coatings of dusting flour and dough and must be cleaned frequently. This process requires removal of the trays, buckets, or belts, and thorough washing of them. The schedule for cleaning is dependent upon the climate, the susceptibility to infestation of the plant, and other factors.

Final Proofers

Conventional and traditional types—Final proofers or pan proofers receive the shaped and panned dough piece and hold it for a time before it is delivered to the oven. During this period, the dough pieces gain most of the volume they will have as finished loaves or buns, though, of course, oven spring will add some additional volume. Fermentation reactions develop flavors and flavor precursors which are essential to the organoleptic properties of the bread. Reactions which both consume and yield sugars will have a strong effect on crust color of the finished product. Water will be lost, and the initial stages of crust formation can be observed. The phenomena which are so important to the quality of the finished loaf, and only part of which have been listed, are affected by the temperature of the dough and this, in turn, reflects to some extent the temperature of the final proofer. Humidity also has an effect on finished product quality, but its action is reflected mostly in crust characteristics.

Final proofers are of several types. The most primitive example in commercial bakeries was a shallow box or drawer which controlled humidity and temperature only by sealing off the dough piece from the bakery air for a short period. Of course, these arrangements are no longer in use, at least, not in developed countries. Eventually, the advantages of better control became apparent and special rooms were designed for final proofing. At first, the dough was taken into the room manually, sometimes by pushing racks of pans into one end of the room and removing them from the other end. Injection of steam from a pipe leading from the boiler was the means of both heating and humidification. Adjustment of conditions was manual and based on occasional reading of the wet-and dry-bulb thermometers, if an instrument of this sophistication was available.

The high labor requirement for conducting the proofing operation by such methods led to the design of several different conveying systems. In the floor-type proof box, racks mounted on casters are used to hold many pan straps. There may be several lanes through the proof box, with a series of racks moved through each. The racks are manually pushed into one end of the box, thereby moving the preceding racks ahead one space, and pulled out of the other end at a rate calculated to give the necessary proof time. The enclosure is a steel frame covered with insulated panels and provided with an automatic air-conditioning system. Overhead rails or floor rails may be used to guide the manually moved racks. Continuous conveying of straps of pans through final proofing enclosures is an advanced method requiring the minimum amount of personnel participation. Figure 5.8 is an example of such a pan proofer.

Controlling temperature and humidity—Conditioning units will include blower, radiators, steam jet, spray units, and eliminators. Figure 5.9 is a diagram of one version of a conditioning unit for a final proofer. In the larger units, stainless steel ductwork will be designed and positioned to distribute the conditioned air in a pattern which is best suited to the conformation of the room and the arrangement of the racks and pans. Air curtains can be installed in loading and unloading openings to reduce loss of conditioned air into the bakery.

The room or cabinet used for pan proofing needs ample heating capacity to quickly counteract infiltrated cold air. This requirement occurs particularly when a door is opened to admit a cold rack with its complement of cold pans and dough. Adding to this burden is the fact that with an open door the outside air rushes into the proof box at the lower part of the open door, and at the same time the conditioned air from inside the box rushes outward. When the number of doors involved and the rate of their opening and closing is considered, it is obvious that a substantial loss of heat and humidity loss will occur. The rate at which the conditioned atmosphere recovers from this abuse is dependent on the sensitivity of the thermostat and the heat capacity of the conditioning equipment.

A proof box does not consume steam continuously; the demand is intermittent and for a great volume at short intervals. If a proof box were a totally inclosed space without air infiltration or ambient heat-subtracting influences, wet-and dry-bulb temperatures could be maintained to fractional degree values. As it is, a proof box has many doors, many cracks, and many cold racks loaded with cold pans and cool dough entering and leaving only a few minutes apart.

If a temperature control apparatus is to perform its assigned duty in

FIG. 5.8. CONTINUOUS RACK PROOFER. AUTOMATIC LOADING AND UNLOADING.
SOURCE: APV (BAKER PERKINS)

a proof box, sensible heat must be available. When steam pressure drops below 15 psig, sufficient heat is not transferred from the heating coil or radiator to the recirculating air to maintain the desired dry bulb temperature. Normally, the wet bulb (humidification) control continues to add moisture to the atmosphere, resulting in an overly humidified condition. Without protective circuitry in the thermostatic complex to avoid creating a dew point, condensation will appear on cold surfaces. Steam pressure is not the critical factor; steam in volume is required.

The pressure of steam indicates its temperature; at 10 psig it is 240°F, at 15 psig it is 250°F, and the higher the pressure the higher the temperature. This is the reason steam from a high pressure boiler must have a pressure reducer in the line and also why bare pipe should be provided for a distance of at least 25 ft to the proof box humidification unit to obtain wet steam.

Pressure of any fluid will be reflected regardless of the diameter of the conveying pipe. To obtain greater volume, a larger diameter pipe

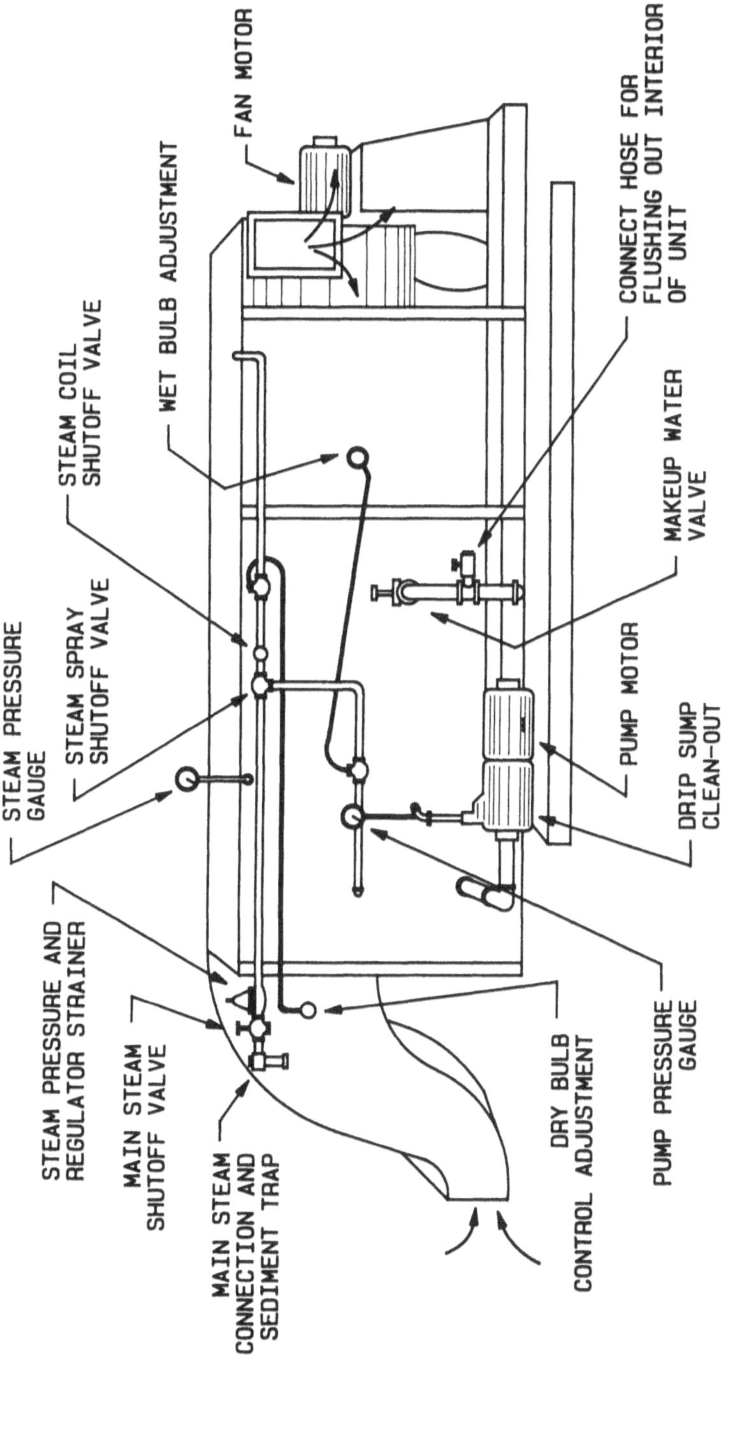

FIG. 5.9. CONDITIONING UNIT FOR A FINAL PROOFER.

and pressure are required. A pound of water, when vaporized, will produce less than 1,000 Btu. If it takes up to 12,000 Btu per hr to maintain 110°F per proof rack, pans, and dough, there will be 12 lb of condensate draining from the radiator. A simple test is to weigh the condensate collected for one hour as it drains through the trap.

Handling the pans—Pan conveying systems somewhat similar to those used in traveling hearth ovens have been built. Other automatic conveying means which have been suggested or designed include belt-and chain-driven mechanisms. Relatively large amounts of floor space are required and construction costs tend to be high. Whatever the method used for conveying the pans, straps, or racks, it should not cause undue vibration, shock, or impact. Toward the end of the pan proof, and perhaps before, the extended condition of the gluten network and the relatively relaxed state of the proteins render the dough structure readily subject to distortion or collapse when subjected to minor shocks.

Modern proofers for large-scale operation automatically load pan straps onto racks which are integral parts of the proofer. In a multitray proofer, pans are grouped automatically and fed to the loading section. The loading conveyor is coordinated with the tray unit, both operating intermittently. As each tray is loaded, the tray unit is indexed upward one tray pitch while the loading conveyor delivers another load. When filled, the tray unit is indexed upward once more and then pushed one space down the length of the proofer by the top ram. This action also causes the movement of one tray unit onto the elevator at the other end of the proofer. The elevator lowers the tray to the lower level where a bottom ram pushes it one space in the return direction. This movement automatically forces another tray unit onto the front-end elevator which indexes it upward through the unloading position and then back to the loading position. Where plenty of floor space is available, the multitray proofer is the simplest, most economical type.

In multi-tier proofers, each tray after it is loaded is raised individually to the top tier of the proofer (see Figure 5.10). Drive chains on each side of the loaded trays convey them back and forth throughout the length of the proofer, dropping them to a lower tier at each turning point until they have reached the unloading station. The trays then continue through the bottom leg to the loading station. In all types of automatic proofers, the dwell time can be varied within predetermined limits so as to fit production requirements.

Another type of automatic proofer comprises eight principal functions: feeding, grouping, and discharge conveyors; automatic loader; unloader; elevator; Lowerator; and double-deck tracks upon which

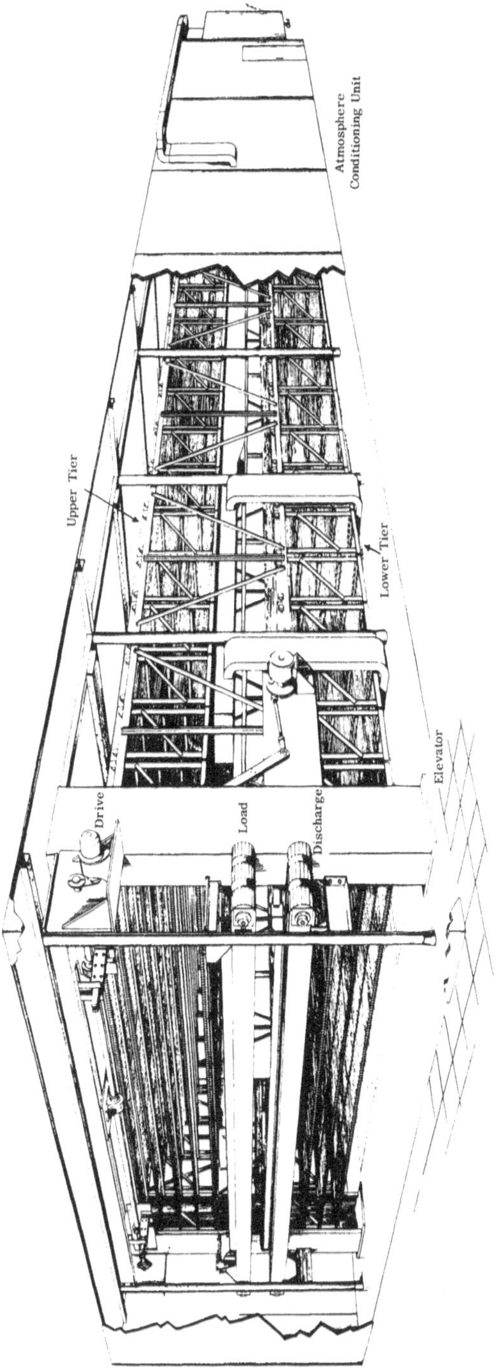

SOURCE: J. W. GREER CO.

FIG. 5.10. AUTOMATIC PROOFER OF THE MULTI-TIER DESIGN.

racks are conveyed by hollow-stud roller chains. The racks are equipped with flanged wheels that roll along the floor rails. Each rack would have, for example, seven shelves, each holding eight straps of five 1.5 lb pans. Four endless conveyor chains are powered by sprockets at the ends of each of the four rails. Uniformly spaced projections engage slots at both sides of the racks and move them laterally on the upper rails, then transfer them to the Lowerator at the rear of the proofer. The Lowerator engages slots at the sides of each rack and moves it to the lower rail where the slots are caught by dogs on the conveyor chains.

The Lowerator and the elevator consist essentially of a set of two endless chains powered by sprockets. The chains operate in the vertical plane and are located on both sides and at the ends of the two sets of rails. Each set of chains is fitted with short lengths of rail, properly spaced to match the trucks. The elevator is more complex than the Lowerator, since it must index the shelves at proper intervals, keeping them level with both the feed and discharge conveyors. The loader and unloader consist of two steel pusher bars on rams mounted at opposite ends of heavy steel rods. As they move in a horizontal plane, the pusher bars simultaneously sweep straps off the feed conveyor and onto a shelf and off a shelf onto the discharge conveyor. A proof box about 18 ft wide, 12 ft high, and 68 ft long could accommodate about 8,120 loaves at a time. Similar equipment is used for rolls.

Valentyne and Hoag (1963) and Petersen *et al.* (1963) described proofers in which the time of transit of a tray of product can be varied. As product passes through the proofer, it travels over several flights of a conveyor, one above the other. Products such as rolls, not calling for as long a proofing time as, say, bread, are not introduced onto the bottom flight but are introduced onto one of the upper flights depending on the proofing time desired.

Bread, rolls, pizza, Danish, and other panned products can be final proofed in rooms provided with two connected spiral conveyors. The pans are placed on the conveyor belt by hand or by automated mechamisms.

Automatic and Integrated Pan Proofing and Baking

Additional advances were made in the late 1970's and the 1980's when systems were developed for continuous conveying of pans from the molder through the oven. For example, panned product is received from the molding machine and grouped. The grouper collects the pans and places a certain number of them in a pre-determined array. The grouped pans are carried in this form to a rack-type proofer, where the loader transfers them onto the proofer shelf.

Grissinger (1973) lists four advances in the mechanical features of rack-type proofers:

(1) Recent types of proofers have only one drive compared to the earlier versions which had four separate drives for moving the racks through the enclosure.

(2) Rack movement is now accomplished with continuous drive motions compared to previously used intermittent motions.

(3) Rack motions were electrically synchronized but are now mechanically synchronized, resulting in simpler electrical circuits and less than one-half the number of limit switches.

(4) Racks were originally free to swing during parts of their travel, causing rack hangups, but they are now fully stabilized throughout their travel.

The pans are unloaded from the proof room (at the same end where they were loaded), and placed on another conveyor which carries them to the oven loader. They pass through the oven, then to the depanner where the pans are separated to return through the system while the loaves, buns, etc., are sent through the cooler, slicer, and packaging line. All without being touched by human hands and, except in the case of emergencies, not even monitored.

These lines originally were developed for loose pans which were carried on various kinds of conveyors, some of which grasped or positioned the pans. Problems in maintaining control of the pans tended to arise if the they were distorted or damaged. Ultimately, these developments culminated in a conveyor to which straps of pans were attached and which carried these pans through a curving path inside an air-conditioned room until pan proofing was complete, then carried them into a room-like oven where they were baked. The attached pans need never leave the conveyor line. This automated system was a genuine innovation in processing. It reduced labor and promoted uniformity. More details will be given in the chapter on ovens.

Expedited Proofing by Microwaves

Conventional bread baking systems include proofers which subject dough to a proofing environment of, for example, 120° to 125°F and 80 to 85% relative humidity for about 30 minutes in order to bring the temperature of the center of the unbaked loaf to about 110°F. Although the dough surface is relatively quickly raised to the desired temperature, considerable time is required before the center of the piece reaches 110°F. Microwave heating is a possible way of speeding up the heat rise and shortening the proofing period. Methods for implementing this process have been developed (Ingram *et al.* 1975). Generally, these methods combine conventional treatment in a conditioned air cabinet with a preliminary microwave heating step. As with most microwave heating methods involving a continuous line, one of the major problems which is confronted is shielding the surrounding areas from escaping radiation.

Microwave heating has been used to speed up the proofing of doughnuts. The entire doughnut can be heated to proofing temperature in a few seconds by microwaves, but about four minutes are allowed for adequate yeast action and dough relaxation. About 30 to 35 min in a conventional proofer would be required for the same effect.

The doughnuts are carried through the microwave tunnel on a canvas belt. This particular system is not applicable to doughs which must be proofed in or on pans because of interference with the microwaves by the metal. A 2.5 kw magnetron tube operating at 2450 MHz is housed in a console at the side of the tunnel and the microwaves fed through a waveguide into the cavity. Air, heated and humidified by conventional techniques, is blown through the cavity to retard evaporation from the dough pieces. The doughnut mix must be specially formulated for this type of proofing to be satisfactory, according to Russo (1971).

LIQUID SPONGES

In the last 20 or 30 years, there has been a considerable shift to the employment of liquid sponges (variously called brews, broths, ferments, etc.) to replace the dough-like sponges with plastic properties on which traditional bread-making processes are based. The most obvious advantage of liquid sponge methods is that pumps, pipes, valves, meters, and tanks can be used to replace the cumbersome dough troughs and space-consuming fermentation rooms of the conventional bakery. Mixing equipment is also simpler and less expensive, and the time required to complete the initial fermentation is often less. The effect of liquid sponges on bread quality is a question which has received a plethora of answers but, from a commercial point of view, it appears to be obvious that satisfactory bread can be made with liquid sponges.

The conditions applied to, and formulas used for, liquid sponges have been discussed in considerable detail in the two previous books in this series. In the present discussion, we will be concerned primarily with the equipment for preparing, holding, and transferring liquid sponges.

Although liquid sponges can be made by manually weighing and depositing the ingredients into mixers, it is considerably more efficient, and more conducive to continuous processing, to automatically meter materials into specialized liquid sponge makers. An example of a liquid sponge maker is shown in the photo reproduced here as Figure

5.11. Typically, such equipment will continuously meter and blend controlled portions of flour, water, yeast, and additives to produce a uniform liquid sponge containing all the formula water and up to 70% of total formula flour. Most often, considerably less flour will be incorporated at this stage since the added viscosity of high flour mixes increases the difficulties of handling the sponge as it ferments. Sponge temperature is regulated by adjusting the water temperature; tanks can be jacketed but seldom are for this type of operation. The finished brew can be put through heat exchangers if it becomes too warm for the final mix. Automatic valving, controls, and programmers are used to obtain a uniform and continuous flow at the rate required. Rates of up to 14,000 lb per hr are readily achievable with fairly small pieces of equipment.

The slurry is pumped from the blender to fermenting vessels where the liquid sponge undergoes those changes which ready it for incorporation into the dough. Fermentation vessels can be batch-type or continuous-type. The batch type will usually be adaptations of vertical tanks equipped with agitators designed to handle liquids of the viscosity expected to be encountered. Positive flow continuous fermentors allow freshly mixed liquid sponge to enter a horizontally aligned cylindrical tank at one end and, after the desired fermentation time has elapsed, to flow out of the opposite end, giving a true first in-first out sequencing. A transfer reel turning slowly within the fermentor imparts positive flow to the liquid sponge through a spiraling in and out pattern. The standard units do not have heat control equipment, but this can be supplied as an option where ambient or other conditions require cooling of the fermentation mixture.

From the fermenting vessels, the liquid will be transferred by an appropriate type of pump through a heat exchange cooler to a refrigerated holding tank. If a continuous fermentor is being used, the speed of the pump must be automatically controlled to deliver fermented sponge at a rate which is synchronized with the fresh mix entering the system, so that all the liquid sponge receives a uniform amount of fermentation.

There are many variations in plant designs for liquid sponge lines. The following is an example of the equipment required for one such plant (Anon. 1987).

(A) Nutrient slurry make-up tank; 500 gallon fiberglass, closed top.
(B) Centrifugal pump.
(C) Nutrient slurry supply tank; 150 gallon fiberglass, closed top.
(D) Nutrient slurry feeder pump; variable speed drive.
(E) Yeast slurry make-up tank; 500 gallon; "B" dome top; jacketed.

FIG. 5.11. Automatic liquid sponge maker.

Source: AMF

(F) Yeast slurry supply tank; 150 gallon; "B" dome top; jacketed.
(G) Yeast slurry feeder pump; variable speed drive.
(H) Continuous liquid sponge mixer.
(I) Positive flow continuous fermentor.
(J) Sponge transfer pump; variable speed drive.
(K) Plate cooler for sponge.
(L) Cold hold tank; 1,500 gallon, "B" dome top; jacketed.
(M) Cold sponge pump; variable speed drive.
(N) Batch weigh tank.
(O) CIP tank and pumps; multi-compartment.
(P) Control panel.

Pumps, pipes, valves, and meters are preferably constructed of stainless steel and must be sized to meet the production needs of the bakery. It is seldom, if ever, necessary to jacket these elements for temperature control. Clean-in-place sanitation is readily adaptable to these installations.

Following is the description of a suggested dual temperature liquid sponge ferment procedure developed for bakeries using conventional mix 50° to 65°F sponge at the re-mix stage and continuous mix 75° to 85°F sponge at the premixer or incorporator (Anon. 1988).

(1) The ingredients for the yeast slurry and nutrient slurry are scaled out and blended with temperature controlled feed water in the slurry make-up tank. The liquid mixes are transferred to the appropriate yeast slurry or nutrient slurry supply tank, each tank being sized to store the requirements for several hours production.

(2) At the start of the production run, centrifugal pumps are used to maintain a flooded condition at the positive rotary transfer pumps which serve to meter the slurries to the sponge blender.

(3) Flour is diverted from the overhead pneumatic flour conveying system into a hopper mounted above the sponge blender. Flour is continuously weighed onto a belt moving at a pre-set rate and allowed to fall into a circular incorporation chamber housing a squirrel cage agitator. Metered amounts of yeast slurry, nutrient slurry, and temperature controlled feed water are piped into the chamber for rapid incorporation with the flour to form a pumpable sponge containing up to 65% of the total flour.

(4) Liquid sponge mix is transferred at an uninterrupted flow rate to the inlet on the positive flow fermentor.

(5) The fermented liquid sponge is transferred by two variable speed positive rotary pumps set to run at a combined flow rate equal to that of the transfer pump feeding the input end of the fermentor. Each pump

transfers the required amount of fermented sponge through the dual temperature plate heat exchanger to the appropriate holding tank.

(6a) Sponge for the conventional mix operation can be transferred by a reversible pump to the batch weigh tank and then back through the pump to a diverter valve and on to the supply header serving the final mixers.

(6b) Sponge for the continuous mix operation is transferred directly to the pre-mixer or incorporator where the dough side ingredients are combined with the sponge prior to the final mixing. The surge or holding tank helps to maintain a uniform metering rate and an uninterrupted operation.

(7) With the exception of the automatic sponge blender, all system components may be cleaned in place with any desired level of automation including optional salvage or recovery-rinse-cycle.

An optional method using heat exchangers follows the above scenario through step 4, then transfers the liquid sponge ferment from the fermentor through a single function plate heat exchanger for cooling to 55°F, then into a jacketed cold sponge holding tank. On demand, a transfer pump moves the fermented sponge to the batch weigh tank, and a second pump transfers the batch to the supply header for the final mixers.

A similar method based on tank-type make-up and positive flow fermentor includes the following preliminary steps.

(1) Temperature controlled water is metered into a dome-top jacketed blending tank. The agitator is started and pre-weighed yeast, yeast food, salt, and other sponge ingredients except flour are added as needed for one batch of sponge. Flour in the scale feed hopper mounted above the make-up tank is incorporated as quickly as possible to avoid gluten coagulation. The agitator is then stopped and the entire batch is transferred to a jacketed surge tank.

(2) While sponge from the surge tank is being metered into the fermentor, a new batch of liquid sponge must be prepared to maintain uninterrupted sponge flow to the fermentor. The transfer flow rate is determined by the desired volume level and the residence time required in the fermentor.

(3) After the established fermenting period, the sponge is pumped through the heat exchanger into the cold sponge tank which is jacketed to inhibit further yeast activity.

(4) The sponge may be transferred into the weigh tank to the final mixer or directly to the premixer or incorporator if a continuous mixer is used.

(5) The entire system may be circulation cleaned and equipped with any degree of automated control.

If a low flour (up to 30%) liquid sponge system is preferred, an alternate method is suggested using tanks for lines of up to 6,000 pounds per hour of finished dough production.

(1) Temperature controlled feed water is metered into Ferment Tank No. 1. With the agitator running, pre-scaled ingredients including yeast, yeast food, salt, butter, sugar, and flour are added, and the slurry is allowed to ferment to the desired pH and titratable acidity.

(2) Ferment Tank No. 2 is then charged following the same procedure described in (1). The procedure is timed to permit the transfer of all the sponge from Tank No. 1 through the heat exchanger to the jacketed cold ferment holding tank by the time the mixture in the second tank has matured.

(3) A rotary pump is used to meter and transfer fermented sponge on demand to the scaling weigh tank or to the mixers, as desired.

The basic principles given in the preceding preparation methods can be used for a water broth ferment system using jacketed horizontal vessels for blending, fermenting, cooling, and cold storage.

(1) Temperature controlled feed water is metered into the first ferment tank. With the agitator running, all pre-scaled ingredients except flour are added as in the Low Flour System.

(2) After the pre-determined levels of pH and titratable acidity are reached, the direct expansion cooling system is activated. Cooling is continued with the agitator running to maintain dispersion of solids in the broth.

(3) After the water broth cools to 40° to 50°F, it is metered to the selected mixer at a pumping rate controlled by the electromechanical timer controlling the running time of the transfer-metering pump.

The dry ingredient incorporator recommended for the above systems is a versatile tank-type blender suitable for batch make-up of yeast and nutrient slurries, and as a flour incorporator. A high speed squirrel cage agitator creates a dual blending action combining an overall swirl with a deep draw vortex to quickly disperse the ingredients while minimizing the risk of gluten agglomeration. It is available in five capacities from 100 to 1,000 gallons, and is designed with a 15° cone bottom to facilitate complete and rapid unloading. It can be provided with a variety of options, including CIP, sensing devices, outer cooling jackets, electric or hydraulic agitator drive, etc.

The automatic sponge blender is designed to continuously meter and blend controlled proportions of flour, water, yeast slurry, and nutrient slurry. It produces a uniform liquid sponge containing all or part of the formula water and up to 65% of the total formula flour. Sponge temperature is regulated during flour incorporation by adjusting the feed water temperature. When used as an integral component of an automated liquid sponge system, the multi-ingredient incorporator can be easily integrated into existing and new liquid ferment systems. The use of metering pumps, automatic valves, and control programmers help to assure a uniform and continuous rate of sponge flow to the fermentor. This piece of equipment is made in two models, one to provide the sponge required for 7,000 pounds of finished dough, and the other double that size. The maker cautions that 65% flour sponges require the addition of all the formula water to the sponge; excessive viscosity will result if less than nine parts water is added to ten parts flour. Optimum viscosity is attained with equal parts of flour and water.

BIBLIOGRAPHY

ANON. 1987. CP Continuous Fermentor, Bulletin L-1-150. Crepaco, Inc., Chicago, IL

ANON. 1988. Liquid Sponge Ferment Systems. APV Crepaco, Chicago, IL

DOWDS, H. M. 1987. Baker's dough proofing and raising unit. U. S. Pat. 4,635,540.

GRISSINGER, G. R. 1973. Automated proofing and baking advances. Proc. Am. Soc. Bakery Engineers *1973*, 128-137

INGRAM, C. E., BRUNK, R. H., and WITKOSKE, E. 1975. Apparatus for making bread and like food products. U. S. Pat. 3,881,403.

MAY, L. 1962. Spaulding's three streamlined production lines for breads, rolls, and donuts. Bakers Weekly *195*, No. 9, 40-49.

PETERSEN, C. W., WITTENBERGER, W. W., and ST. JOHN, J. M. 1963. Bread handling apparatus. U. S. Pat. 3,101,475.

PFENING, F. D. 1972. Personal communication. The Fred D. Pfening Co., Columbus, OH

RUSSO, J. R. 1971. Microwave proofs doughnuts. Food Eng. *43*, No. 4, 55-58.

SCHIFFMANN, R. F., STEIN, E. W., and KAUFMAN, H. B., JR. 1971. The microwave proofing of yeast-raised doughnuts. Bakers Digest *45*, No. 1, 55-57, 61.

VALENTYNE, P. H., and HOAG D. H. 1963. Dough processing apparatus. U. S. Pat. 3,101,143.

CHAPTER SIX

FORMING AND MOLDING EQUIPMENT FOR BREAD-LIKE PRODUCTS

INTRODUCTION

The subject of dough forming and shaping is so complex that it was necessary to divide the discussion into three chapters, this one which contains material about bread and bread roll equipment, pizzas, English muffins, etc., chapter eight, which includes descriptions of cookie and cracker equipment, and chapter seven, which describes all other forming equipment. In the present chapter, the greatest amount of space has been devoted to bread molding equipment, not only because bread and bread rolls are the products processed in greatest volume by bakeries, but also because the principles involved in their manufacture are often applicable to the processing of other kinds of bakery foods. Special forming equipment for several other bakery products bearing a general similarity to bread in their composition and method of manufacture will be discussed, but not in the detail used for bread and roll molding. The main emphasis in this, as in the other two chapters on forming equipment, will be on machines suitable for wholesale or large retail bakeries. There are several publications on retail shop practices available to technologists and entrepreneurs who are interested in that aspect of the bakery arts, and there would be little justification for duplicating such coverage in the present volume. On the other hand, the lack of availability of organized discussions of mass production equipment dictates the fullest possible use of this space for consideration of such machinery.

Current procedures and machinery are emphasized, but some historical material is included when it is considered necessary for a better understanding of the present state of the art. Trends and the reasons for trends in equipment design are discussed in relation to the past.

Examples are chiefly confined to U. S. practices. It is recognized that these are not the most advanced in every case, but they are likely to be the most important to the reader of this volume. Where important and instructive deviations from U. S. practices are thought to be readily applicable to situations existing in this country, they are brought into the discussion.

In general, types of equipment described in other chapters—sheeting rollers, laminators, proofers, dividers, rounders, etc.—will not be discussed in this chapter even though they form parts of the lines used to mold and shape products. It was necessary for clarity and completeness, however, to give brief descriptions of these auxiliary devices when describing complete make-up lines. Otherwise, the reader might have been left with a mistaken concept of the whole processing system.

LOAF MOLDERS

Function of the Molder

In bread-making plants, the molder receives pieces of dough from the intermediate proofer and shapes them into cylinders of loaf size ready to be placed in the pans. There are several types of molders, each of which will be described later, but all except the extrusion type have four functions in common: sheeting, curling, rolling, and sealing. Some writers consider the last two as a single function, since they are performed simultaneously.

Figure 6.1 is a schematic diagram of a simple type of molder which performs these functions. The dough as it comes from the intermediate proofer is a flattened spheroid. The first operation of the molder is to flatten this piece out still more into a thick sheet which can be properly manipulated in the later stages of molder operation. This effect is usually achieved by two or more (usually three) consecutive pairs of rollers, each succeeding pair being set closer together than the ones which precede it. The first pair of rolls, called the head rolls, exerts only a relatively slight pressure on the dough piece. The second set of rolls, the center rolls, presses the dough more strongly, while the last set—which may be called either the sheeting rolls or the lower rolls—exerts the greatest pressure on the dough sheet. In the case of a reverse sheeting molder, there are two single rolls, called the reversing roll and the receiving roll, which intervene between the center rolls and the lower rolls.

The gradual reduction in thickness effected by this multiple roller system minimizes the punishment received by the dough so that tearing and similar problems are reduced. Sometimes a single "flattening" roll is located above the infeed conveyor which leads to the molder. This flattening roll performs a slight initial reduction in thickness which facilitates engagement of the dough piece by the first pair of rolls.

It has become fairly common to encase some or all of the sheeting rolls with Teflon sleeves in order to prevent them from adhering to the dough pieces. When the molder sheeting rolls have not been treated in this manner, scraper blades are placed so as to assure the separation of dough sheets from the rollers.

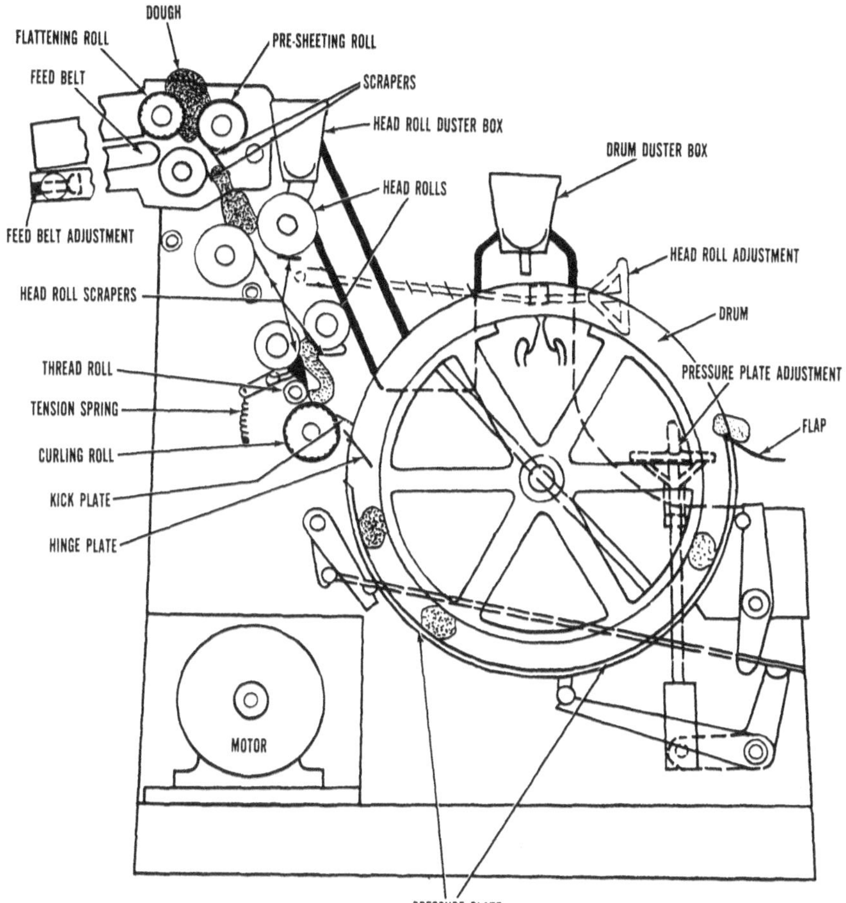

FIG. 6.1. DIAGRAM OF A SIMPLE DRUM MOLDER FOR BREAD LOAVES.

After the dough has been sheeted out, it is curled up into a loose cylinder. This operation is conventionally performed by a special set of rolls, as indicated in the figure previously cited. Alternatively, it is accomplished by a pair of canvas belts. The lower conveyor belt moves the dough piece forward until the upper curling belt or mat engages the front end of the piece and holds it while the rest of the sheet continues forward, causing the sheet to curl into a loose cylinder. A more advanced development substitutes a short length of woven metal mat or linked thin metal bars for the upper curling belt. In these machines, the metal curling device is affixed just above the conveyor belt, with one end resting on the belt. As the dough piece passes under the curling device, the weight of the latter creates enough drag to pull the forward end of the dough piece up and delay it while the conveyor belt rolls the piece into a cylinder.

The layers in the cylinder of dough are not tightly adherent when it leaves the curling section. The next function of the molder is to thoroughly seal the dough piece so that it will expand into the typical loaf shape when it is proofed. In addition, entrapped air between the dough layers is expelled and the cylinder of dough is lengthened so that its axial dimension is somewhat greater than the inside length of the pan. One way of achieving these results is by rolling the dough cylinder between a large drum surfaced with canvas and a semi-circular compression board having a smooth surface. Clearance bewteen the drum and the board is gradually reduced along the route of dough travel so that the piece is constantly in contact with both surfaces and gradually becomes compressed.

A more common arrangement is a flat pressure board and a powered belt which gradually squeeze the dough cylinder into shape as a result of the decreasing distance between them. At high speeds, say 80 pieces per minute, dusting flour is usually applied just before the dough piece goes under the pressure board. Pressure boards range in length from 24 to 48 inches. The surface of the pressure board is often altered by some kind of "buildup" to give better control over the shape of the dough piece. A layer of sponge rubber covered by canvas is a common expedient. Angles made of metal or wood strips or half-cylinders are sometimes attached to the pressure board, with the points of the angles pointing toward the intake, and the whole structure covered with canvas. Thick pieces of canvas belting are also used to build up the pressure board. These pressure board modifications are intended to permit use of a loose pressure board without getting pointed ends on the dough pieces. They leave the ends of the dough pieces rather loosely sealed so as to allow the gases to escape, the end result being a more even grain throughout the loaves.

An integral component of most modern molders is the automatic panning device. Empty pans are carried by a conveyor past the end of the molder, and the loaves are transferred from the molder and positioned in the pans by an apparatus operated by compressed air.

Types of Molders

As the dough piece passes through the sheeting rolls, there is a tendency for the moisture content of the trailing edge to become increased at the expense of the leading sections. This redistribution of moisture results from the effects of compression on the dough structure. In the normal course of events in a conventional molder, the trailing edge, which is of relatively high moisture content, ends up as the outside layer of the cylinder. It has been thought for many years that it would be preferable, from the standpoint of loaf performance during proofing and baking, to have the wetter portion of the dough sheet folded into the center of the loaf. Observations on hand molded doughs formed in this manner seemed to confirm the superiority of "reverse molded loaves."

Some of the major modifications which have been made in molder design are the result of attempts to avoid folding the dry end of the dough sheet into the center of the loaf. Commercially accepted developments have included the cross-grain molder (Figure 6.2) and the reverse sheeting molder. The former type curls the dough sheet at right angles to its direction of travel through the sheeter rolls. As a result, the wetter edge of the dough forms one end of the loaf rather than the outside layer. The crossover effect is achieved by changing the direction of travel of the dough 90° after it leaves the sheeting rolls. The first of these machines developed used a turnover or "flipflop" method of transfer, while a slide transfer or "shootover" method later came into use.

Reverse sheeting molders were devised to curl the sheet of dough so that the wet end of the piece would be folded into the center of the loaf. This is accomplished by turning over or reversing the dough piece between the second or third set of rolls (i.e., between the center rolls and the sheeting rolls) thus placing the original trailing end (the wet end) in the leading position (Figure 6.3).

Another molder that was developed primarily to give loaves a more uniform cell structure twists the dough pieces after they have been rolled into cylinders. Also, the unique appearance of these loaves is evidently attractive to some consumers. For whatever reason, twist bread has been very popular in some sections of the country for many years. The twisting operation was formerly performed entirely by hand, an obviously uneconomical practice in this highly competitive industry. Machines were made available to perform the twisting at a rapid rate and with results which are as good as, or better than, those achieved by hand twisting. In these molders, dough cylinders fall into U-shaped cups, each end of a cylinder being supported by a separate cup, and the cups are then twisted in a rotary motion.

In a typical twist bread molder, the sequence of operations is as follows. As the dough pieces leave the molding conveyor, they first trip a flap which actuates the dough traps and the twisting mechanism of the molder and then drop into the selecting trap. The latter device opens each time the flap is tripped, and deposits pieces successively into each of two pockets of the collecting trap. When the collecting trap

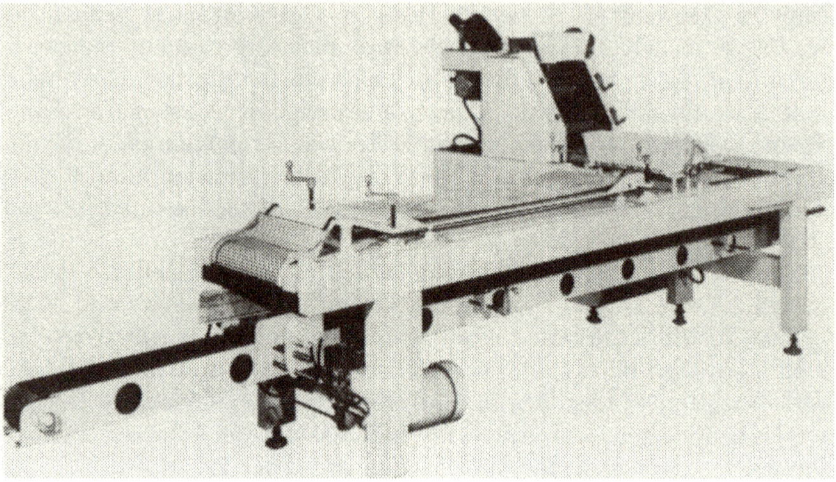

FIG. 6.2. A CROSS-GRAIN MOLDER-PANNER FOR BREAD LOAVES.
SOURCE: STICKELBER

has accumulated two dough pieces, they are released into the first set of twisting cups, which then rotate 180°, but in opposite directions, to apply the first two twists. These dough pieces are now deposited in the second set of cups, where the same kind of twisting action is again applied. After this final operation, the loaf forms are deposited into pans for final proofing and baking by the usual means.

Controlling and Adjusting the Molder

The first adjustment (in terms of sequence of the processing) possible in the molder is setting the head rolls. It has been reported that most steel head rolls are run with an opening of 0.140 to 0.180 inch. With plastic covered sheeting rolls of equivalent size, it is possible to reduce the opening to 0.060 in. Many authorities seem to feel that the closer these rolls can be set without tearing the dough, the better will be the grain in the finished loaf. In any case, the optimum setting will be determined by the conditions existing in each particular plant. It is evident that the lower sets of rolls should be set more closely than the upper set, and that the difference between the settings of the first two pairs should be greater than the difference between last two pairs.

Adjustment of the scrapers on the head rolls (if plastic covered rolls are not used) should be just sufficient to release any dough which has a tendency to stick. Scrapers set too close tend to create friction and may even result in scoring of the roll surface.

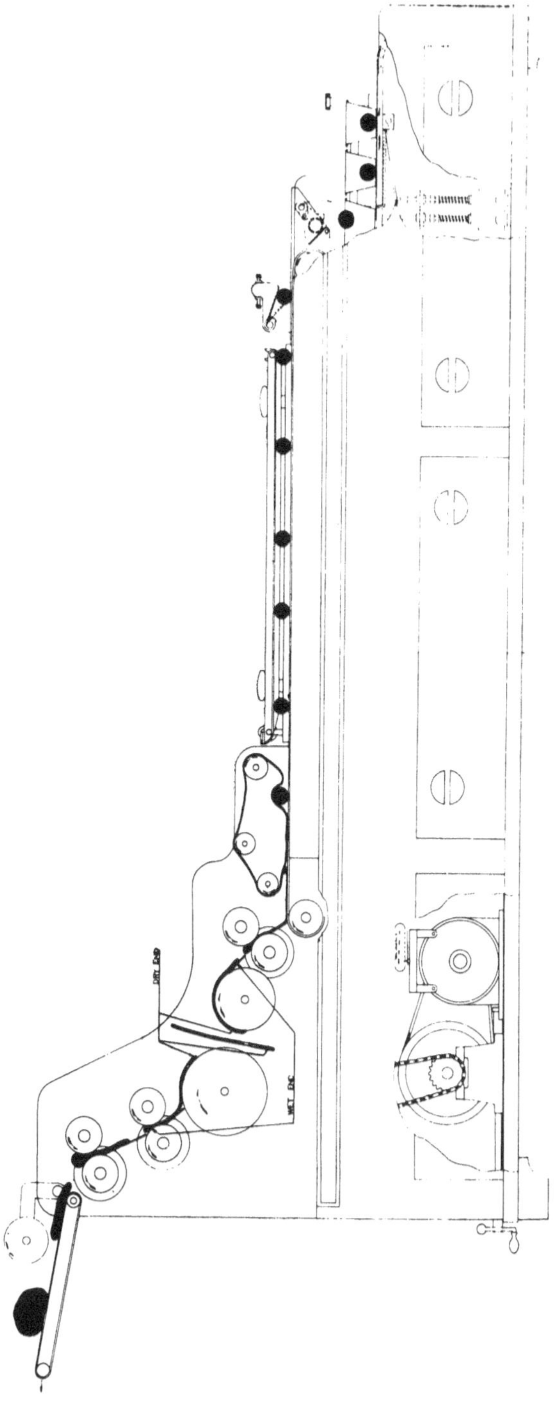

FIG. 6.3. Diagram of a Reverse-Sheeting Molder Panner for Bread Loaves.
Source: Read Standard, Inc.

The speed of the molder can be adjusted, and should be sufficient to allow adequate space between the dough pieces arriving from the intermediate proofer.

The length of the dough cylinder discharged from the molder is governed by adjusting rails or guides lying between the compression surfaces. These guides should be far enough apart so that the dough piece is just slightly longer than the pan.

Clearance between the compression surfaces is another variable which can be controlled in molders. Enough pressure should be exerted on the piece to thoroughly seal the contacting layers of dough. Too much pressure will result in dumbbell shaped pieces and consequent misshapen loaves of bread. Too little pressure, in addition to causing inadequate seals and holes in the bread, may give dough pieces an oval longitudinal section causing them to bake into a misshapen loaf. Gradual application of pressure is necessary to efficiently remove air bubbles from between the dough layers.

Dusting flour applicators should be controlled to give the minimum rate of addition consistent with the prevention of stick-ups and other malfunctioning of the machinery. Too much flour added at this critical stage will result in spots and streaks in the bread, holes due to poor sealing, and other defects.

Maintenance requirements include lubrication of the mechanical parts, cleaning away dough particles and flour from the working surfaces, and application of divider oil to the rolls after cleaning.

Extrusion Molders

Extrusion molders for bread loaves operate on a totally different principle from the molders described previously. They are used on continuous systems where the dough is pumped to the extrusion head as a homogeneous mass not divided into pieces.

In one widely used system, the extruder immediately follows the dough developer. The dough is continuously mixed as it progresses downward to the exit of the developer, where it passes through a narrow slit opening to the extruder. After passing this point, the dough is extruded as a cylinder which is cut off by opposed blades. The diameter of the cylinder is normally kept constant, but the blades can be programmed to cut off lengths which give the needed weight. The knives are actuated synchronously with panner action, so that a pan is properly placed when the severed dough piece falls into the panning area (Figure 6.4).

BREAD ROLL MOLDERS

Bread rolls and buns are formed on equipment which differs in several important respects from the loaf molding equipment used for sim-

FIG. 6.4. THE DIVIDER-MOLDER-PANNER STATION IN A BAKER DO-MAKER CONTINUOUS BREAD LINE.

ilar types of doughs. In some cases, dividers, rounders, proofers, and molders are part of an integrated unit which maintains dough piece registration throughout processing. Frequently, an automatic panning unit is also part of this complex. There will often be specialized shaping, cutting, and molding devices following the rounder or proofer to form kaiser rolls, split rolls, Parkerhouse rolls, etc. Applicators for poppy seeds, salt crystals, and the like may also be included in the line.

The bun divider is somewhat similar to the dividers for bread loaves previously discussed. A horizontal cylinder containing cavities rotates beneath a hopper filled with dough. Hoppers generally have a capacity of 250 to 500 lb of dough. Since a trough will contain enough dough for at least 15 min running time, and usually longer, changes in dough properties with time can create serious problems in maintaining uniform weights. Attempts have been made to overcome the density change by constant working of the dough during the time it is being held, or by placing the dough under pressure while subjecting it to a slight mixing action. Many bakeries now use degassing pumps to feed dough into the divider hopper.

Dough is sucked into the divider cavities by the retraction of pistons which then cut off and expel the dough pieces onto the rounder conveyor as the cylinder completes its rotation. A replaceable plastic wear insert between the hopper and cylinder protects the cylinder against damage from foreign objects. There are two types of bun dividers available for high speed operation, a four pocket divider yielding up to 400 pieces per minute and a six pocket divider putting out up to 600 pieces per minute. Although the six pocket divider is normally the most efficient, there are certain panning patterns for frankfurter and hamburger buns which cannot accept this kind of output and it is the practice in such cases to block off one or two (depending on the pan configuration) of the pistons. The description which follows refers to a recent model of a six pocket divider (Broaddus 1978). The divider head unit has a vacuum system that draws dough from the hopper into the cavities formed by variable stroke pistons having a scaling range of about 1 to 4.5 oz. The vacuum pump is a 7.5 hp unit. In older models it is mounted under the divider bed, but in the newer versions it is mounted externally to avoid heat build up in the divider and to make servicing and repair easier.

A pressurized oiling system provides positive and uniform oil dispersion for the divider head. Individual needle valves lubricate the cylinder bases and plunger, insuring the quick and complete release of dough pieces. A wall-mounted pump is furnished as standard equipment and can be mounted near the oil supply.

The most common type of bun rounder in this country consists essentially of a conveyor belt and a set of concave bars set at an angle to the line of belt travel. The contacting surfaces of the bars are lined with Teflon or are made of polyethylene stiffened and supported by metal strips. As they drop from the pockets of the divider, dough pieces are positioned on the belt so that they are quickly pulled under the proximal end of the one of the rounder bars. If it is a six pocket divider, there are six rounder bars. The compound motion imparted by the straight ahead movement of the belt and the sidewise force exerted by the rounder bar causes the dough piece to rotate. If the dough has suitable physical properties and the piece size is in the correct range, this is a very effective and reliable rounding method in spite of its simplicity. Such rounders do not have as positive a placement control as certain other types and their action may result in a imperfect registration at the proofer intake, but on normal runs this is seldom a major problem. Production rates of up to 320 pieces per minute can be attained with these divider and rounder combinations. Figure 6.5 shows the divider, rounder, and proofer entry area of a roll line—in this case, an English muffin line.

At the end of the rounder belt, the dough balls fall into Teflon-lined chutes which release four or six dough pieces simultaneously into the trays of the proofer, the number depending, of course, on the divider output.

Dough pieces leave the intermediate proofer and enter the molder, where they are formed into the desired shapes, such as hamburger buns, hot dog rolls, club rolls, salt sticks, etc. A typical molder will consist of (1) a molder chute having a gate operated by a cam mounted on a cam shaft synchronously driven by the proofer depositer, (2) a set of adjustable molder rollers, (3) a molding belt with an auxiliary curling mat, (4) a standard auxiliary pressure board unit for forming hot dog rolls, (5) a set of hot dog dough gates, and (6) an indexing mechanism for synchronizing the release of the formed dough pieces with the cavities in the pans. Attachments can be obtained for making twin rolls, clover leaf rolls, hard rolls, pull-apart bread, etc.

Proofed dough balls can be discharged in rows onto a phasing belt that indexes them under a multi-head stamper, which will have designs characteristic of the product being formed, such as the five or six curved radial cuts applied to Kaiser rolls. The stamped dough pieces are then conveyed via a right angle conveyor into a reciprocating transpositor which is sometimes referred to as a "to and fro" device. The transpositor spans the width of the platens of a final proofer, which can be as much as 12 ft wide. After final proofing, the dough pieces are

FIG. 6.5. THE MOLDER IN A BREAD ROLL LINE. DIVIDER AND INTERMEDIATE PROOFER ARE ALSO SHOWN.

SOURCE: CLOCK ASSOCIATES

removed from the moving platens by a stationary crawler and transferred onto an oven band, where the rows are condensed in order to obtain the maximum pieces per bake (Cummings 1987).

CROISSANTS

Production of croissants includes, among other processing steps, laminating butter into a yeast leavened dough similar to bread dough, reducing the finished laminate to a sheet of the proper thickness, cutting triangles from the sheet, and rolling the triangles into the typical croissant shape. Any and all of these steps can be mechanized, partially or completely. Specialized devices are not required for mixing the dough, laminating it, and sheeting it. Cutting can be done by rotary dies which sever triangles from the sheet. In the small retail bakeries, simple fabricated rollers moved by hand can be used for cutting the triangles. For rolling, curling, and panning the croissants, manufacturers offer equipment of almost any degree of automation and rate of production which could be desired.

The manufacturer of one semi-automatic line describes the process as follows (Anon. 1986). The dough is laminated into a sheet slightly thicker than required, thus saving considerable preparatory time. The dough sheet is rolled up on a wooden dowel which is placed on holders situated above the infeed belt. An automatic device continuously feeds the sheet of dough to calibrating rollers which reduce the dough to a thickness required to insure desired weight of the croissants. The calibrating rollers perform the additional function of joining the end of one sheet of dough to the beginning of the next sheet. This allows continuous operation and eliminates any waste (operator participation is needed in the joining function). The triangles are cut in two stages, by two rotary dies, in order to prevent any sticking to the dies. Triangles are automatically conveyed to the rolling machine with their bases parallel to the rollers. The rolling machine can be adjusted to give the dough as many turns as desired. The finished product has a raised spiral design and is delivered to the pan ready for proofing. About 4,000 croissants per hour can be made with this equipment.

A somewhat more automated line of considerably higher capacity has been described in detail by Rijkart (1984). Vertical high speed mixers provided with 40 hp motors and having a capacity of 300 lbs of dough are used. Ingredients are weighed and metered into mobile bowls which can be hydraulically lifted into place. The bowls are closed with a lid during the mixing period of about one minute. Minimal development is given in the mixer since the laminating process will provide adequate development to the dough. In the completely automatic laminating system, a dough feeder elevates the dough into a sheeter head consisting of two feed rollers and a sheeting roller. One of the top rollers has deep grooves and the other has fine triangular grooves. The bottom roller has a smooth surface which gives the dough sheet a perfectly smooth surface. Speeds of the feed rollers and the bottom roller are separately adjustable. This adjustability is considered to be important to prevent overworking the dough during sheeting.

The sheet of dough from the sheeter head is fed onto a supply table which has been covered with a thin layer of flour. At this point, fat is applied by an extruder which transforms blocks of butter or other shortening into a continuous layer of fat without destroying the plasticity. The operating part of the shortening extruder consists of two transport rollers with retractable knives. After the extruder deposits the fat sheet on a layer of dough, the fat-dough composite is wound into a spiral by a curling roller. There are usually six layers of shortening and seven layers of dough in the composite, but the number of layers can be adjusted to achieve the desired results.

The roll of dough and fat is now fed onto a second supply table from which it is fed to a pressing roller for the purpose of preliminary sheeting. Then, cross rollers sheet the dough across the width of the belt, reducing its thickness in the process. Rollers on the cross pinners move across the dough at 100 strokes per minute while rotating at a speed such that their circumference travels about 10% faster than the crossing motion. Up to about 40% reduction can be obtained at this point. Following the cross pinner is a reduction roll stand containing rolls of relatively large diameter and having an adjustable speed infeed conveyor.

The additional laminating which is necessary to obtain the desired finished structure can be done by manually folding the dough strip and sheeting it by any convenient method. The fully automatic laminating devices described elsewhere in this volume can also be used for this process. A brief description of the laminating device recommended in Rijkart's article is included at this point for the sake of completeness. After passing through the two roll sheeter, the dough strip is carried by a conveyor belt to a guillotine which cuts pieces about 20 inches long from the 24 inch wide strip. One reason for this cutting is that it releases some of the tension built up in the dough by the repeated sheeting.

The relaxed dough pieces are stacked by means of a retracting conveyor which drops the dough squares onto a lower conveyor belt moving in a cross-wise direction to provide a layer of overlapping pieces. Total layers at any one point can be adjusted to give from 16 to 48, considered to be the optimal range for finished croissant doughs. The dough array at this point is ready for final sheeting, or for a rest period if the latter is required. Final sheeting is preceded by dusters which apply flour to the top of the dough and to the conveyor belt. First, the dough pieces are slightly pressed together by a reduction roller and then fed to a cross pinner which reduces the thickness considerably. Then the strip passes through the side guide or cross pressing rollers which keep the dough sheet in line and help make the edges straight. In the final

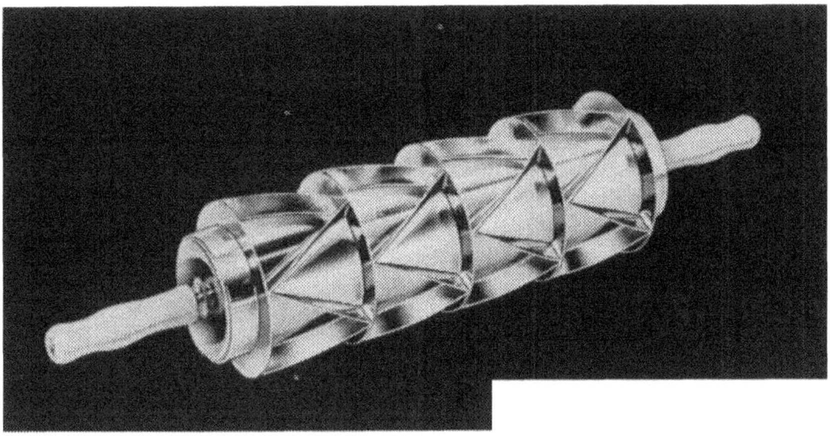

FIG. 6.6. ROTARY HAND CUTTER FOR CROISSANTS.
SOURCE: MOLINE DIV. PILLSBURY

sheeting operation, the dough passes through the multi-roller which consists of six smooth rollers rotating in a circular orbit above the lower, large diameter roller. The dough is pressed on the lower roller while it is being contacted by one of the small rollers, so that the dough is not continuously pressed and stretched. A reduction of thickness of up to 10 to 1 is said to be obtainable with this equipment.

From the multi-roller, the dough strip passes through a final gauge roll set to give it the thickness required at this stage. Before the last step, the dough has an opportunity to relax as a loop is formed by a loop control which speeds up or slows down the whole line in front of the last reduction roller. After the last two roll pinner, the dough is cut into blocks which are placed in retarders to rest for about 12 hours.

The dough block, having undergone the final relaxing step, must then be sheeted to a thickness suitable for the croissant cutting and rolling machine. In Europe, many manufacturers use a block processor which consists of a supply table, flour dusters for the belt and the dough, a cross pinner, a multi-roller, and a final gauging roller.

The final shaping process consists of cutting triangles from the sheet of laminated dough, rolling the cut pieces into tapered cylinders with the point of the triangle on top, and curving the cylinders into crescent shapes. Cutting triangles is done with a hand roller knife in the smallest bakeries, by a fabricated rotary die cylinder (see Figure 6.6) for manual applicatiom in larger retail bakeries, and by cutting machines (usually combined with the curler) in wholesale bakeries. The rotary cutting machines usually employ two cylinders, each of which makes one or two of the line cuts, to avoid the hang-up of dough which is all too likely in a triangle die. There are a number of croissant machines which satisfactorily roll the croissant, according to Rijkaart. Croissant rolling machines are rather simple devices, consisting of two conveyor belts traveling at different speeds. Dough triangles are placed between these belts with the point trailing. The leading edge is pulled back by the slower traveling upper belt, starting the wrapping motion which continues until the entire triangle is curled up into the typical horn shape. Incidentally, the same type of machine can be used to form salt sticks, except the latter are generally rolled into a tighter cylinder and are not curved into crescents after rolling.

The rolled dough is curved, usually manually, and deposited in trays for the final fermentation step. Proofing is normally at 98°F and 80% relative humidity for 60 minutes. Baking for a croissant of the usual size will be at 430° to 450°F for 14 to 15 minutes. Steam is added to the oven during the first minute of baking so that the dough does not crack.

Rowe (1985) described a continuous automatic process for croissants which completely eliminates the retarding stage for finished dough blocks but does start with retarded dough. Dough that has been retarded is fed into the dough hopper of a laminator and shortening of the roll-in type is fed into fat hopper. This type of laminator extrudes a hollow tube having an outer layer of dough and an inner layer of fat. The tube is flattened by a conventional roller and then fed to the folding unit which forms up to eight layers. The loosely laminated dough structure is then conveyed to a so-called stretcher unit consisting of a series of small rollers which revolve on their axes and at the same time are carried in an oblong path which places them in contact with the dough for only part of the rollers' travel. As the rollers contact the dough, their surface is moving in a direction opposite to that of the dough which is passing beneath them. As the dough passes under the rollers, it is also passing over three different conveyors operating at different speeds. The net effect of these actions is that the dough is continuously stretched without encountering excessive pressure which could break it down.

The stretched dough sheet is conveyed to the next folding and laminating table and then to another stretcher. From here it goes to the make-up section. Rowe describes a croissant molder which fits on tracks over the make-up table and cuts the dough sheet to the proper triangular size, curls the triangle, and deposits the dough pieces on the make-up table. Speed of the unit described is 15,000 croissants per hour and it requires one operator. Croissants are manually curved and panned for proofing, requiring another four persons. Proofing at 85° to 90°F and a relative humidity of 80% for 55 minutes is recommended, as is a baking time of 15 minutes at 400°F.

ENGLISH MUFFIN LINES

The first fully automatic lines for producing English muffins were installed about 1960 (Noel 1962). They performed the functions of: dividing; rounding; dusting dough pieces with corn meal; loading to a proofer; proofing under controlled time, temperature, and humidity; dusting trays with corn meal; 360 degree transfer of dough pieces to griddle; grilling; turnover of muffins; finish grilling; elevating; conveying to cooler infeed; cooling, discharge to packaging line; carton setup to packers; and automatic feeding to wrapper. Since this time, a vast expansion of English muffin production has occurred, and pre-engineered lines are available from small semi-automatic operations (see Figure 6.7) to high speed fully automatic plants.

A more complete description of the earliest fully automated lines can be deduced from the illustrations accompanying Noel's article. Dough for the muffins is scaled and rounded by a vacuum roll machine similar to that used for hamburger buns and the like. The pieces pass under a corn meal duster conveyor before entering the proofer. Excess corn meal is returned to the dusting equipment. Dusted dough pieces fall into depositing cups which put them into a 12 pocket tray in an overhead air conditioned intermediate proofer. Bottoms of the proofer trays were previously dusted with corn meal. Proofing time can be regulated by a variable speed drive. The overhead proofer has a fully enclosed, insulated cabinet and the temperature and humidity are controlled. Texture and grain characteristics of the finished product can be varied by adjustment of the proofer controls. Proofed dough pieces are turned over as they leave the proofer and drop into the cups of the grill. These cups can be varied in diameter and height to suit preferences of the targeted market. They are Teflon-coated so that muffins will not adhere to them. The traveling hearth plate of the grill moves into the oven chamber, which is heated by direct gas firing. At the end of the top run, the grids make a 180° flip for the return passage, at which time the muffins are turned over and deposited on a second traveling hearth where they receive grilling on the other side. When they reach the end of the second grill, the finished pieces discharge to an elevator which lifts them to the overhead cooler conveyor system.

Additional details of English muffin production, especially as adapted to conventional roll equipment, were published by Pfefer (1976). Mixing of doughs can be successfully accomplished in several

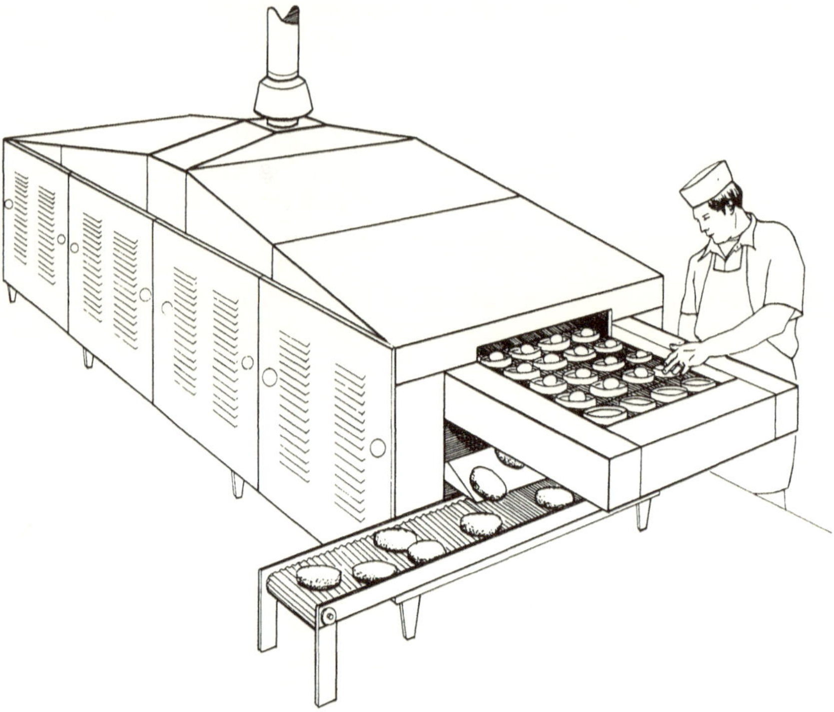

FIG. 6.7. SMALL SEMI-AUTOMATIC GRIDDLE FOR ENGLISH MUFFINS.
SOURCE: PRODUCTION LINE EQUIPMENT, INC.

kinds of mixers, including horizontal dough mixers and vertical mixers, but not on the ultra-high speed mixers. A cool dough with high absorption is preferred. The floor time should be kept short, with a suggested maximum of 15 minutes between the end of mixing and the beginning of dividing. Any dough divider capable of cutting and dropping a soft dough can be used. The dough pieces can be rounded before proofing or dropped into the proofer trays without rounding. Muffins produced by the first alternative are more regular in outline, which may or may not be desired by the consumer. Before the pieces go into the proofer, they are dusted—generally with corn flour—by conventional dusting devices. Proof time is approximately 30 minutes at 110°F wet bulb and 96% relative humidity.

From the proofer, the dough pieces are deposited into griddle cups by a turning mechanism which maintains the top of the proofed dough piece as the top of the dough piece in the grill cup. At this point, griddle cups are open at the top and are not at full baking temperature. After approximately 2.5 minutes of travel time, top covers are brought down into contact with the griddle cup. About this time, the cups have moved into the burner zone of the griddle where temperatures of 475°F are established. After four minutes grilling time, the muffins are dumped from the griddle cups and inverted by means of a slide mechanism. They travel an additional 3.5 minutes on a flat hearth without rings. After leaving the griddle, the muffins are cooled in, for example, a multi-tiered atmospheric cooler. Muffins may be split by mechanisms employing stainless steel needles which perforate the roll on both sides as it passes down a narrow conveyor belt.

An additional variation was described by Thompson (1981), who had developed a proprietary "no-time base" for the product. The dough is brought out of the mixer at 68° to 70°F and given a floor time of 5 to 30 minutes. It is divided in the usual way at about 75 cuts per minute on a four-pocket divider. Rounding is done in the same equipment as used for buns. The dough pieces are dusted with corn meal or a mix that contains corn meal and panned in a special pan containing cups which are one inch deep and four inches in diameter. Cups are double-walled to provide a dead air space and a heat sink; the usual pan has 24 of these cups rigidly affixed and the bottoms of the cups are slightly raised. Proofing is carried out under typical bun conditions; that is, in a wet box at 115°F for about one hour or until the dough rises to about one-eighth inch from the rim. Special lids having about twenty diamond shaped perforations are placed over the proofed muffins and they are baked for about nine minutes at 510°F. The lids are removed, and the muffins depanned (preferably by a vacuum depanner).

The importance of high absorption coupled with faster mixing, a cool dough, and a three-stage mixing process for efficiently producing high quality English muffins was emphasized by Juers (1982). Special mixer agitators also aid in the development of these doughs, it was said. Scaling of the dough is not a particularly critical factor, although a slightly higher level of divider oil may be required to insure a trouble-free run. Good coverage of the dough piece with the right granulation (not too coarse) of corn meal, corn flour, and/or rice flour is essential to proper handling in and out of the proofer. Poor coverage can result in cripples. Dusting powder which is too coarse can cause uneven baking, carbon buildup on the burners, excessive consumption, and possibly flash fires. Proofer design can affect dough response and product quality. To obtain the desired conformation, dough pieces should be proofed on a fairly flat surface which is coated with a blend of corn flour and corn meal. This permits the dough to flow without rising excessively. If the cups in the proofer trays are too deep, proper flow will not be possible. To overcome this problem, various types of sleeves can be placed over the trays. If the material is properly chosen, sleeves can also reduce sticking.

Correct adjustment of temperature and humidity is critical. The proofer should be run at a dry bulb setting of 110° to 115°F. Relative humidity will depend on the system, the dry heat setting, and the dough temperature. The latter factor has an important effect on the humidity. The lower the dough temperature and the higher the proofing temperature, the more humidity is created by the dough itself. This can be corrected by using large temperature differentials between the dry and wet temperature settings or by installing a live steam injection system controlled by a wet bulb thermometer in the duct. Too much humidity in the proofer will make the dough pieces too wet to deposit easily and uniformly in the griddle. A relative humidity which is too low will decrease dough flow and result in small, non-uniform muffins with a dull outer appearance.

The depositor is another critical area where production problems can occur. The depositing mechanism is hourglass shaped and acts to turn the dough piece between its drop from the proofer tray and its landing on the grill. To improve performance of this mechanism, the depositor surface should be coated with a release compound and it should be powered by a variable speed motor. Most of the time, the depositor is operated at a speed slow enough to just lay the dough pieces evenly into the griddle cups. Slight sticking to the depositor can often be corrected by slightly increasing its speed. The slide which, along with the depositor, forms the transfer complex, can be improved by installing a piece of fiberglass cloth on its surface. The fiberglass should be allowed to extend slightly beyond the bottom end of the slide.

The griddle is probably the cause of most production problems. Its top section consists of cups in rows of 12 across. Cups are usually 3.875 inches in diameter and one inch high. Some griddles do not have cups, but just a flat band. There are usually 18 burners located at the top section of the griddle just below the traveling cups. It is important to have good heat at the beginning of the bake. Top flight plates are flat plates that travel above the cups for a specific period of time. They are not present on all ovens. Adjustment of these plates has a definite influence on the finished product. To obtain a porous muffin, the plates should be situated above the cups just enough to permit the dough piece to expand without touching them. The last 18 inches of top flights should be placed so that they barely touch the top of the proofed dough piece and make a flat area about the diameter of a half dollar.

If the plates are set too low, the dough piece will be restricted in its rise and the finished products will be very uniform but not porous. If the top plates are too high along their entire length, steam from the muffin will escape, sometimes causing the dough to set too quickly and resulting in non-uniform but very porous muffins. If the last 18 inches of the top flights are not lowered sufficiently to flatten the muffins very slightly, the dough pieces will peak when coming out from under the top flights, but will be uncooked under the surface. When these muffins travel down the slide, their skin will rupture and they will stick to the slide causing wedge-shaped muffins and many cripples. Also, if small pieces of dough are seen clinging to the flights, this may be an indication that more heat is needed.

According to Juers, the height of the top flights is very critical to the production of porous muffins. When equipment is new, it is easy to adjust the top flights above the cups and obtain uniformity, but with time, it is possible for the metal frame of the griddle to warp under the extreme temperatures. There will usually be 5 to 6 feet between the adjustments for the top flights; to maintain a uniform height and decrease warping, bolts should be welded to the lower frame every 18 inches and adjusted upward to stop the flight plates from warping. The bottom flight plate is not as critical as the top cup section since the muffin is nearly completely baked by the time it reaches the bottom flight. Its main function is to cause the desired coloring of the muffin. Through much service, the rails on which the cups travel can wear to the point where the clearance between the cup and the top plates will vary along the length or width of the conveyor. Also, the chain to which the cups are attached may stretch, causing pinching of the muffin when it is being deposited onto the bottom plates.

The cups should be blown out with air as they are emptied to prevent carbon buildup and black specks on the finished product.

A commercially available automatic English muffin proofer has the following specifications and features (Lupo 1987):

(1) Doors and fixed panels bonded and sealed with stainless steel inside and No. 4 finish stainless steel outside. All side, end, and bottom doors have refrigerator style heavy duty chrome plated hinges and safety door latches.

(2) All side, end, and bottom doors have edge mounted extruded double lip vinyl gasketing.

(3) Stainless steel shafts and plated sprockets throughout.

(4) All cross supports over product zone are of non-corrosive and non-painted construction.

(5) Cross flow air distribution system in proofing cabinet. Distribution manifolds and air plenums are damper controlled. Cabinet air is replaced three times per minute.

(6) Steam injected air conditioning system with electronically controlled wet and dry bulb recording controller. Temperature and humid-

ity are electronically modulated to insure straight line control of the proofing chamber atmosphere.

(7) Mill finish aluminum supply and return ductwork between floor mounted or ceiling hung conditioner unit.

(8) Automatic spring-loaded tensioning for the main proofer chain, to insure slack free operation.

(9) Proofer drive synchronized with dough divider and griddle. When the electrical control system is in the automatic registration mode, the proofer drive masters both the divider and griddle into proper registration.

(10) Automatic product loader which accepts dough pieces from a six wide dough divider and deposits twelve wide into the pocket proofer trays.

(11) Automatic product unloader for accepting dough pieces from inverted proofer trays and depositing them into the griddle cups. This is accomplished with a motorized two-stage variable speed twelve pocket rotary cylinder.

(12) Motorized variable speed twelve spot corn meal dispenser in the load/unload leg to meter spots of corn meal on to the proofer trays prior to product loading.

(13) Aluminum proofer trays with twelve pockets and smooth invert tray ends. Trays have washable sleeve type tray covers secured with stainless steel retainers.

(14) Solid state adjustable switch assembly for controlling the automatic loader, automatic unloader, corn meal dispensers, and automatic registration.

(15) Load/unload leg is guarded by removable acrylic panels, magnetically interlocked for personnel safety.

An automatic English muffin griddle or oven for large scale production is described and specified as follows:

(1) Heavy gauge integral cup flights, twelve cups wide. The cups are Teflon coated and 3.8 inch wide by 1 inch high. Other cup sizes can be provided.

(2) Heavy gauge hearth and cover flights are pan glazed.

(3) A high efficiency combustion system consisting of ribbon band burners, individual burner premixer, external manual throttling valve for matching heat to product size and production rates, combustion blower with air inlet filter and monitoring pressure switch, pre-piped IRI block and bleed-type valve train manifold, solid state electric ignition, and delayed zero speed solid state switch to prevent nuisance shutdown of the system.

(4) Fixed primary cover flight frame with adjusting screw supports.

(5) Secondary cover flight assembly with two worm gear type manual adjustments for control of product shape prior to transfer from cups to hearth.

(6) Replaceable wear rails for flight chain.

(7) Synchronous griddle drive to insure registration with the proofer.

(8) High temperature ball bearings with silicone fiber glass seals.

(9) Automatic spring-loaded take-up on cup, hearth, and cover flight chains to compensate for chain expansion and wear.

(10) Teflon turnover chute for cup to hearth transfer.

(11) Exterior stainless steel enclosures with side-hinged swing out doors on both sides and the end.

(12) Counter rotating cleaning brushes for the cup flights, with self-contained drive and delayed shut down when griddle is stopped.

(13) Cleaning brush for the hearth flight, with self-contained drive and delayed shut off.

(14) Griddle exhaust hood.

Another manufacturer of large scale production units for English muffins has provided the following description of a typical line (Clock 1987):

(1) Standard models are made for 500, 750, 1,000, 1,250, 1,500, and 2,000 dozen muffins per hour. Equipment is made in modules that allow for future expansion.

(2) Dough is mixed in a horizontal mixer and pumped to the receiving hopper of the divider.

(3) A Model K roll divider and rounder drives the proofer, oven loader, and oven. All elements are synchronized regardless of the production rate. This is necessary because, for example, some bakers run sourdough muffins at 250 pieces per minute and whole grain at 235 pieces per minute.

(4) The proofer has stainless steel trays and stainless steel insulated panel doors. Also, a stainless steel conditioning unit.

(5) A corn meal recovery system is located between the Model K and the proofer and at the oven loader. There is a spot sifter at each loader position.

(6) The unique single conveyor, double-plate oven has top and bottom zone modulating controls. The cam track allows the baker to close the lid as desired in the top zone. Exterior panels are hinged stainless steel type.

(7) Scorer/splitters and penny packers are optional equipment.

A small semi-automatic muffin griddle for use by small wholesale bakeries, in-store bakeries, and institutional bakeries is being offered by one manufacturer. The units are all 14 ft long; the different sizes are from 44 to 70 inches wide and produce 1,200 to 3,600 English muffins per hour. These machines accept rounded and proofed dough pieces, which are normally hand fed into the cups. Muffins bake for 6 to 9 min and are then dumped onto a discharge conveyor. These griddles use ribbon burners which can be individually adjusted. Other features include (Lupo 1987):

(1) The Teflon coated cups can be provided in sizes up to 5.38 inches in diameter and 1.5 inches high.

(2) The hearth and cover flights are silicon pan glazed.

(3) Built-in cup to hearth turnover chute.

(4) Combustion system consisting of pipe ribbon burners, main heat control valve, solid state electric ignition on each burner, industrial style combustion system control, monitoring switch for low fuel gas pressure, low air pressure monitoring switch, and adaptability to natural gas, propane, or manufactured gas.

(5) Griddle drive and combustion system are electrically interlocked, preventing unwanted heat buildup on stationary flights.

(6) Griddle drive speed adjustable to allow variable grilling times.

(7) Cover flight assembly can be adjusted while it is in motion.

(8) Sides of the griddle have removable sheet metal enclosures.

(9) Integral hood to collect hot gases.

(10) High temperature ball bearings with silicon fiber glass seals.

(11) Replaceable flight chain wear rails on complete griddle.

The layout for a complete production line for English muffins will be found in Figure 6.8.

PRETZEL EQUIPMENT

There are two basic types of pretzel forming equipment, the kind which cuts a thin slice from an extruded dough tube having a cross-section like a pretzel and the type that first forms a thin strand of dough and then twists it to form the pretzel shape. There is also a process for cutting pretzel shaped pieces from a flat sheet of dough (Figure 6.9), but this is used mainly in Europe and probably mostly for cookie doughs. These preceding descriptions apply to the traditional three lobed pretzels. So far as the author knows, all rod-shaped or stick pretzels are extruded. The following discussion applies to the small dry or cracker-like pretzel, not to the pretzels which are often sold hot.

Pretzel doughs can be mixed on horizontal dough mixers or most other types of mixers suitable for developing a stiff dough. Pretzel doughs tend to be very stiff, so the mixer must either be constructed to accept strong resistances or smaller dough batches must be made. A

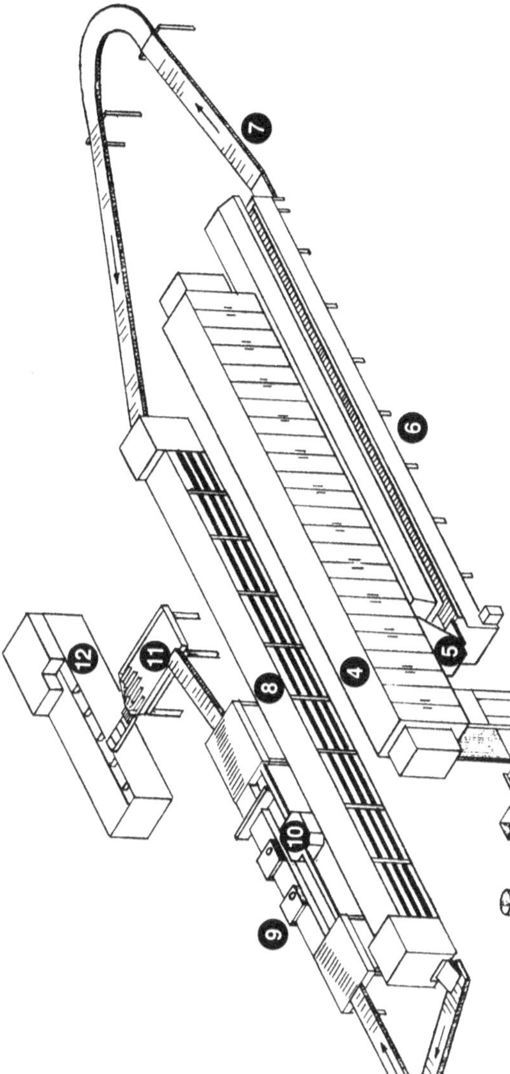

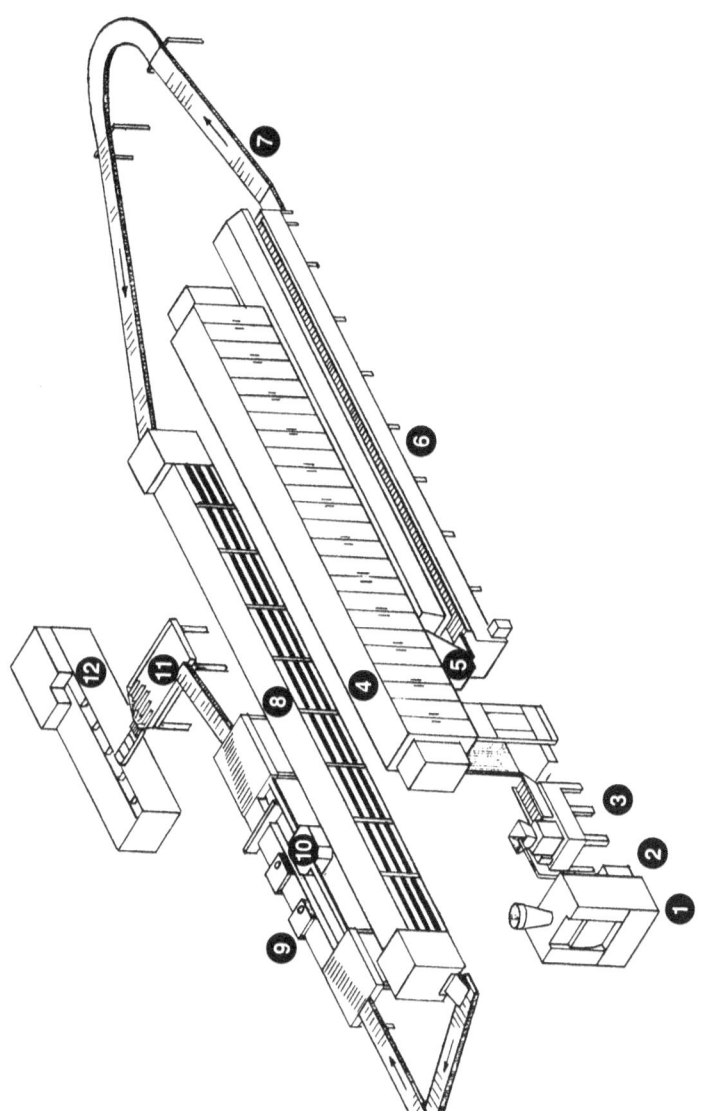

FIG. 6.8. A COMPLETE LINE FOR LARGE-SCALE PRODUCTION OF ENGLISH MUFFINS. (1) MIXER. (2) DOUGH PUMP. (3) DIVIDER-ROUNDER. (4) PROOF CABINET. (5) OVEN LOADER. (6) DOUBLE PLATE OVEN. (7) OVEN DISCHARGE CONVEYOR. (8) COOLER. (9) WEB SLICING. (10) SCORER-SPLITTER. (11) MUFFIN STACKER. (12) PACKAGING.
SOURCE: CLOCK ASSOCIATES

FIG. 6.9. Two cutting machine roller dies forming a pretzel shape. The first roller cuts out the interior spaces and the second roller cuts the pretzel outline.

Source: Hecrona

patent by Blain and Zabrodsky (1987) describes the continuous manufacture of pretzel dough by method in which the ingredients are continuously metered to a mixer appearing to be, from the inventors' description, an auger type mixer conveyor.

In Reading Pretzel Machinery Co. equipment, dough is placed in a hopper from which an auger forces it through a slot in the face plate of the extruder. The dough emerges from the extruder in small strips of circular cross section. A knife automatically cuts the strip into pieces of predetermined length. This segment drops on to a canvas belt that carries it under a second belt. As the dough is gradually squeezed between the belts, it is rolled and formed to the desired thickness. At the end of the rolling process, the string of dough has its ends clipped so that all lengths are uniform. The dough strip then enters a twisting machine. As the shaped pretzel leaves the twister, it passes under a roller that exerts a slight pressure, thus fastening the knots. AMF equipment uses twisters that operate differently (see Figure 6.10).

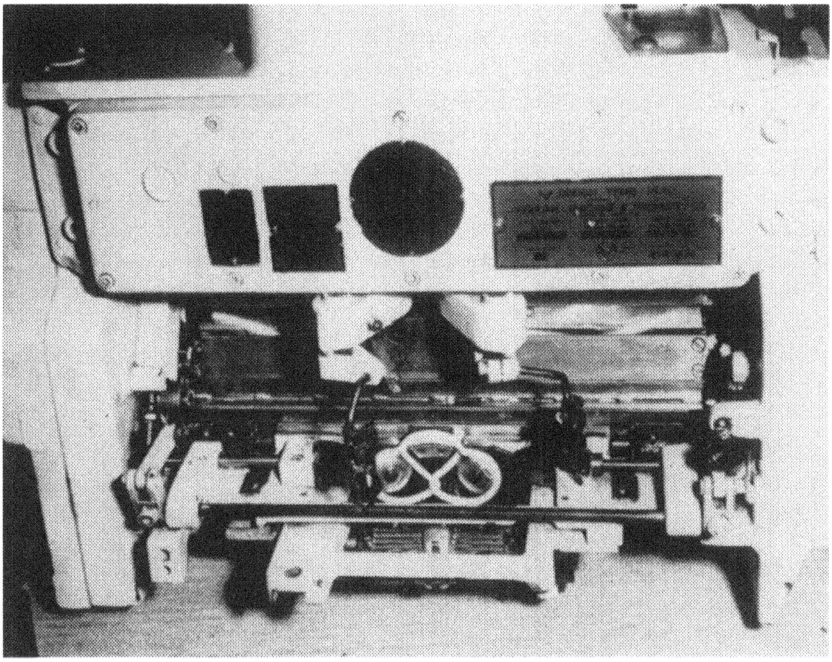

FIG. 6.10. PRETZEL TYING MACHINE.

SOURCE: AMF

The other type of pretzel former, now used for most of the pretzels made in the U. S., is similar in several respects to the wire-cut machines used for making cookies. A stiff pretzel dough is pushed by two rollers into the casing of a screw conveyor. The auger forces the dough under high pressure into a chamber having the lower surface penetrated by tubes leading to orifices shaped like pretzels. The dough continuously extrudes through these orifices and is sliced off in appropriate thicknesses by a blade held in a frame which is moved back and forth. The blade moves forward for the cutting action while it is held very close to the bottom surface of the die, and it is dropped downward a slight distance for the return stroke so that it does not interfere with extruding dough. As the dough pieces are severed, they drop on to a conveyor belt which carries them to the dipping and baking sections (Sterrett Campbell 1988).

In either type of equipment, the subsequent treatment is about the same. By means of a reciprocating conveyor, the raw pretzels are placed across a proofing belt approximately 40 ft long. From the proofing belt the dough passes through a caustic bath. The caustic dipping section consists of two tanks. There is a smaller tank through which the unbaked pretzels travel, and a larger make-up tank (usually at a lower level). The caustic solution is pumped from the make-up tank to the upper tank in which its level is maintained by adjusting the overflow pipe. This system keeps the volume in the upper tank constant. The caustic solution of 1.0 to 1.5% sodium hydroxide is held at 186° to 195°F. If the caustic concentration becomes too high, there will not be a complete conversion to sodium bicarbonate in the baking and drying cycles and the pretzel will be hot to the taste due to residual sodium hydroxide. There appears to be no FDA regulation on the amount of sodium hydroxide in the caustic solution or on the pretzel.

Immediately after the pretzels leave the caustic solution, they are salted. The salter consists of a supply hopper from which the salt is sprinkled by a grooved roller. Salt slides down a chute until approximately two inches above the pretzels, then drops the rest of the way. The salted dough pieces then enter the oven, which is often about 50 ft in length. One supplier offers oven modules 5 ft long which can be assembled to give a baking chamber any multiple of that size; as supplied, the ovens require manual lighting and adjusting of each individual burner. The bake section is the top portion of the oven and it has burners over and under the band that carries the pretzels. Time in the bake section is controlled by a variable speed drive and is between 4 and 5 min. The moisture at the end of the bake period should be about 15%.

Baked pretzels go down a slide to the drying section, which is underneath and separated from the baked section by heavy insulation. Pretzels that cling to the baking belt are removed by a scraper blade. The belt in the drying section travels in a direction opposite to that of the belt in the baking section. Speed of the drying belt is variable, but it will always travel at a much slower speed than the baking conveyor. Pretzels form a bed several inches deep and remain in the drying section from 25 min up to 90 min. Temperature is held in the range of 225° to 250°F. There is much debate over the drying time and its total effect. In addition to reducing the moisture to the desired 2.0 to 2.5%, the long drying time is said to be necessary to temper the pretzel so that it will not break too easily during packaging and transport.

Stick pretzels are extruded using, for example, a group of five extruding heads containing 10 to 12 holes per extruding head. The dough is forced through the extruding head by means of a pressurizing helical conveyor and falls on to the proofing belt. As the dough strip nears the end of the proofing belt, it is cut into the desired length by a group of circular knife blades. These blades travel across the belt, cutting the dough strands. When the knives reach the edge of the belt, they rise and return to their starting point. The raw stick pretzels pass through caustic and salting applicators similar to those used for twist pretzels. Baking temperature is usually kept at a constant 420°F and the time is between 4 and 5 min. The drying section is run at 225° to 250°F, with the sticks exposed for approximately 55 min.

Log and nugget pretzels are made much as the sticks are, except that they are cut off at the extruder head.

BAGELS

It would seem at first glance that bagels could be formed in the same way as doughnuts. As a result of the use of very strong flour (or of flour supplemented with vital wheat gluten), however, regular bagel dough is too stiff to be cut or extruded as doughnuts are, and the cut edges do not round off completely during proofing. Bench preparation involves hand rolling a cut piece into a cylinder and pressing the ends together. Automatic forming devices generally follow this pattern.

The basic principle of most bagel forming machines is that a cut, extruded, and/or rolled strip of dough of the proper weight is gradually reduced in thickness and bent into a circle between a sleeve type conveyor and a tapered mandrel. The narrowing channel eventually forces the ends of the strip together and seals them, forming the completed doughnut shape.

In a machine patented by Patchell and Goldberg (1963), a piece of dough is forced onto the tip of a cone and then rolled up the cone by mechanical fingers until it assumes the proper shape and size. A later patent (Thompson 1974) describes a machine in which a partially pre-shaped batch of dough of fairly uniform width is supplied on a conveyor to a dough dividing apparatus which forms the dough into two strips and then cuts them into rectangular pieces. These cut pieces are shaped between a stationary mandrel and moving forming cups into dough toroids suitable for bagels. A further development (Thompson 1984) was a machine which took bulk dough and divided and formed it in somewhat the same manner as described in the earlier patent. According to the inventor (1988), the latter patent can reasonably be regarded as representing the current state of the art and will be discussed in greater detail.

The machine in question includes a dough divider of drum-like configuration having around its circumference rectangular dough receiving channels into which a pre-shaped dough batch is pushed by compression rollers. The intermittent movement of the drum can be adjusted to control piece size. Sensors are positioned above the drum to detect the buildup of dough at the divider drum before it passes beneath the compression rollers and to prevent flooding the drum with excess dough. A rotating cut-off knife and a dough wiping means are provided to complete the separation of a measured volume of dough from the divider cavity. The dough strips so formed are propelled by gravity and a rotating wheel through a chute and guide means to the dough forming location. The dough forming apparatus includes a series of articulated, toroid forming, segmented, tubular cups mounted on a cup chain. These segmented cups open under the force of gravity as they move along the bottom pass of their conveying chain. They are closed by side guides as they approach the upper lay of the conveyor. A toroid forming mandrel is mounted on a single vertical support arm which is pivoted to facilitate pivoting of the mandrel away from the forming cups. This permits the use of an alternative forming apparatus.

The patent describes a pair of top and bottom preshaping and dough moving means before the dough strips are engaged between the moving cups and the stationary mandrel so as to roll and squeeze the strips into a cylindrical form. The cylinder is gradually elongated and narrowed as the channel between the mandrel and cup is tightened. Finally, closure of the two ends of the strip occurs as they meet at the top of the mandrel; the hinged sides of the cups at that point have been brought together by guides along the sides of the conveyor.

After forming, bagels are proofed. Provisions for proofing take many forms. In the simplest, bagels are hand placed on wooden peels which are slid into racks. The proofing area may be simply a warm place in the bakery. More advanced procedures use intermediate proofing cabinets similar to those described in Chapter 5. Proofed bagels are often refrigerated to retard fermentation until the final steps of boiling and baking.

Immersion in boiling water for a short time forms the basis for characteristic shiny brown skin of the bagel which is completed during baking. Hot water treatment can be performed in almost any kind of steam heated kettle. Agitators are not required. For large scale production, the dough pieces can be dropped into a rectangular vat and removed by some kind of skimming apparatus after they float to the top. The same kind of apparatus used for this purpose in doughnut frying vats could doubtless be adapted for use in a boiling water bath.

After a brief drying step, the bagels are baked in conventional types of ovens (Marracini 1986). A recent development is the baking of bagels without prior boiling; rack-type steam ovens are used to get a glossy crust. Opinions as to the quality of the non-boiled bagels differs among experts (Petrofsky 1986).

GRISSINI AND BREADSTICKS

There is a considerable production of breadsticks, the thin cylindrical pieces of bread sometimes baked almost to dryness but in other varieties retaining some of the soft crumb. The very small pieces found in individual serving packs, and having a crisp texture throughout, are baked from dough which has been extruded through circular orifices by the usual types of simple extrusion equipment. For more conventional doughs, bench top devices are available from Italian manufacturers to sever a strip of sheeted dough passed under a grooved roller. More elaborate automatic machines are available to perform essentially the same function at a much higher speed and with the ability to vary the diameter and length of product. They include automatic and continuous dough kneaders and sheeters. A typical machine incorporates the sheeter itself with a loading hopper in the gauge roll section, to form the dough into a ribbon of the required thickness. Dough from the final pair of gauge rolls is allowed the necessary recovery time on the machine band, and, after this, is gang-slit and finally chopped off by a reciprocating knife.

PITA BREAD

The distinguishing characteristics of pita bread, and variations which are sometimes called Arab bread, pocket bread, etc., are that it is a very thin piece of baked dough, circular in outline, and the top crust is separated from the bottom crust except at the circumference. Superficially, it somewhat resembles a flour tortilla and like flour tortillas these breads are generally made from very lean doughs—flour (usually strong flour, sometimes supplemented with gluten), water, yeast, salt and (sometimes) sugar.

Most of the steps in the processing of pita bread are similar to those used for other types of bread. A flow diagram would show mixing, a relaxing stage, dividing, rounding, an intermediate proof stage, sheeting, a final proofing stage, baking, and cooling (Cooper 1986). Straight dough procedures yield satisfactory products. Dough balls, rounded and intermediate proofed in the normal manner, are passed through a series of sheeting rollers which flatten each piece to a thickness of about 0.0625 inch. Size, conformation, and skin characteristics are controlled so that the final "loaf" is as near a perfect circle as possible and the crust has no breaks or cracks. The amount of dusting flour should be kept to an absolute minimum. The flattened rolls proceed to a final proofer which, in large scale operations, may be a series of many conveyor belts stacked one above the other, so that the dough pieces can be carried back and forth several times before exiting the proofer. No pans or other supports are used. With a relative humidity of about 65%

maintained in the proofer, the crusts become tough and rather dry, features which are important in obtaining the proper response during baking. The proofed loaves are baked on a mesh belt in a very hot oven with a high contribution from radiant heat. The temperature should be at least 700°F and at some points perhaps as high as 1,000°F (Cooper 1986). Within about 30 sec, the dough expands almost to a ball shape, the gas bursts through at some point, and the dough contracts somewhat. Due to cooking and dehydration of the crust, complete collapse does not occur, and an interior space continues to exist. Further contraction occurs during cooling, but it may be necessary to compress the loaf further by hand or roller to get it fitted into the package.

Some recent installations use a sheeting and cutting method to form the loaves, an obvious way to get the perfect circles that are seldom achieved in the sheeted ball operation. It seems that the cut edges do not prevent the necessary oven expansion.

PIZZA CRUSTS

In describing pizza crust equipment, the two main types of production have to be differentiated. The local pizzeria will normally use some kind of pressed dough equipment, if it makes its own crusts and does not hand form them. There are several different types and sizes of pizza crust presses, including some that are suitable for individual pizzerias (see Figure 6.11). The vast numbers of crusts that are made for frozen pizzas, and the fairly large number of crusts that are made for sale to pizzerias and some other outlets, are made either by a sheeting and cutting process or by pressing lumps of dough into thin disks. Nearly all commercial pizzas are yeast leavened, though some also include a little chemical leavening to give a final burst of gassing in the oven.

Stamping methods for forming pizza crusts were described by Lehmann (1986), with some details of the processing equipment commonly used for this purpose. He stated that shells prepared on automated lines by the stamping method are generally made by a no-time dough process. The dough is mixed 3 to 5 min in a conventional horizontal bread mixer. Strong development is not intended, the desire being to produce a soft, smooth and very relaxed dough which will flow out and exhibit minimum shrinkage at the stamping station. Batch size should be adjusted so that the entire mixer load can be put through the divider in not more than 12 min. Dividing and rounding are also done in conventional equipment. The dough balls are passed through an oil spray and enter an overhead proofer maintained at 95° to 100°F.

After 10 to 15 min of proofing, the dough pieces are deposited into pans which are specially designed to be compatible with the pizza

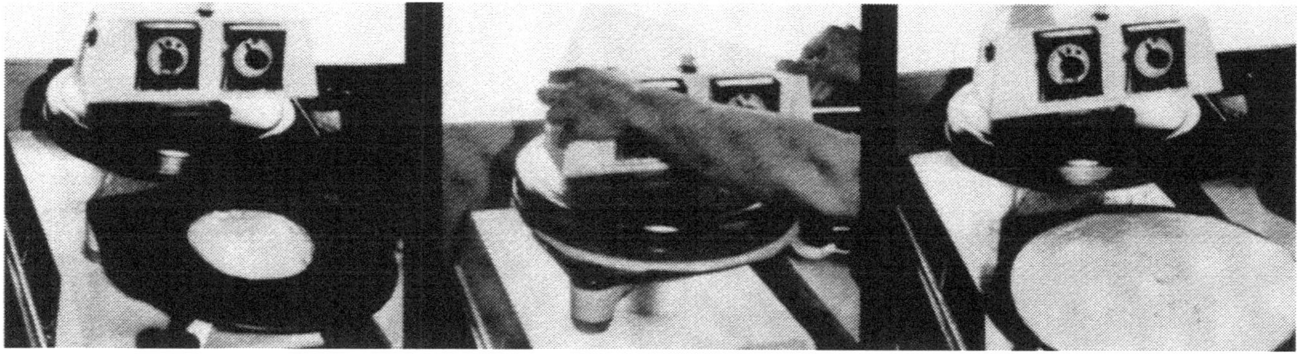

FIG. 6.11. MAKING PIZZA CRUSTS ON A SMALL PRESS. LEFT: THE LOWER PLATEN IS SWUNG OUT AND A WEIGHED PIECE OF DOUGH PLACED ON IT. CENTER; THE LOWER PLATEN IS RETURNED TO ITS SELF-ALIGNING POSITION UNDER THE UPPER STATIONARY PLATEN AND THE HYDRAULIC PRESSING MECHANISM ACTIVATED. RIGHT: THE BOTTOM PLATEN IS LOWERED AND THE PRESSED DOUGH REMOVED.

SOURCE: PROPROCESS CORP.

press. These pans are then carried directly into the first stamping station, where the dough is pressed out to a specified size. In some cases, the initial stamping step extends the dough only part way, and the crust is extended to its final diameter by a second pressing step after the dough undergoes a short relaxation period. For best results, the dough should rest at least two minutes between the stamping stations, but some bakers make do with only 45 seconds. The shorter time will often lead to excessive shrinkage when the dough is baked.

Thin crusts will go directly to the oven after they are pressed. Thick crusts may either go directly to the oven, or receive an additional 5 to 10 min of final proofing at 90°F to promote greater crust thickness. Some bakers will press the freshly baked crusts under a compression belt or floating roller to reduce their thickness slightly and achieve greater uniformity in crust height. Texture—eating quality—suffers as a result of this treatment, the crusts taking on the cardboard-like characteristics that are frequently remarked upon by consumers.

Further information on preparing pizza crusts by large scale equipment was published by Fischer (1981). For pressing operations, a straight dough with as much as five percent of yeast and one or two percent of shortening is divided and rounded by conventional methods such as those used for loaves and rolls. Dough pieces are given some floor time and then sent through an intermediate proofer, proofing being carried to the point which will enable the dough to color well when baked. Proofed dough is placed on an oiled pan and, usually, more oil is applied. Pressing can be done once or twice, or with heated platens. Pressed pizza crusts are baked in a pan and, usually, in a tunnel or tray oven. A major negative of the pressing method of pizza crust manufacture is the large number of pans that are in use at any one time.

Preparation of pizza dough by sheeting and cutting methods is fairly straightforward. For large scale commercial operations, horizontal dough mixers are perfectly satisfactory methods of assembling the ingredients and developing the dough. Almost any other kind of mixer which can develop the dough and can economically process the rather stiff dough that is characteristic of pizza crusts can be used—vertical planetary mixers, for example. Dough temperatures of 78° to 82°F are considered desirable. The dough is allowed to rest for 5 to 10 min after mixing. It is then extruded on to the mixing line. The extruded strip of dough will pass through a series of reduction rollers to reduce its thickness to one-eighth to three-sixteenths of an inch for thin crusts or one-fourth to five-sixteenths of an inch for thick crusts. The dough then passes under a docker and, finally, is die cut to produce the desired type shell. Formed dough pieces proceed through a final proofer maintained at 90° to 95°F and 80 to 85% relative humidity. Proof times are about 8 minutes for a thin crust and about 30 minutes for thick crusts. Immediately after proofing, the shells are baked at about 425°F.

For sheeting and cutting lines, the dough may or may not be proofed in the trough. Sheeting rolls may be arranged in a cascade so that the dough falls from one to the other before going underneath the cutter or they may be arranged in the the more common way, on a horizontal line. Rollers may be sheathed in Teflon and pick off rolls, rather than knives, may be used to transfer dough from the gauge rolls to the conveyor. Docking is considered to be necessary for crusts baked without topping. In some factories, the dough is baked on the same band it is proofed on, to eliminate a transfer of the soft, sticky disc of dough. In other installations, the proofing is handled by transferring the cut pieces to a special carrier. This carrier allows the pickup nosing conveyor to get beneath the cut and proofed pieces and transfer them to the oven with no distortion. The sheeted and cut crust may be proofed before baking. Baking times for these crusts are on the order of 3 to 5 min. Band ovens are frequently used.

A method of preparing sheeting and cut pizza shells on a large scale is described in a patent of Moline (1977). The production line described by Moline is composed largely, if not entirely, of conventional dough processing equipment. Dough is mixed in either a batch or continuous mixer and then transferred to a dough extruder which can process it into a relatively thin continuous dough web. The web is then sheeted by conventional pairs of reduction rollers until it assumes the width and thickness deemed suitable for further make-up. The continuous web is transported on an endless belt conveyor for a time required for proofing. After proofing, the expanded dough is sheeted again to a predetermined width and thickness. Circles are cut from the web by conventional type roller cutters or other means. Docking rollers, which punch holes in the dough to prevent excessive rise, may be utilized at this point, although they are probably not necessary if filling is to be applied immediately. Trimmings generated in the cutting operation are collected on belt conveyors which carry them back to the beginning of the line where they are blended into new dough.

A patent obtained by Groth (1968) describes a type of apparatus which can be used to make pressed pizza crusts. A divider, presumably of any standard type, forms dough balls of the weight appropriate for one pizza crust. Each one of these balls is deposited on a separate aluminum plate having the approximate diameter of the pizza shell. The aluminum plates are conveyed on drive chains through a pressing mechanism. When a plate is centered in the press, the drive mechanism stops temporarily, and a hydraulic drive forces a circular mold the size of a pizza crust on top of the dough ball, causing it to assume the thin disc shape of a typical pizza shell. Subsequently, the aluminum plate with its raw crust is moved to a docking machine which perforates the outer surface of the shell to prevent excessive expansion in the pre-bake heating which follows.

Some inventors have concentrated their efforts on devising machines which duplicate the actions performed on the pizza sheet during hand preparation. The rather quick straight downward pressing motion of the usual pizza press machine has been thought by some pizza experts to be damaging to the textural properties of the crust, making it somewhat cracker-like. One approach to improvement was revealed by Pacilio (1987), who patented a machine which simultaneously compressed and applied centrifugal force to a lump of pizza dough. Dough is placed on a support surface or pan which can be rotated. While it is being rotated, the press slowly descends and at the same time pushes outward; excessive spread of the dough is prevented by the pressing surface. Other inventors (DeChristopher 1987) have concentrated their efforts on modifying the design of the surface which contacts the dough. In this case, the stamping platen is provided with many downwardly projecting protuberances—bumps—which press the dough into a relatively flat configuration but with a dimpled surface that leaves thick portions between indentations for entrapment of gases.

Most pizza crusts are probably topped immediately with sauce, cheese, and other adjuncts, rather than being baked separately. Much of this preparation is done in pizzerias where topping application is essentially a hand operation, except, perhaps, for some simple metering devices for sauce. This obviously labor intensive step has been the object of considerable development work by equipment manufacturers. An integrated system for preparing crusts, applying toppings, and baking was patented by Triporo and Eckels (1987). This invention describes a "tamping" press having a knobbed contacting surface which performs thrusting motions at successively different positions relative to the pan. This gradual and varying pressure is said to yield a crust similar in texture to those made by pressing with the fingers, i.e., similar to handmade pizza crusts. A conveyor carries the pan with its portion of dough through the press, under outlets which automatically meter topping ingredients on to the crust, and into the oven portion. The conveyor moves continuously in an oval path so as to bring the pan back to the dough extruding area after the finished pizza has been removed.

Some of the largest manufacturers of frozen pizza have changed to frying, rather than baking, the crusts before applying topping. The patent of Totino *et al.* (1979) is exemplary. There is no doubt that frying, under proper conditions, can improve crust texture and flavor, and—most important—lead to better retention of quality during frozen storage. In a typical line of this type, the dough would be mixed in a conventional horizontal mixer, dumped into a trough and given some floor time, pumped through an extruder to form a fairly thick layer, passed through about three stands of reducing rollers (usually preceded by flour dusters), cut into circles by rotating die cylinders, proofed by passing for a few minutes through a tunnel into which steam is being injected, dropped into a vat of hot oil, carried through the oil by a wire mesh conveyor, scooped up at the other end by another wire conveyor, shaken and blown to remove excess oil, cooled, and topped. After the circles are cut, the trimmings are removed and conveyed to the trough for recycling. Docking is absolutely essential to prevent the dough pieces from expanding into a ball shape in the fryer. Proper choice of docking pattern prevents the formation of large bumps on the surface. These bumps are almost as objectionable as the ball shape since they interfere with application of topping and may make it difficult or impossible to fit the pizza into its package. Different kinds of docking apparatus have been used. It appears that thick pins rather than thin ones, and more rather than fewer pins, represent desirable approaches. The pins should penetrate the crust, or nearly so. A mere dimple will not serve the purpose. It has also been suggested to use heated bars or punches to tie the top and bottom surfaces together.

Preparation areas have also been developed for delivery vehicles, so the pizza can be assembled and baked while the truck is in transit to the consumer. One of these arrangements is described in a patent granted to Abbott *et al.* (1986). The truck is divided into a driver's station and a pizza preparation area. The latter, also described as the kitchen area, includes a pie case, a pizza preparation station, and a chair for supporting a cook while in transit. Pizzas are made from uncooked but proofed dough loaded in pans, and cooking is done in a conveyor oven which is part of the equipment.

BIBLIOGRAPHY

ABBOTT, M. T., STREEPY, G. S., PAULUS, J. R., BARRERA, R. and BREWER, D. E. 1986. Pizza preparation and delivery system. U. S. Pat. 4,632,836.

ANON. 1985. Le Croissant Dore. Tecknomatik, Padova, Italy.

ANON. 1986. New wrinkle to fully automated process. Prepared Foods *157,* No. 8, 201.

BLAIN, W. A., and ZABRODSKY, J., III. 1987. Continuous pretzel dough manufacture. U. S. Pat. 4,691,625.

BROADDUS, M. R., JR. 1978. Soft roll equipment. Proc. Am. Soc. Bakery Engineers *1978,* 122-132.

CAMPBELL, S. 1988. Personal communication. Atlanta, GA

CHIAO, T. T., and CHIAO, C. C. 1986. Apparatus for forming wrapped food products. U. S. Pat. 4,574,690.

CLOCK, T. Q. 1987. Personal communication. Clock Associates, Portland, OR

COOPER, I. 1986. Pita/pocket bread. Proc. Am. Soc. Bakery Engineers *1986,* 151-156.

CUMMINGS, R. P. 1987. Automation of specialty roll products. Proc. Am. Soc. Bakery Engineers *1987,* 145-153.

DECHRISTOPER, E. L. 1987. Method of making a pizza-type product of dough. U. S. Pat. 4,696,823.

FISCHER, H. A. 1981. Pizza crust production. Proc. Am. Soc. Bakery Engineers *1981,* 170-174.

GROTH, F. A. 1968. Method and apparatus for forming pizza shells. U. S. Pat. 3,379,141.

JUERS, A. A. 1982. English muffins. Proc. Am. Soc. Bakery Engineers *1982,* 46-51.

LEHMANN, T. A. 1986. Pizza crust. Proc. Am. Soc. Bakery Engineers *1986,* 167-177.

LUPO, J. 1987. Personal communication. Production Line Equipment, Rockaway, NJ

MARRACCINI, M. 1986. Bakery's growth attributed to bagel forming machine. Baking Industry *153,* No. 1868, 22.

MOLINE, R. V. 1977. Method for forming dough shells. U. S. Pat. 4,046,920.

NOEL, E. M. 1962. Fully automatic production of English muffins. Bakers Weekly *194,* No. 9, 32-34.

PACILIO, V. C. 1987. Pizza dough spreading apparatus. U. S. Pat. 4,690,043.

PATCHELL, H., and GOLDBERG, I. 1963. Apparatus for automatically forming dough rings for making bagels. U. S. Pat. 3,080,831.

PETROFSKY, R. 1986. Bagels. Proc. Am. Soc. Bakery Engineers *1986,* 143-151.

PFEFER, D. N. 1976. English muffins. Proc. Am. Soc. Bakery Engineers *1976,* 51-55.

RIJKAART, C. 1984. Croissant production. Proc. Am. Soc. Bakery Engineers *1984,* 137-145.

ROWE, C. S. 1985. Croissants. Proc. Am. Soc. Bakery Engineers *1985,* 154-162.

THOMPSON, D. T. 1974. Machine for forming bagels. U. S. Pat. 3,792,940.

THOMPSON, D. T. 1984. Compact dough dividing and forming machine. U. S. Pat. 4,478,565

THOMPSON, D. T. 1988. Personal communication. Thompson Bagel Machine Corp., Los Angeles, CA

THOMPSON, J. B. 1981. English muffins. Proc. Am. Soc. Bakery Engineers *1981,* 141-145.

TOTINO, R. W., BEHNKE, J. R., WESTOVER, J. D., and KELLER, R. L. 1979. Fried dough product and method. U. S. Pat. 4,170,659.

TRIPORO, P. R., and ECKELS, S. 1987. Apparatus for making pizza. U. S. Pat. 4,634,365.

CHAPTER SEVEN

FORMING DEVICES FOR OTHER PRODUCTS

INTRODUCTION

This chapter contains discussions of the forming equipment for sweet doughs, pies, doughnuts, cakes, pancakes, tortillas, biscuits, and a few other items. Some of these products are made from yeast leavened doughs, others from chemically leavened or unleavened mixtures. Dough pieces for frying (honeybuns, doughnuts, etc.) can also be made on these lines. Machines for applying adjuncts (e.g., icing depositors, oil and wash sprayers, chocolate enrobers, filling injectors, and salt sprinklers) will be discussed in a subsequent chapter. When necessary, the auxiliary equipment which operates in conjunction with the forming devices will also be described, but such digressions will be held to the minimum consistent with clarity and accuracy.

FORMING DEVICES FOR SWEET DOUGHS

General Considerations

The multiplicity of shapes and dough types found in bakery dessert goods leads to the use of many different combinations of the relatively few automatic devices available to the baker. Some of the most common pieces of equipment required for shaping sweet dough products are sheeters, applicators, curlers, and cutters. These units are often combined in pastry benches, automatic sweet goods machines, etc., which process the dough after it has been sheeted or extruded. Of course, any bakery which has been in operation for a considerable time will have developed gadgets, equipment modifications, and totally new devices which are unique to their plant. Large operations may have machines specifically designed for their particular needs and which are are kept secret for competitive reasons. This article concerns the types of equipment offered as stock items (more-or-less) by equipment manufacturers in the U. S. and abroad.

Dough extruders for sweet doughs move dough from a hopper through a large orifice, either rectangular or circular in outline, by means of a screw or auger. The dough strip falls on to a conveyor belt or into a sheeting roller for further processing. Since the dough is not forced through a small opening, little harm is done to its cell structure.

In some cases, a cut-off mechanism is controlled by a variable electronic timer to give pieces of a desired weight. This equipment is particular useful for extruding large (e.g., 15 lb) dough pieces to be sheeted out as Danish pastry. In other cases, a continuous sheet is made of the extruded dough.

Many of the final forming devices—the curling rollers, guillotine cutters, longitudinal cutters, etc.—can be used for any of the sweet goods doughs, such as Danish, roll-in sweet doughs, puff pastry, and regular sweet doughs. The following description includes a sequential listing of the equipment which would be found in a mechanical bench suitable for a medium-sized retail bakery offering a wide variety of sweet dough products:

(1) An introductory conveyor to bring the dough to the sheeting rolls.

(2) Three sets of sheeting rolls to reduce the dough in gentle stages to the proper width and thickness.

(3) Optionally, a cross rolling sheeter could be installed at this point. For most doughs, it would not be needed.

(4) A make-up conveyor on which the dough is transferred as the subsequent operations are being performed.

(5) A dispensing tank for automatically applying oil or any other kind of liquid.

(6) A cinnamon and sugar dispensing unit, adjustable for rate of application and for width of application.

(7) One or two curling rolls, for coiling the dough into an endless cylinder.

(8) A sealing and guide thimble which moves the coil back to the center of the conveyor belt.

(9) A cutting device which may be a rotating die or guillotine cutter.

(10) A panning table with drop leaves on each side. The sweet rolls, coffee cakes, etc., are removed from the belt here and placed into baking pans by hand.

(11) Variable speed electric motors for driving the conveyor belts, curling rolls, and various dispensers.

The layout of a sweet goods line suitable for a large wholesale bakery is shown in Figure 7.1. A sweet goods machine equipped for a fairly large retail shop or small wholesale operation is shown in Figure 7.2.

Curling Rollers

The master baker, working at his bench, will sheet out a piece of dough in an approximately rectangular shape, and apply cinnamon "smear" to the sheet (avoiding both edges). He will then roll the dough into a cylinder by grasping one of the longer edges near the end and

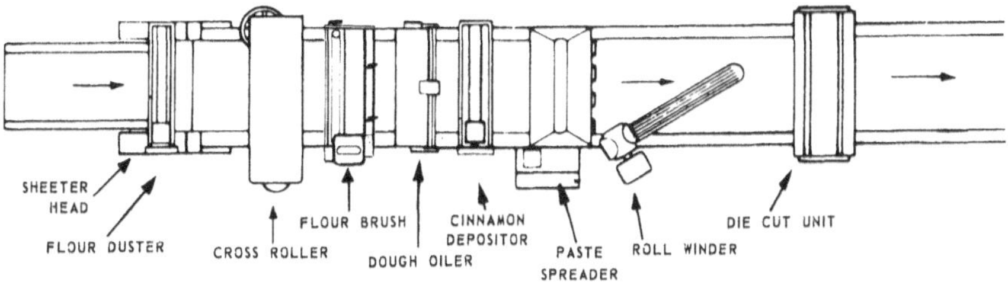

FIG. 7.1. ARRANGEMENT OF PROCESSING DEVICES ON AN AUTOMATIC SWEET GOODS MACHINE.

SOURCE: ANETSBURGER

rolling the sheet over on itself to form a cylinder of spiral cross section. The curling process will be continued until a long tube of fairly uniform diameter is formed. During this manipulation, he will apply slight pressure on the upper surface to prevent entrapment of air and to assist the sealing process. When the curling has been completed, the outward edge is usually sealed into the side of the roll to avoid "unraveling" and to provide an obstacle to leakage of the filling. He will then cut sections of, say, one-half to one inch from the dough cylinder and place these sections, with a cut side up, on the tray, pan, or peel which will hold them during the proofing and baking process. When baked and iced, this will be a cinnamon roll. There being an insufficiency of master bakers to supply the needs of the trade, there is a requirement for some mechanical method to automatically form cinnamon rolls and the like.

If a steel cylinder is placed slantwise of a conveyor belt carrying a strip of dough to which filling has been applied, with the proximal end of the cylinder near the edge of the belt and the distal edge somewhere near the center of the belt, and the cylinder is rotated so that the motion of its lower edge is opposite to the motion of the belt, the cylinder will tend to pick up the edge of the dough and then continue to pick up and roll over the remainder of the strip so that, if conditions are correctly adjusted, a continuous roll of dough will be formed. Crosscuts of this roll will show alternating spirals of dough and filling, the same pattern as found in the common sweet roll described in the preceding paragraph.

These curling rollers (also called roll winders) are found on nearly all mechanical benches for sweet dough products and are used for many varieties and styles of dessert items. By placing a curling roller on each side of the belt, two dough strips can be curled at the same time. Some of these machines have segmented rollers so that one or

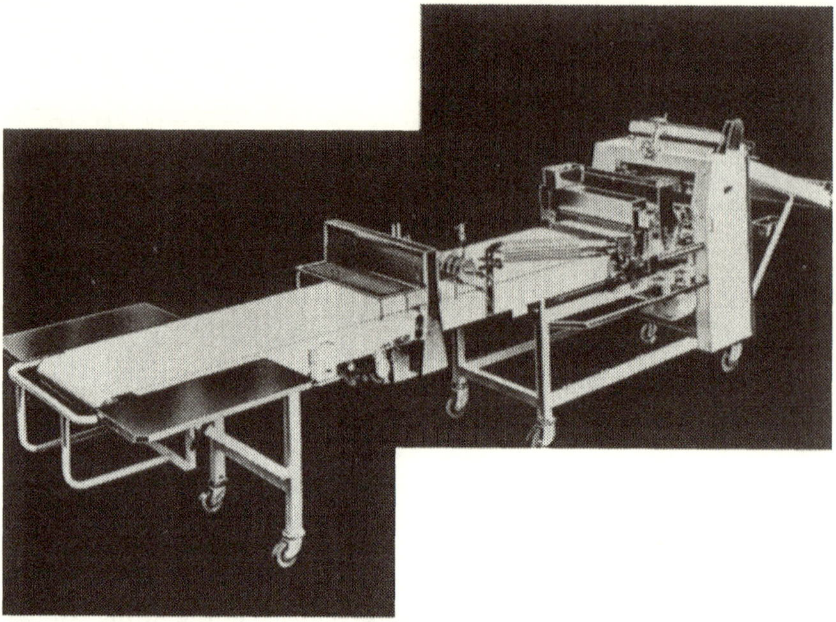

FIG. 7.2. A SWEET GOODS MACHINE WITH SHEETING ROLLS, DUSTER, CURLING ROLL, SLITTING KNIVES, GUILLOTINE, AND PANNING TABLE.
SOURCE: MOLINE DIV. PILLSBURY

more can be removed and the others moved to different locations on the axle, leaving a free space above the belt. This allows two of the roll winders to be put on the same side of the belt; the earlier formed dough cylinder passes under the empty space on the second roller. If very thin strips are being formed (as for braided coffee rings) several small roll winders (generally, not more than four) can be aligned parallel to each other to multiply the capacity of the line.

Most of these cylinders have shallow and narrow grooves alternating with broad ridges placed lengthwise to assist the roller in grasping the dough sheet. They are rotated by separate motors (not necessarily adjustable in speed) and are, usually, mounted so that their angle to the belt can be adjusted. The distance from the belt should also be adjustable. It is not a good idea to have the roller touch the belt, since it rotates in a direction opposite to belt travel and the friction which would result from contact would not only cause much wear on both belt and cylinder, but would heat the belt surface, cause tracking problems, and interfere with production in other ways. It is important to have the roller close to the belt, however, since there is a possibility the dough will not be picked up and will jam under the roller if the distance is too great. This is not likely to happen if the dough is rather dry and extensible and is fairly thick, but is a distinct possibility if the dough is wet, sticky, weak, and thin.

Cutting Devices

In the preparation of individual pieces from a continuous strip of dough, there comes a time when the strip must be cut. Cutting can also form part of the design of an individual piece. For lengthwise cutting, rolling disc knives (one to many on a shaft, powered or not) are suitable. For crosswise cutting, there are several options. The up-and-down motion of the guillotine has been found useful in many applications; disc knives rolled across the belt offer a more complicated solution; rotating cylinders with cutter blades are often used. For more complicated cutting, various types of rotating dies can be applied, as in the forming of doughnuts. Cutting with high pressure jets of air or water has been proposed and some experimental equipment tested.

Guillotines—These devices are used to cut across the belt, severing the dough strand, strip, or sheet into segments. They have a reciprocating, up and down motion (see Figure 7.3). Since the conveyor belt will normally move the dough continuously beneath the cutter while the blade is going through the dough, there will obviously be some restraint on the dough's forward motion, possibly causing distortion of the piece. This effect is minimized by making the cut a rapid one and by advancing the guillotine blade a little as it completes the cut. As a result, cuts can be made without retarding the dough sheet enough to cause observable distortion.

The other problem sometimes observed at the guillotine is sticking of the dough to the blade, so that the strip is brought up and back as the blade retracts. If not corrected, this can cause serious repeated jamming. Suggested cures for the problem include stretching a thin rubber sheet beneath the cutter so that the blade does not actually touch the dough, and passing the blade through oiled wipers before each cut. The blade can also be sheathed in Teflon.

Rolling blades—The simplest of all automatic cutting devices are disc blades rotating freely on an axle. Usually, several blades are arranged along the axle at equal distances from each other. Such devices are used to make lengthwise cuts in strips of dough passing down a conveyor belt. More effective, though more complicated, are the knives which are rotated at the same speed as the belt. Much more complicated are the disc cutters which are rolled across the belt by a mecha-

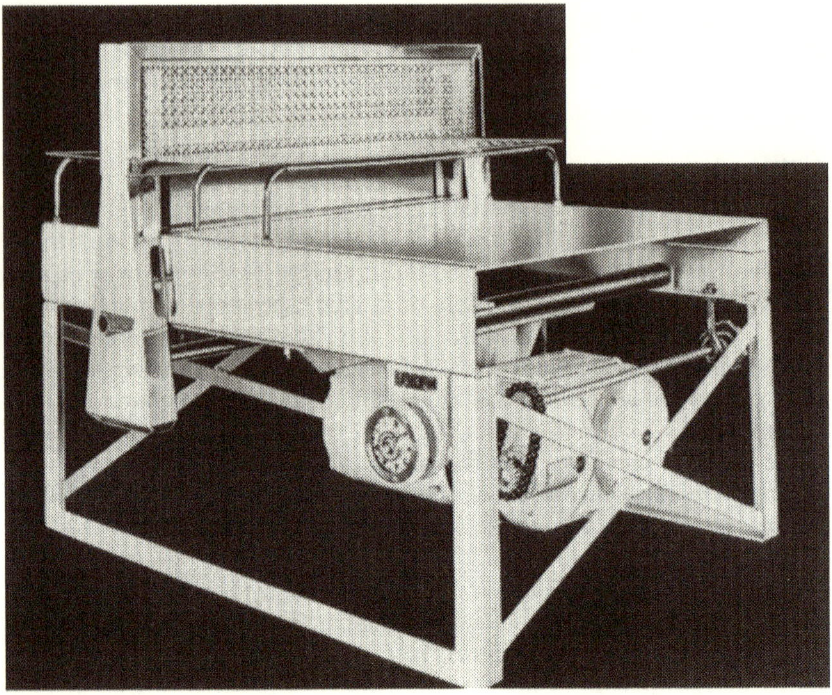

FIG. 7.3. GUILLOTINE CUTTER.
SOURCE: MOLINE DIV. PILLSBURY.

nism which must take into account the forward motion of the product. The discs can complete their circle by being brought back to the starting side of the belt from above or below the belt. These cutters are used on products such as baked fig bars which are difficult to cut by guillotines because of their sticky or tough qualities. The blades can be washed as they travel below the belt on their return path.

Rotary dies—Many kinds of products can be formed of dough blanks cut from continuous sheets by rotary dies. Doughnuts (hexagonal or circular cuts) are sometimes cut this way and other types of fried foods such as long johns and bismarcks are usually made with these devices. Very many kinds of baked products go through a die cutting step during their preparation. Croissants, pizza crusts, tortillas, and cookies are just a few of the items which can be formed in this way. Of course, there are alternate ways of preparing some of these items, extrusion for doughnuts and pressing (stamping) for pizza crusts, for example. Rotary dies can be supported either on powered axles or rotated passively by their contact with the dough or belt. The former arrangement is more suitable for mass production, and it is necessary that the speed of the circumference of the die be well coordinated with the speed of the conveyor belt.

The dies themselves are usually constructed of thin metal strips welded, soldered, or otherwise affixed to the surface of a drum or cylinder connected by circular end pieces to an axle. It is also possible to construct such dies by machining cavities in a solid metal cylinder or, rather, around the surface of a thick metal tube. When thick sheets are being cut, the reduction in dimensions as the walls of the cutters go toward the center of the cylinder should be taken into consideration. The difference in dimensions from top to bottom of the cavity is not necessarily a disadvantage; it can even help to separate pieces from one another or pieces from scrap as a result of the slight compression applied to the upper part of the dough blank. If problems do arise, distortion can be minimized by increasing the diameter of the cutter.

More complicated arrangements of rotary cutting dies include using two cylinders to complete one design, using a second cylinder to emboss a design or dock the sheet, adding ejector plungers to the cavities, and heating the blades.

One problem with rotary cutting cylinders is that thin doughs or sticky doughs tend to hang up in the cavity, especially if the cavity is small. Even a momentary clinging to the cutter can be unacceptable if it causes distortion of the piece. Still worse is the situation where dough pieces remain in the die for a complete revolution. This is sometimes observed when cutting croissant triangles with a single cylinder, the three narrow points and the thin dough combining to cause delayed or imperfect release. In this case, the problem has been solved by using two cylinders, one of which cuts two sides of the triangle while the other cutter severs the base. Similar principles can be applied to many other designs.

Where docking is required, as in some cracker doughs, pizza crusts, etc., the sheet will often first be docked by a cylinder with pins affixed to its surface and then cut into squares or circles by a second cylinder. Since the web is still continuous as it leaves the docking station, the dough is pulled off the pins by the forward part of the strip. In some cases, the docking pins make the sheet stick to the belt more firmly, assisting release from the die at the cutting station.

Using ejector plungers to push out the dough pieces is an alternative which has been applied in the production of some difficult small pieces.

Though expensive, it is a positive solution to the problem of sticking in the die.

Equipment for Laminated Goods Such as Danish Pastry

If a layered dough or roll-in dough is to be prepared, as for puff pastry or Danish pastry, repeated sheeting and folding of the dough and fat combination is required. Dough brakes, which have been used for preparing dough and fat laminations for many years, were described in a preceding chapter. Reversible sheeters, also previously described, can be used for preparing layered doughs. Large scale operations will generally use automatic laminators which replace substantially all the manual labor needed by dough brakes and reversible sheeters. To briefly recapitulate the laminating process used in Danish and puff pastry, the following steps are found in conventional versions: (1) a dough, preferably undermixed but of good stability—that is, resistant to breakdown when overmixed—is sheeted out to give a fairly thick layer, (2) fat, either in sheets or chunks, is applied to part of the dough layer, (3) the part of the dough sheet which is not covered with fat is folded over the fat, (4) the fat and dough combination is sheeted out and again folded, and (5) the folded fat and dough combination is again sheeted out and folded, often after a rest period. Step 5 is repeated any number of times as necessary to yield the type of product wanted by the baker. Usually, a rest period in the refrigerator is given to the dough piece after each sheeting and folding operation so that the dough can relax and the fat can congeal. Factors affecting finished product quality in laminated dough products have been covered in great detail in "Formulas and Processes for Bakers."

Fully automatic laminators will be discussed in the chapter on cookie and cracker equipment, since they are essential parts of soda cracker lines. There are two major types of automatic laminators used for quantity production of Danish pastry. In the first of these, dough is extruded in two continuous belts with shortening extruded in a layer between them. This three-layered sandwich is reduced in thickness by one or more roll stands, and the thinned sheet is cut into pieces about 18 inches long. Cut pieces are transferred in a shingled, or partially overlapping, fashion to a belt running at right angles to the original conveyor. After passing along a conveyor for a time sufficient for the dough to relax somewhat, the shingled sheets are pressed into a continuous strip by another reducing roller, or set of rollers. Again, the strip is cut and shingled before being reduced to about one inch in thickness. This final dough strip is cut to pan size and retarded for make-up perhaps 12 to 16 hours later.

In a more recent type of equipment, the dough strip is kept in continuous layers—not cut—which is overlapped on itself in a diagonal pattern before it is reduced in thickness. The original dough-fat-dough sandwich described above is either formed by the same three layer extrusion method or is formed by extruding a cylinder of dough inside of which is a layer of fat. In the latter case, the cylinder is flattened to make the three layer complex which starts the whole process. Several different devices have been developed to make the overlapping layers.

Danish products can be made by a continuous flow system which starts with extrusion of dough as a strip onto a conveyor belt. The dough strip is thinned out and widened by one or more pairs of reduction rollers. Subsequently, a number of narrow bands of shortening, covering perhaps one-half of the surface, are extruded onto the dough. The roll-in fat should equal at least 25% of the dough weight. A curling roller forms the dough and fat into a cylinder which is then flattened out by other sets of rollers. This is similar to a three-fold operation done manually. A cutting wheel divides the strip of dough in lengths to fit a bun pan, which is then put in a retarder. Some operators give the dough a little floor time (room temperature rest) before sending it to the retarder. After 1 to 4 hr rest in the refrigerated room (about 34° to 38°F) the dough pieces are folded in four layers, sheeted out, then folded and sheeted again. The dough can now be retarded until it is scheduled for make-up, or it can be cut and shaped immediately. The pieces are filled, formed, and topped as required for the particular item being made, and then proofed. Room temperature Danish should be proofed for about 50 min to 80 min at approximately 95°F dry bulb and 85°F wet bulb temperatures. Baking time and temperature depend on the piece size and, to some extent, the shape.

Once the laminated structure has been formed, the Danish dough can be made into sweet rolls and coffee cakes by the same kinds of forming equipment described elsewhere.

Semi-automatic Sweet Goods Machines

These "mechanical benches" are assemblages of several types of devices connected by conveyor belts. They may either be specialized for the manufacture of one type of product or be convertible for making several products. They are used primarily for yeast-leavened sweet goods such as cinnamon rolls, coffee cakes, Danish pastries, doughnuts, bismarcks, and long johns, but may also be used for baking powder biscuits, shortcakes, soft rolls, etc. Production rates are intermediate between fully automated lines and hand work. A description of a typical bench suitable for general production is given below.

The infeed conveyor carries the dough to three sets of sheeting rolls by which it is brought to the desired width and thickness. This dough strip is then delivered to the make-up conveyor where oil or any other kind of liquid can be applied to it by a spray, drip, or roller applicator. A powder dispenser then applies cinnamon sugar or any other kind of pulverulent spice or seasoning. Next in line is a curling roll which can automatically wind the strip into an endless helical coil. Adjustments can be made so that the straight seam lies along the bottom. A sealing and guide thimble (also called a bell) moves the cylinder back to the center of the belt. The endless roll of dough may now be flattened by rollers, slit by disc cutters, and chopped into segments by a guillotine cutter. Finally, the pieces emerge into the panning section of the conveyor where they may be hand panned, automatically panned, or dropped into a proofer or fryer.

Cannon (1987) described a modern high volume plant for sweet yeast raised bakery products. The dough, which is continuously mixed, is delivered in chunks to a hopper leading to a dough pump. Dough is forced through extruders onto the productiom belt. In this plant, sweet doughs are run on a laminating line (more common for Danish doughs) because this type of processing gives a better cell structure, a more uniform grain, and better shelf life. In the laminating section, a first extruder lays down a continuous strip of dough. On this the fat pump deposits a continuous layer of fat, and finally the fat is covered by another dough strip. A set of compression rollers squeezes the three layers together, then a reciprocating belt laminator folds the dough in several layers on the belt. Dusting flour is applied and the dough layers are sheeted by reduction rollers. One more set of laminating and sheeting devices are applied, to give the final dough the thickness needed for the forming and shaping operations.

Several different products are made on this line, as is usual for a bakery offering a wide variety of sweet dough products. To make apple turnovers, separated deposits of filling are discharged on to continuous strips of dough which are then cut in squares and hand folded and sealed in the triangular form. Apple strudel is formed by extruding filling continuously on dough strips, then plowing the strips into closed cylinders which are finally severed into relatively long pieces by a guillotine cutter. A fruit and cheese Danish is made similarly by extruding strips of dough, depositing a cheese filling continuously, using a roller to make a lengthwise indentation in the filling, depositing a fruit filling in the indentation, plowing the combination strip of dough and filling to give a continuous cylinder which can be hand formed into circles or other forms. Cinnamon buns are made by applying the cinnamon filling to the dough strip, running the strip through curling bars to form the spiral cross-section expected in this product, and severing the continuous cylinders into short sections to make the individual rolls. Puff pastry is also made on this line, though with a different dough formula and different type and quantity of interleaving fat. The manufacturer claims a satisfactory product is obtained without the usual rest periods in the refrigerator between sheeting and laminating steps.

Example of a Production System for Sweet Rolls and Coffee Cakes

Sweet yeast raised doughs can be mixed as straight doughs, remixed straight doughs, or sponge doughs. Shaffer (1977) recommends the use of remixed straight doughs as producing the eating qualities of a straight dough and the machining and keeping qualities of a sponge dough. He also believes the machining quality is superior to that of a sponge or pre-ferment type of dough. For the initial mixing operation, the dough is mixed in a regular 1,000 lb. horizontal mixer for about 15 min, or to the pickup stage. At this time the dough temperature will be 82° to 85°F. A 75 to 90 min bulk fermentation in a dough trough is allowed. The fermented dough is dumped into a hopper and pumped through a continuous mixing developer with a capacity of 3,600 lb per hr. Speed of the developer can be varied from 40 to 100 rpm, and the normal range for processing sweet doughs is 65 to 80 rpm.

The developed dough is extruded through a four inch stainless steel tube tapered to a three inch tube and provided with sanitary fittings to deliver a single dough strip. Since there is no nozzle on the end of the three inch tube, the dough will come out round and with smooth edges. The dough is delivered onto a floured belt about nine feet long. Dust is applied to the top, and the dough transferred to an incline belt about 18 ft long with a roller about one foot from the lower end of the incline. This roller flattens the dough and helps to form a thin skin on it. The dough is now relaxed on another conveyor for 6 to 8 min and then transferred to the make-up equipment.

The dough strip passes under three floating rollers which spread and shape the dough for consistent feeding into the sheeting rollers. After passing through the sheeting rollers, the strip is dusted again and then goes under a cross-roller which rolls it to the desired width, say 24 inches. The sized strip goes under another floating roller to smooth out the dough before it passes through the slitter and trimmer. Slitters divide the dough into two strips and the trimmers remove any excess dough. The strips, now of uniform width and thickness, pass to a lateral switch conveyor which carries them to the make-up belts. They next pass under a series of several floating rollers (5 to 8 in the system described by Shaffer) which make the dough strips thinner and wider.

A powered flour brush removes excess flour from the two thin strips of dough, each about 15 inches wide at this point.

Fillings such as wet cinnamon, oil, preserves, jelly, etc. are pumped on to the dough strips. Rollers or spreaders are used to spread the filling to the desired width. Curling rollers are used to form the strips into cylinders. The dough-and-filling cylinders are transferred to an automated cutting and panning device.

When the dough cylinders reach the end of the dough belt of the cutter-panner, they are cut into bun size pieces by a blade which cuts horizontally against the dough belt. The dough pieces are dropped directly into foil containers of appropriate dimensions, and the containers are automatically dropped on to the pan belt.

A second make-up table is used for producing hand-panned coffee cake. A single large dough-and-filling cylinder is rolled for this application. There are panning aprons at the discharge belt so that personnel can hand pan the finished dough pieces.

PIE MACHINES

In discussing equipment for preparing pies, it is necessary to distinguish between baked pies, pressed crumb crust pies, and fried pies as there is a considerable difference in the way these three types of dessert items are formed.

Forming Equipment for Baked Pies

There are two popular methods for forming pie crusts, (1) sheeting the dough and cutting out circles (discs) and (2) pressing the dough into shape between two dies. Satisfactory crusts are being made by both methods. The sheeting and cutting method is said to permit the achievement of better texture but it generates a considerable amount of scrap which has to be remixed, re-sheeted, and re-cut. The stamping method should lead to only a few percent of rework, at least as far as the forming operation is concerned.

Very simple bench-scale machines can be obtained for press-forming pie and pastry crusts in aluminum foil tins or in regular pie pans (see Figure 7.4). An upper die, shaped to form the upper surface of the bottom crust, is hydraulically actuated to press the dough into shape. This upper die is electrically heated to increase the plasticity of the dough mixture. A bottom die or form holds the pie tin which functions as a mold for the outer surface of the bottom crust. The manufacturer claims this press can produce up to 700 pie shells per hour and has a range of sizes from 1.5 through 10 inches diameter. The hydraulic drive is powered by a built-in electric motor.

FIG. 7.4. A SMALL PRESS FOR PIE SHELLS.
SOURCE: EKCO

The sequence of operations in a commercially available pie production line using the sheeting and trimming approach is:

(1) Metering the dough. An automatic divider converts the bulk dough into two rectangular dough blocks which will ultimately become the bottom and top crusts. Weight adjustments can be made while the divider is running.

(2) Conveying the dough blocks from the divider to the crust rollers.

(3) Sheeting the bottom crust. One dough piece is automatically

cross-rolled and deposited over a moving pie tin. Size and thickness can be varied by adjusting the roller settings.

(4) Docking the bottom crust. "Docking" in this context means pressing the dough sheet into contact with the pan. As this step is completed, the dough sheet closely fits the sides and bottom of the cavity.

(5) Wetting the rim. A simple dispenser and applicator wets the upper rim of the bottom crust so that the top crust will adhere to the lower dough piece after the trimming operation.

(6) Forming the rim. Planetary spinning heads form and trim the edge of the dough sheet.

(7) Filling the shell. A measured quantity of fruit filling is deposited into the bottom crust. Adjustments of delivery rate can be made at this point.

(8) Sheeting the top crust. The second dough piece is automatically cross-rolled and deposited over the moving pie pan containing the bottom crust and filling.

(9) Sealing. The bottom and top crusts are sealed together around the rim and excess dough is trimmed off. Plain or fancy crimps can be formed at this point.

(10) Spraying. A wash or glaze can be automatically sprayed on the top crust if desired.

(11) Recovering trimmings. The excess dough from the shaping operation is returned to the dough divider.

It is said that production rates of 600 fruit pies per hour can be achieved with a labor force of three people when using equipment of the type described above.

One of the leading manufacturers of pie making equipment recommends a line consisting of the following machines after the mixer. The line described below forms a system capable of producing 6,000 ready-to-bake two crust meat or fruit pies, 16,000 tarts, or 3,600 fancy rimmed pie shells per hour:

(1) Automatic pan dispenser—an optional feature that is synchronized with the pie machine and automatically drops pie pans into the pan holders.

(2) Automatic dough divider—a heavy duty, ruggedly constructed machine that converts bulk dough into two rectangular dough pieces which are ultimately rolled into the bottom and top crusts. Many of its operations are readily adjustable.

(3) Dough ball delivery conveyors—two conveyor belt systems that automatically transport the dough pieces from the divider to the crust rollers so that they can be cross-rolled into the top and bottom crusts.

(4) Bottom automatic crust roller—automatically cross-rolls the bottom dough piece and deposits it over the moving pie pan. Adjustment controls allow for a wide variety of size and thickness variations.

(5) Docker—a flexible plunger which automatically presses down the bottom crust dough into the pan so as to fill out the pan's contours.

(6) Automatic wetter—a simple electromechanical device that wets the upper rim of the bottom crust so that the top crust can be sealed to it in the trimming operation.

(7) Automatic rimmer—through the action of its planetary spinning heads, this device forms the dough sheet in the pan into fitted and trimmed pie shells, with fancy built-up rims.

(8) Fruit filler—automatically deposits a precisely measured quantity of fruit filling into the bottom crust as it sits in the pie pan. Widely adjustable.

(9) Top automatic crust roller—automatically cross-rolls the top dough piece and deposits it over the moving pie pan containing the bottom crust and filling.

(10) Trimmer—designed to seal the top and bottom crusts around the rim and neatly trim off all the excess dough from the top and bottom dough sheets. Forms plain or fancy crimps.

(11) Finger-bar take-off conveyor—a take-off mechanism that gently removes the completed pie in its pan from the plateholder and deposits it onto a pie sprayer or take-away conveyor belt.

(12) Pie sprayer—sprays the top of the pie with a special wash prior to baking so as to give the baked pie a glossy, finished appearance.

(13) Take-away conveyor—an open mesh or flat belt conveyor that receives pies from the sprayer unit and transports them to the unloading station.

(14) Trimmings return conveyor—this conveyor, mounted in the main frame, returns the dough trimmings from the trimming operation back to the starting end of the pie machine for transfer back to the vicinity of the dough divider.

(15) Trimmings elevating conveyor—two belt conveyor units which combine to take the dough trimmings off the trimmings return conveyor and deposit them into the dough divider hopper for reprocessing into more bottom crust dough pieces (Pluta 1988).

Additional descriptions of the various work stations and pieces of equipment in lines of the above type may be helpful in visualizing the operations performed in a sheeting and docking line. These follow.

The operator loads pie dough into two feed troughs where endless belts aided by flush-mounted flanges feed the dough into individually adjusted reduction rollers so it can be gently formed into a continuous stream. The dough is divided by rotary cutters into 3 by 4.5 inch blocks of proper thickness to give the required weight. The blocks are then automatically flipped over and deposited on separate belts leading to the bottom and top crust rollers.

A trimmings take-away and elevating conveyor picks dough trimmings off of the trimmings return conveyor and transfers them to the cross-elevating delivery conveyor. From the delivery conveyor, they will be automatically dropped into the trough feeding the bottom crust roller to be mixed with fresh dough before being used for the bottom crust.

Each dough block is swept off its belt onto a cross belt which automatically conveys it through two successive pairs of cross rollers. These reduce the dough in thickness and convert it to an oval shape. Each pair of cross rollers has a thickness adjustment setting. Crust position is controllable in four directions by a "joy stick" control—forward, backward, right, or left—while the machine is running.

The rolled dough piece is swept into a final set of cross rollers for a fourth reduction step to give the final crust thickness and shape it into roughly rounded form. A powered discharge belt then deposits the finished crust onto the moving pie plate. An orbital docker gently presses the crust into the plate. An automatic wetter sprays a fine water mist to insure that the bottom and top crusts seal together when the finished filled pie passes under the orbital trimmer/crimper/sealer.

Filling is deposited into the bottom crust by a spiral fed filler unit which operates off the line shaft. The filling will normally be transferred by an air-powered pump into the filler hopper from a mixer bowl or drum. From the filler hopper to the filler spiral feeder, the transfer is by a sanitary pump. Pump and hopper are often combined on a detachable dolly holding the positive displacement sanitary pump, volume controller, magnetic trap for tramp iron, pump disconnect handle, and Tygon tube. Advantages claimed for the spiral filling unit include minimal breakage of fruit pieces, superior filling accuracy, ease of cleaning, and simplicity of operation.

At the top-crust rolling station, there is a fine tuning thickness adjustment wheel on the final set of cross rollers, hardened aluminum alloy carrier plates, stainless steel removable plateholders, and a revolving brush to remove any flour from the top crust. An automatic rimmer section can be added to make fancy crimped rimmed shells. Two rimmer heads, each powered by a 1.5 hp motor, simultaneously engage and follow two plate-holders for a short distance. Three crimped trimming wheels revolving around the perimeter of each rimmer head trim excess dough from the edge of the plate, and a packing wheel gently packs this dough into the head die. Rimmer heads then are lifted up, return to the rear of the station, and engage two more plateholders to repeat the operation.

The discharge end of the pie machine has level take-off rails and fingers to avoid the spillage that might occur if the rails were inclined. A cam operated device lifts the pie from the plateholder and transfers it to the take-off rails. Filled shells are carried away and transferred by the finger bar conveyor to the stainless steel link belt take-away conveyor.

If desired, pies can be sprayed with milk and egg wash (or any other kind of wash) by portable airless sprayer units feeding sprayer nozzles located in a hood above a conveyor attached to the discharge end of the pie line. The flexgrid conveyor has a stainless steel recovery container to catch the excess wash.

Another optional accessory is the automatic lattice topper. This unit takes the round rolled pie crust as it is discharged from the crust roller dough slide and carries it into a lattice top cutter (a rotary die) which cuts offset slits in the crust as it passes between the cutter and the spreader belts. Spreader belts stretch the cut crust sideways as it moves forward, thus opening the slits to form a lattice top. The crust then is dropped onto a filled pie as it passes under the discharge plate of the lattice topper.

Basic change parts enabling the operator to change from one size pie to another include the orbital trimmer head, a set of stainless steel plateholders, an orbital docker pad, and a set of take-off rails with stainless steel backing plates.

The frame of the automatic pie machine consists of welded two inch square tubing of 11-gauge stainless steel, mounted on leveling legs (Pluta 1988)

A pie-making machine which does not roll and sheet the dough but press forms it into the desired shapes was described by Gageant (1964). A metered amount of the blended ingredients is fed through a series of stations where it is formed and placed in a pan. Various filling materials can be dispensed into it. After filling, a die-formed pastry cover is placed over the pan and its edges crimped to the bottom crust. The porous metal dies release dough from the forming surface by slight air pressure without the need for dusting flour. In some cases, the crust molding die may be heated, say to 250°F to facilitate the dough flow (Atwood 1964). Since the shortening is substantially liquefied by this technique, air-injection or other means must be used to release the dough from molds.

Figure 7.5 is a view of a linear pie production line which uses the

FIG. 7.5. A LINE FOR MOLDING AND FILLING TWO-CRUST PIES. THREE ACROSS SIZE.
SOURCE: COLBORNE MANUFACTURING CO.

pressing method of forming the crusts. This particular model can mold up to three 10-inch crusts with each pressing. Shown are two dough hoppers and, in the center, a filling hopper. Material flow is from right to left.

Packaging and chilling or freezing would normally follow the above operations. In some cases, the pies would be baked for fresh distribution routes or, occasionally, for freezing.

Fried Pies

The usual fried pie is a single serving pastry weighing 4 to 6 oz. Consumers eat the pie out of hand, rather than with a fork. The conformation may be semicircular, rectangular, or approximately triangular.

Doughs can be mixed successfully on vertical mixers equipped with a dough hook, on double-armed kneading mixers, or on slow speed horizontal mixers (Burris 1979). In one operation, thought to be representative of general practice, batches ranging from 250 to 1,000 lbs are mixed in open bowl double-arm mixers. Doughs are transferred, either manually or automatically, to a twin screw extruder that feeds a ribbon of dough through a series of sheeter rollers which compress the dough into a specified thickness. To make crescent shaped pies, the dough is sheeted into a thin, narrow, and continuous ribbon which is fed over a series of open clam shell pie formers. A Teflon coated roller cuts the excess dough from the molds. Filling is metered and deposited at intervals. Once the filling is deposited, the clam shells close, enveloping the filling in the dough. Fluted edges on the clam shells crimp the dough and seal the filling within the pie, after which the pies are transferred by a conveyor to a submersing fryer. Conveyors within the hot fat hold the pies under the surface and carry them to the discharge end. While still warm, the pies are glazed with a sugar icing. They are then cooled and wrapped. Fat-based chocolate coating is sometimes applied to fully cooled pies (Anon. 1986).

Processing lines seem to vary considerably between pie manufacturers. The following equipment sequence is one alternative. Bulk dough is conveyed under a corrugated roller which produces a strip from 0.5 to 0.75 inch thick. Three pairs of sheeting rollers reduce the dough sheet to its final thickness of 0.125 inch. The dough sheet is carried over aluminum molds mounted on conveyor chains. These molds have the overall configuration of a square with rounded corners but are hinged in the middle. A measured amount of filling is deposited on one side of the dough, then the opposite side is brought up and over by raising that half of the mold. This process also cuts off the excess dough at the edge and crimps and seals the edge. The general principle is shown in Figure 7.6. Trimmed dough is returned to the mixer.

Another approach to forming is to shape the pies on the conveyor belt. One such system folds the dough over the filling after which aluminum molds mounted on the circumference of a wheel seal and trim the pies (Havighorst 1976). Still another method applies a wide strip of dough across a conveyor belt and adds filling in continuous strips. The unfilled side of the dough strip is lifted up and over, then dropped down to cover the filling. Disc knives cut and crimp the pies into several continuous lines. A second cutter squeezes the filling apart, then cuts and seals the ends of individual pies. Mechanical transpositors can be used to line up the pies on the infeed belt of the fryer.

Fillings for fried pies are mixed and cooked in steam kettles of standard type. Slow sweep type agitators are needed in order to avoid excessive breakage of the fruit pieces. "Cream" or pudding fillings of chocolate, vanilla, lemon, buuterscotch, and banana are also used and they are prepared either in steam kettles or in continuous jet cookers.

Frying is normally done in continuous deep fat fryers of conventional type with submersing conveyors. The pies are often glazed in a waterfall type coating device which applies about 0.5 oz to each pie. After this treatment, air is used to blow off excess glaze. Coated pies are cooled for about 90 min on a spiral cooler or equivalent equipment.

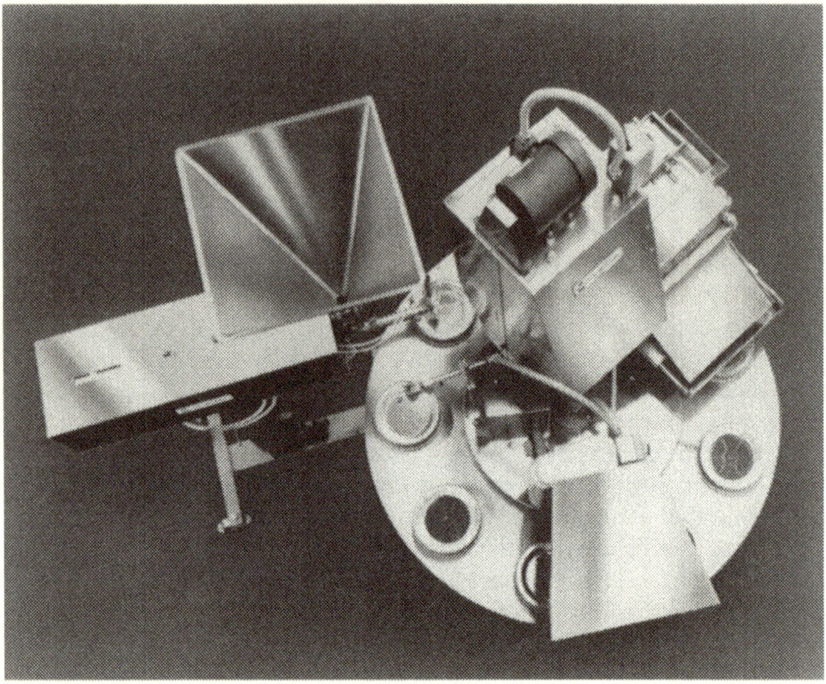

FIG. 7.6. NINE-STATION CIRCULAR MACHINE FOR MAKING CRUMB CRUSTS, FILLING THEM, AND TOPPING THEM.
SOURCE: COLBORNE MANUFACTURING CO.

Pressed Crumb Crusts

Graham cracker crusts or other types of crumb crusts, such as those used for some cheese cakes and for some cream pies, are usually made by pressing a crust mixture consisting of graham cracker crumbs, vanilla wafer crumbs, gingersnaps, or chocolate cookie crumbs, sugar, shortening, and flavoring ingredients. The shortening provides a binding action and, since it is the only structural element in these crusts, they tend to be very fragile. The pressing action is accompanied by a spinning motion of the upper mold which distributes the crust mix up the sides of the pan which forms the bottom mold.

The basic steps in preparing a pressed crumb crust are these:

(1) The graham crackers or cookies are milled or crushed between rollers. Milling must be performed with equipment giving very gentle action, since the production of large amounts of fines must be avoided.

(2) The ground crackers or cookies are sieved to remove excessive fines or pieces which are too large. A sifting device with gentle action is

preferred, so that additional breakdown of the crumbs is minimized. The sifter should be fitted with screens having fairly large openings.

(3) The sieved material is gently mixed with melted or softened shortening, sugar, flavors, salt, and the rest of the ingredients.

(4) A weighed portion of the mix is deposited in a pie tin.

(5) The pie tin is placed in the plateholder of the pie press.

(6) The top mold descends with a spinning action, forcing the crumbs across the bottom and up the sides of the pie tin, and finally compresses the mixture into a compact layer.

(7) The top mold is raised.

(8) A measured amount of filling is deposited in each shell.

(9) Whipped topping is applied in a pattern to cover the filling.

(10) The finished cream pie or cheesecake is automatically lifted from the plateholder and carried by a conveyor to the packaging area.

If the product is to be sold as an unfilled crust for home preparation, the above steps 8, 9, and 10 are replaced by the application of a close-fitting, semi-rigid plastic cover bearing a paper label. The plastic cover is secured by crimping the extended rim of the aluminum foil pan over its edges.

Machines suitable for preparing and filling crumb crusts can be of either linear or circular configuration. Linear machines tend to be the higher speed equipment, and are often used for preparing the tart shells and cookie crusts that are sold in supermarkets for preparing cream pies in home kitchens.

The circular machines are of variable capacities and are widely used by wholesale bakers as well as by some institutional bakeries. They are probably more versatile than the high speed linear units and certainly occupy less space. A typical unit would have nine stations, or plates, and produce 25 to 30 finished packaged pieces per minute with two or more operators. These units include, among other devices, a pan depositer, a crumb dispensing station, an automatic spinner head, an orbital cream filler head, and an orbital topping head. The topping head can be fitted with different dies to give different topping patterns. Figure 7.7 shows equipment of this type.

DOUGHNUT EQUIPMENT

Generally, conventional horizontal or vertical mixers can be used for making doughs and batters for yeast raised and cake type doughnuts, but efficient production requires specialized equipment subsequent to the mixing stage. There are two methods of forming doughnuts, the sheet-and-cut method and the extrusion method. Automatic doughnut production in a linear operation involves seven basic operations: (1)

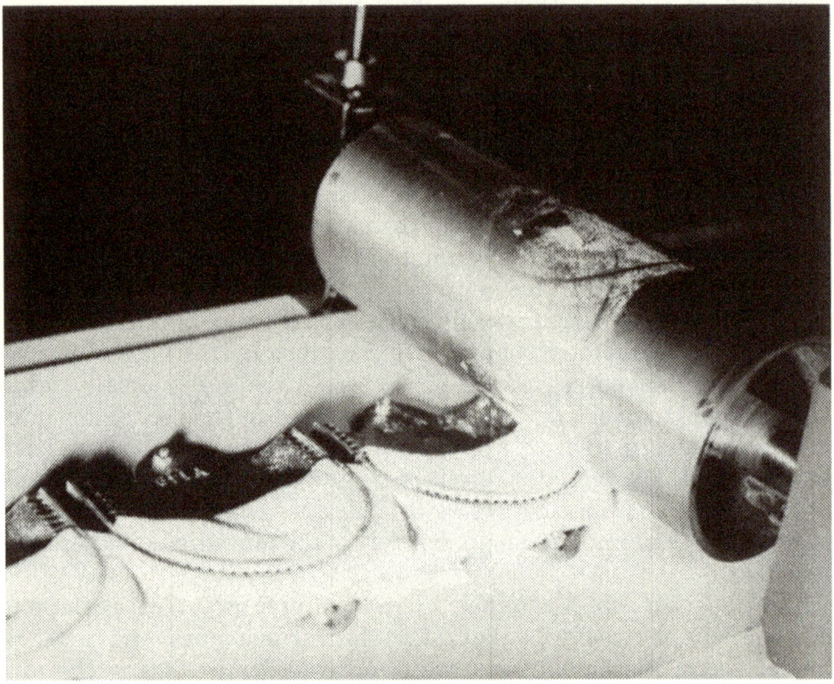

FIG. 7.7. FORMING AND SEALING FRIED PIES.

scaling ingredients, (2) mixing, (3) either extrusion forming or sheeting and cutting, (4) proofing, if the dough is yeast leavened, (5) frying, (6) finishing, and (7) packaging. Some of the equipment alternatives available to the commercial producer of doughnuts will be discussed below. Fryers will be covered in more detail in a subsequent chapter.

Yeast raised doughnuts of the conventional hole-in-the-center type can be formed either by cutting hexagonal or circular pieces from a dough sheet or by extruding the dough through a circular orifice. Very soft doughs and batters are obviously not adaptable for sheeting and cutting methods. Doughs varying in consistency from very soft to moderately tough, as well as batters, can be handled on extruding equipment.

Sheeting and Cutting Processes

The preparation and cutting of a sheet of doughnut dough varies little from the generalized system described for sweet doughs in previous sections of this chapter. Ingredients are scaled, the dough mixed (straight doughs, with few exceptions), the dough passed through sets of sheeting rollers and gauge rollers or hand rolled, and pieces cut from the dough by die cylinders. The devices found on a complete sweet goods bench would be adequate for all these operations, although a special cutter is needed.

The amount of trim or scrap dough generated during the forming operation should be a major consideration when making a choice of equipment. Cutting a sheet of dough with hexagonal cutting dies will reduce the scrap; these cutters are often called "scrapless" although, of course, there is some scrap generated along the edge of the strip of dough. The wider the dough sheet, the less scrap is generated in both the hexagonal and the round cut, provided that the rotary cutting die is sized to take full advantage of the width of the strip. Using a strip so narrow that the central part of a hexagon or circle falls on the outer edge of the strip is a sign of poor processing. Round cutters will generate perhaps 60% doughnut and 40% scrap in the average well-run line, while hexagonal cutters can approach an 80% doughnut and 20% scrap ratio. Rectangular cutters (as for long johns) and square cutters (as for some Bismarcks) would, of course, generate very little scrap.

It is important to keep the amount of scrap returned to the mixer at a minimum, not only because the production capacity is reduced by reprocessing a material which has already gone through most of the line, but because scrap dough is more developed than freshly mixed dough and therefore changes the dough's consistency and its response to the whole series of processing steps. Since scrap cannot be avoided, it is important to keep the amount added to each new mixer batch at the same level.

The center circles, or holes, can be removed by a rotating "picker" attachment (see Figure 7.8). These constitute a few percent of scrap, or they can be fried separately to give a specialty item liked by many consumers.

After yeast leavened doughs have been cut into the finished form, they are usually allowed to proof. This is a critical step, and unusual conditions can be involved, sometimes with a view to developing a dehydrated layer on the dough piece so that a characteristic type of crust will appear during frying. Specialized types of proofers, often resembling intermediate proofing chambers or rack type cabinets, are used.

The three types of automatic proofing systems for yeast raised doughnuts are:

(1) The proofing cloth system in which the pieces are cut from a sheet of dough on the bench or make-up table and then transferred to a proofing cloth, screen, or proofing board. The dough pieces are then transferred on their supporting material to a conventional proof box.

FIG. 7.8. HEXAGONAL DOUGHNUT CUTTER WITH HOLE PICKER.
SOURCE: MOLINE DIV.

After a proof time of 25 to 35 min, the proofing cloths are taken to a feed table from which the dough pieces are transferred to a conveyor-type fryer.

(2) A second type of production system involves the manual transfer of raw doughnuts from a conventional make-up table to the flights or baskets of an automatic proofer. Humidity and temperature are maintained by automatic controls. The uniform proof imparted to the product by the controlled conditions in the cabinet permits close control of product size and quality. Since a constant load is being sent to the proofer, and there are no doors to open or close, uniform temperature and humidity conditions can be maintained.

(3) The third type of proofing system uses microwave heating to speed the leavening action. When irradiated by microwaves, there is a substantially uniform temperature rise throughout the dough piece, eliminating the slow transfer of heat by conduction or convection which must occur in conventional proofers. This process is discussed in more detail in the chapter on proofing enclosures.

Extrusion Equipment

The extrusion method is certainly the most efficient type of processing method since it is truly scrapless, at least so far as the forming operation is concerned. It is also less labor intensive. If it were not for the fact that the consumer, that fly in the ointment of all efficiency experts, often demands raised doughnuts, there would be no sheeting and cutting operations in existence.

Simple, inexpensive devices for extrusion forming are available for the retail baker. These doughnut cutters, one of which is shown in Figure 7.9, deposit the batter directly into a fryer. Extruders for yeast leavened doughs, mostly used in large scale operations, will deposit the scaled doughs directly into the infeed conveyor of a continuous proofer. Extrusion depositers do require that doughs have certain characteristics in order to operate satisfactorily, however.

Although it is possible to sheet a chemically leavened dough and cut doughnut shapes from it (much like soda biscuits are made in large scale plants), cake doughnuts are generally extruded from a mass of soft dough (or thick batter) held in a hopper. Both vacuum-mechanical or pressure-extrusion systems are in use. In pressure extrusion, a rotary valve delivers batter from the hopper into a chamber where it is subjected to 4 to 10 lb of air pressure. Several tubes lead from this chamber and are closed at the exit end by cutting valves. When these valves are opened, pressure forces batter down the tubes and around the cutting dies. Closing of the valves severs the ring of dough, allowing the raw doughnut to drop. Product weight is a function of the size of the opening, batter viscosity, air pressure, and length of time the cutter is open. The vacuum system creates a negative pressure ("vacuum") by retracting cutters or plungers to draw batter from the hopper to the cutter cylinders. Each cutter has a separate cylinder for measuring the proper amount of batter. At a predetermined point, the cylinder is closed off, fixing the amount of batter which will be extruded in the form of a ring through the die opening.

Extrusion methods drop the cut dough pieces directly into the hot fat. If a continuous fryer is involved, as there would be in all large-scale operations, the drop must be registered with the movement of the flights or pusher bars which move the doughnuts through the vat. Although this adjustment is critical, it is not difficult to maintain on modern equipment, assuming normal operator vigilance and a dough which does not hang up in the cutters. If the conveyor timing is improperly coordinated with the dough drop, distorted products can result from the pusher bars contacting the piece before it is firm enough to maintain its shape. Even worse is the dropping of the dough directly on top of the conveyor bar; this causes hangups in the fryer and a large amount of scrap product (Fischer 1976).

In the pressure methods, product weight is affected by batter viscosity, pressure on the batter, and length of time the die is open. Product size cannot be varied at the individual orifices because a single pressure chamber feeds all the cutters. The vacuum method is less

PILLSBURY FIG. 7.9. HAND OPERATED DOUGHNUT EXTRUDER WITH BATCH FRYER.
SOURCE: DAWN EQUIPMENT CO.

sensitive to the viscosity of the batter, which tends to increase as the batter ages, and the shape of the extruded ring is the same from all cutters since the opening does not change in size or shape.

Worn or damaged cutters cause scaling problems and appearance defects in the finished products. Worn cutters leave strands ("whiskers") protruding from the cut surfaces and they drop small pieces of dough into the fryer. Badly worn cutters will produce out of shape doughnuts.

By modifying the cutter plungers in a vacuum-mechanical system, different shapes of doughnuts can be produced. The star center doughnut cutter is equipped with a piston seal on the forming piston and there is a row of degassing pins around the upper surface of the forming piston. The degassing pins change the rate of expansion in certain areas of the deposit, causing a "star" to form in the center, instead of a round hole. Star center doughuts are surface fried. Old fashioned doughnut cutter plungers have piston seals and there is also a degassing plate above the plunger pistons to assist in releasing gas. The latter action leads to a denser batter so that the frying doughnut stays longer beneath the surface. In addition, the shape and position of the forming piston is changed to give a tubing effect to the cut. Lower frying temperature and a special formula are generally required to give the so-called old-fashioned doughnut, which is characterized by a very rough surface with strong color contrasts.

Miniature doughnuts can be made with cutter plungers that are very similar to the plain and star center cutters. They have adjustments for increasing the seal pressure, and the cutter cylinder is modified to provide a cleaner cut and drop. Miniature doughnuts are usually surface fried but are sometimes fried submerged.

Cutters for crescent, or semi-circular, doughnuts are made with attachments placed on the surface of the forming pistons for plain doughnuts. They are available in a variety of sizes for closing off a portion of the ring of cut batter. When the dough deposit fries, its expansion causes the cut to partly straighten out, so that a crescent is formed. A stick dougnut cutter plunger is equipped with a piston seal and sleeve in place of the usual forming piston. The sleeve has a slotted extrusion orifice through which the batter is forced during the cutting cycle. The rod of batter is flat, and should not round out during its expansion in the fryer, for this would lead to turning in the oil with generally poor cooking results.

A ball doughnut cutter plunger is equipped with a piston seal and sleeve instead of the usual forming piston. During the cutting cycle, batter is forced out of round holes, forming small balls which drop into the oil. Cutters for crinkled doughnuts use a cutter plunger having a forming sleeve and forming piston designed to give the characteristic shape of this variety. The form of the fried doughnut allows the consumer to break the doughnut into bite-sized pieces.

Plunger cutters for French crullers are designed to include a counter-rotating forming die and forming piston. The rotation of the die and piston impart a twisted shape to the batter deposit. This shape is held through the frying and finishing processes. It is either surface fried or submerge fried. A special formula is required (Belshaw 1976).

A further development in vacuum-mechanical equipment has been the introduction of multispaced cutter heads. Since both the die shapes and the center-to-center spacings can be changed, several varieties and sizes of doughnuts can be formed simultaneously. When smaller pieces are being made, more pieces can be extruded on the same belt width, eliminating empty proofer spaces and increasing utilization of fryer capacity.

Extruder systems have been developed for depositing doughnuts with a ring of jelly inside (Moyer 1986). Thus, instead of filling the doughnut after frying, it is fried with the jelly inside. A modified feed system draws jelly from a reservoir into a ring shaped outlet placed in the same region where the dough is being extruded. The jelly must be a starch-based, lower moisture formula since the conventional fruit jelly based on a pectin system does not perform satisfactorily during extrusion or during frying.

A completely automatic production system for doughnuts would usually include equipment for making both extruded and cut doughnuts from both yeast leavened and chemically leavened doughs. The line would require, as a minimum, an extruder for yeast raised dough, an automatic proofer, an automatic cake doughnut cutter, an automatic fryer, a fat melter and leveler, a sweep conveyor, an automatic glaze applicator, and a screen loader. A typical linear doughnut layout is shown in Figure 7.10.

Frying, Cooling, and Decorating

Extruded doughnuts are almost always dropped directly into the frying vat. Sheet and cut methods require a transfer from the forming station to the frying vat. Often a proofing step intervenes. This requires extra labor and some extra equipment but allows considerably greater flexibility in scheduling the frying process.

As is well known, nearly all varieties of doughnuts must be turned over a little more than halfway through the frying process in order to get a symmetrical shape and a uniform crust color. In batch fryers, the doughnuts are turned over manually, but continuous vat fryers (which are available in very small scale versions as well as in very large capacity models) are made with conveyors which automatically turn the dough pieces over at the proper time.

The point at which the doughnut is turned over during its travel through the fryer is a significant factor affecting the shape and color of the finished product. If the dough piece is turned too soon, volume will be low. When it is turned over too late, it may be misshapen, absorb too much shortening, and lose too much moisture.

Doughnuts are customarily cooled before they are packaged. This prevents condensation from forming in the container. Cooling is usually conducted by conveying the fried products through a room temperature area, possibly with fans directed upon the conveyor to assist heat transfer.

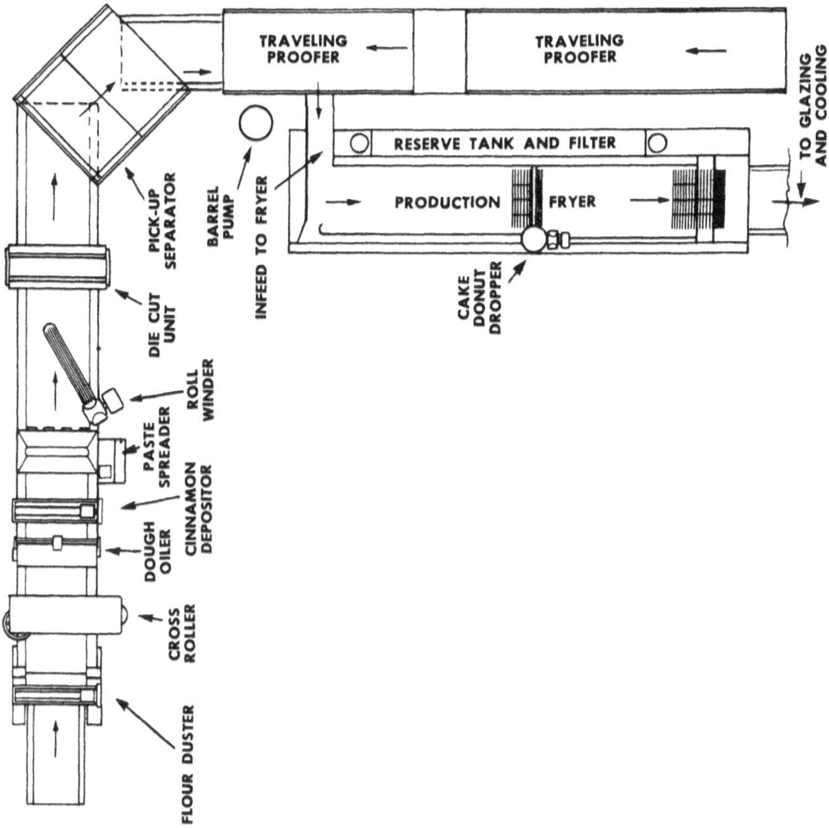

FIG. 7.10. LAYOUT OF AN AUTOMATIC PRODUCTION LINE FOR DOUGHNUTS.
SOURCE: BELSHAW

Most doughnuts are decorated and modified after frying by one or more of the following operations: (1) coating with powdered sugar, (2) covering wholly or partially with glaze, icing, or frosting; (3) sprinkling with nut pieces, coconut shreds, candy bits, or other particles; or (4) injecting with fillings of the jelly, pudding, or jam type. Some of the equipment used for these operations will be described in a subsequent chapter.

CAKES

For all practical purposes, cakes and other products made out of fluid batters assume the contours of the container in which they are baked (except for the top crust), so the forming or shaping machinery can be regarded as the pans themselves and the pumps, pipes, and metering devices which deliver the batter to the pans. The major pieces of equipment required in a continuous mixing and depositing line for cake are shown in Figure 7.11. Finishing equipment which enrobes, deposits, injects, spreads, or otherwise adds fillings, toppings, etc., have very important effects on the appearance, flavor, cost, and nutrition of the final products.

Batter depositors may be either manifold-type or volumetric hopper types. Accuracy is a prime consideration. Ability to handle batters containing pieces, such as raisins, nuts, and chocolate chips, without pulverizing them is important. The depositor should not markedly change the specific gravity established at the mixer. Whether deposited on the oven band or in open pans, the top surface of these cakes will be flat or slightly domed. Generally, the smaller the pan the more prominent the domed shape, so that top contours of cup cakes and muffins are significant factors affecting consumer acceptance. Differentiation in the larger cakes must depend upon post-baking operations such as cutting, rolling, layering, and decorating.

Of course, fancy pans to make cakes shaped liked Easter eggs, or for that matter bunnies and lambs, are traditional. It is possible to bake cakes in pans shaped like Texas, and there are even people who want to do such things. The chief difficulties encountered when using these fancy pans are getting the batter to fill the cracks and crevices so that the finer details of the pattern can be seen, and getting a relative uniform bake in the thick and thin portions of the cake. The design problem is largely a matter of avoiding fine detail when doing the artwork, and adjusting the batter formula so that the viscosity is low enough to allow the mixture to flow into all parts of the pan. Vibrating the pan strongly during and after depositing the batter is of considerable help, although it may tend to bring all the larger gas bubbles to the surface of the batter with unfortunate effects on appearance of the finished cake. Leavening should be on the low side, since substantial expansion during baking will tend to lift the batter out of small crevices. Large bubble size also tends to blur the finer lines, making a fine grain desirable. A slow bake will help some in avoiding a combination of unbaked areas in the thick parts and burnt points and stripes.

Lining fancy pans with Teflon or some other nonstick coating improves detail and also reduces the possibility that some critical part of the design will remain in the pan when the cake is dumped.

Baking cakes as a continuous ribbon on an oven band has become common in large installations. Individual cakes are formed by cutting pieces from the baked strip and combining them in various ways. Of course, a somewhat similar process was followed when the smaller shop used sheet pans for baking layer cakes and the like. The obvious disad-

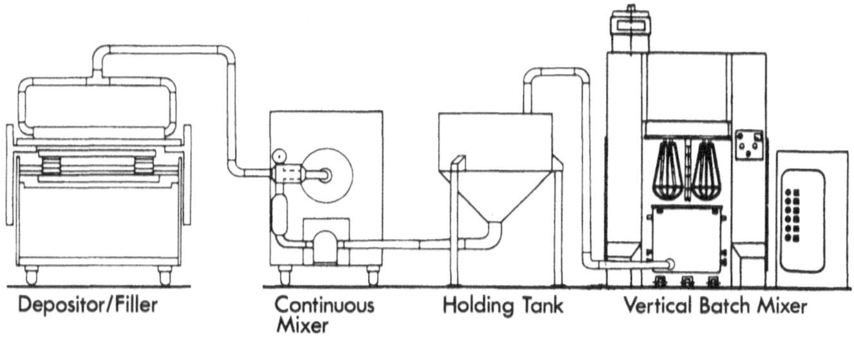

FIG. 7.11. AUTOMATIC DEPOSITING LINE FOR CAKE.
SOURCE: WESTERN BAKERY IMPORTS

vantage is that any shape which is not rectangular results in the generation of considerable amounts of scrap which is difficult to utilize. Even when observing this limitation, however, many varieties can be obtained.

In manufacturing band cakes, batter is prepared in a continuous mixer and pumped to a manifold, or batter distributor, positioned above the oven band. The manifold spreads the batter in a uniform depth, although more than one strip may be used if baked edges are preferred to cut edges, as they may be when preparing cake for Swiss rolls (Freihofer 1985).

In addition to restricting the shape of finished products, baking cakes on a band also establishes limits on formulation. It is important that the batter does not spread significantly during baking. Sponge cake batters of high specific gravity are preferred. The finished cakes are often lower in moisture than is considered satisfactory for conventional cakes. Fillings, icings, frostings, and other adjuncts of relatively high moisture content are liberally applied to furnish the textural characteristics the consumer expects.

After the baked cake has been separated from the oven band by a knife, it is transferred in a continuous ribbon onto the cooling conveyor. When some cooling has occurred the cake is cut into strips by disc knives and then, for the preparation of Swiss rolls, filling is deposited in a continuous stripe. Lateral cuts are made, often by rotating disc knives which move across the band. The individual pieces are then rolled mechanically or by hand to form the finished rolls. In preparing layer cakes, two iced or filled strips can be plowed up and over to form a double layer before the horizontal cuts are made.

Some very successful small cake manufacturers have found that many varieties can be produced from the same basic type of batter. Differentiation between the varieties is established by changing the fillings, toppings, icings, and enrobings. Adding an automated machine for injecting fillings is one of the easiest and most efficient ways to expand varieties in a line of small cakes. Newer automated finishing techniques include depositing creme fillings into dessert shells before enrobing. More personnel are required and some additional equipment is necessary for each variety, but the specialized equipment is generally considerably less costly than a new line. Consumers seem to appreciate the opportunity to choose something different, and greater sales volume results.

Totally oven-finished items are efficient to produce (Hokes 1977). These involve the deposition of materials on top of cake batter or beneath the cake batter before baking. Among the materials which have been used in this way are jelly, streusel, raisins, blueberries, nut pieces, cheese pastes, coconut crunch, flavored crumbs, and cinnamon pastes. Equipment is available to do this automatically. Several oven-finished varieties can be produced in the same size pan and with the same batter, or with minor variations in the batter.

THE RHEON ENCRUSTER

This device, the basic design of which is now more than 20 years old, is specified here by its trade name since it appears to be unique in the way it operates. Many kinds of devices have been suggested, and used, for enrobing a filling in a kind of dough, or one kind of dough or batter in another. Things like ravioli, apple dumplings, soft and crisp cookies, and wafer shells with filling are some examples. What appears to set the Rheon encruster apart from the earlier equipment is its adaptability to many different kinds of coatings and fillings, as well as its method of operation. It is somewhat less versatile in the shape of product formed, however, being limited to ball shaped items for all practical purposes. No doubt a limited range of shaping procedures could be applied after the balls are formed. It can also be adapted to continuous extrusion of cylindrical products such as tamales. There are some limitations on the consistency of the filling and dough, and of piece sizes in the filling. Production rates are fairly low, as compared to some of the other equipment designed for making filled products.

The unusual aspects of this machine lie in the actions by which the dough is transferred and shaped following its extrusion from the dough hopper. Dough is gently forced from the dough hopper by a set of two rotating screw conveyors and a stability roller. The screw conveyors are designed to be without a center shaft for the final two-thirds of their length. When the dough reaches the compound nozzle assembly, it is formed into a ring or doughnut shape. In the forming or ring assembly the dough is reformed in a continuous strip to completely enrobe the filling that has been directed by another extrusion method to the center of the dough forming ring. The machine is said to operate at a top speed of 3,000 pieces per hour.

PANCAKES, CREPES, BLINTZES, AND FRITTATEN

Automatic and semi-automatic devices have been made available for the quantity production of pancakes, crepes, and frittaten. There are simple low-output devices suitable for making crepes for restaurants or small retail bakeries which consist of a motorized and heated conveyor serving as the griddle and a gravity dispenser for batter. A rectangular crepe is prepared in this equipment. The same manufacturer makes an industrial machine capable of making 700 to 1,000 pieces 33 cm in diameter. This device includes a cylindrical rotating hearth bearing circular depressions for containing the batter. Cooked crepes are stripped off on a conveyor belt leading to an automatic receiving table (Anon. 1972). Other devices more suitable for restaurant use than for retail or wholesale bakery production include an electrically heated machine making circular crepes 6.5 to 12 inches in diameter and of adjustable thickness. It operates at the rate of 180 to 360 per hour (Anon. 1973).

Blintzes could be formed on most kinds of crepe machines with simple modifications of formulas and processing conditions.

Another manufacturer describes the procedure for using his equipment, which is based on a circular horizontal rotating hearth. The hearth is divided into an inner and an outer section. Crepes or pancakes are prepared in the following manner. During a pre-heating period of about 20 min, the containers for the cooking oil, the batter including a reserve supply, and the various sweet and savory fillings are made ready. When the baking temperature is reached, the oil spray and the batter spreader are started by merely pressing a button. The batter flows from the batter dispenser onto the baking plate in a continuous stream at a rate determined by the setting. The batter starts to bake immediately.

After one revolution of the inner baking plate, the batter is automatically turned onto the outer baking plate, where final baking takes place. The speed of rotation of both baking plates can be set to obtain the best baking time for the batter. The baked pancakes, either flat or rolled, are lifted from the discharge conveyor and transferred to the filling area where various prepared compositions are applied, if the the finished product requires them. In this way, the following ready made foods are produced: plain pancakes, rolled pancakes filled with jam, cheese, meat, vegetables, etc., Kaiserschmarren, pizzas, and frittaten. The latter are thin strips (like noodles) cut from pancakes and used as ingredients in soup and the like. Accessories include a jam or cheese filling device for pancakes, a raisin spreader for Kaiserschmarren, filling and rolling devices for meat or vegetable rolls, and a shredding device for frittaten. Claimed production rates include 2,500 pancakes of 4.75 inch diameter, 1,700 pancakes of 6.6 inch diameter, or 175 lb of frittaten per hour. Baking times will be 0.5 to 2 min (Anon. 1982).

Ruhdorfer (1974) patented a method and equipment for making frittaten which involves feeding a liquid pancake batter to the upper roller of a pair of large heated rollers. The baked batter is removed to a conveyor belt and dried with a hot air blast. Cutting of the partially dried dough into strips is accomplished by cutting discs. The strips are severed lengthwise by a knife revolving with a roller and acting against a fixed knife. The half-dried frittaten are immediately guided to a vat of hot oil which further dehydrates them. The resultant product measures about 30 to 40 mm in length and 2 to 3 mm in width, and it has a moisture content of less than 7%. Advantages of the frying step is that the frittaten have improved flavor and much longer shelf life, as compared to the usual product.

TORTILLAS

Traditional tortillas are made of a paste (masa) formed by grinding corn kernels which have been soaked in hot lime water. The masa is patted into a thin cake, similar in size and shape to a pancake, and baked (originally) on a heated rock or pottery surface. This procedure has been mechanized, in steps, so that restaurants serving Mexican food could reduce the labor requirement for making tortillas, tacos, enchiladas, etc. Until fairly recently, however, the equipment was small in scale, simple in design and operation, and relatively high in labor requirement.

Mostly as a result of the popularity of fried corn chips and, later, the explosive growth of chains of restaurants serving food based more-or-less on Mexican cuisine, as well as the rise of central commissaries for these foods and the popularity of frozen versions of the foods, the preparation of tortillas was fully mechanized.

Havighorst (1971) and Clark (1981) described mechanized plants used to prepare corn tortillas. The cooking cycle begins when the operator charges a kettle with 2,000 lb of white corn kernels and a measured amount of water. The kettle contains an upper perforated ring from which water is sprayed over the corn. A steam injection ring in the bottom part of the kettle maintains the contents within a desired temperature range and agitates the corn.

The corn is first hydrated with water maintained at about 120°F. Then, the temperature is raised to 165°F to gelatinize the corn, after which the heat is shut off. During the cooking cycle, hot water is sprayed over the corn and continuously recirculated by a pump connected to the discharge valve. When the temperature drops below 140°F, recirculation is discontinued and the corn is allowed to steep undisturbed for about 12 hr. The steep water contains 12 oz hydrated lime for each 100 lb corn.

After completion of the steeping period, the corn is flushed with fresh water into a draining conveyor. This conveyor discharges the kernels into a screw-type elevator in which the corn is again washed to remove free starch and other debris before it is milled. The mill consists of a stationary lower stone disc and a rotating upper stone, both of 16 inches diameter and 4 inches thickness. This buhr mill of generally conventional design is powered by a 30 hp motor and grinds about 3,000 lb per hr of corn.

The ground alkalized corn dough, or masa, is conveyed to the hopper of a tortilla cutting head by a screw-type extruder. A thick sheet of masa is extruded between sizing rolls, and the resultant ribbon is cut into discs of appropriate size by the usual rotating dies.

Gas-fired ovens held at 600°F bake the raw tortillas in 30 to 32 sec. Tortillas are at about 175°F at the discharge but are cooled to 85° to 90°F in a multi-tiered conveyor through which ambient temperature air is circulated.

In a recent development, whole corn kernels are ground and blended with water and about 0.2% calcium oxide. This mixture is fed to a cooking extruder where it is worked and heated until the masa reaches proper consistency. It is then extruded and cut into tortillas, which are said to compare favorably with conventional tortillas.

Machines have been developed to feed balls of masa to a heated compression unit which forms them into tortillas of uniform weight and dimensions. Also, a compact tortilla press and oven has been patented. This device divides the masa into pieces of weight suitable for one tortilla, presses the unit into disc shape, and toasts each side. Presumably, the latter machine would be particularly suitable for restaurants, small tortilla shops, institutions, etc.

There is also a white flour tortilla based on wheat flour which differs in other ways from the traditional tortilla. This wheat, or flour, tortilla has a blander flavor than masa tortillas and, probably just as important, is better adapted to wrapping around fillings without cracking. The commercial forming process for flour tortillas is similar in many respects to that used for corn tortillas. The usual composition appears to be low grade wheat flour, salt, a small amount of shortening, enough water to prepare a dough, and, often, a little baking powder. A very soft dough is formed into flat, thin sheets which are toasted on a baking surface until part of the starch has been gelatinized and the moisture reduced to acceptable levels.

Schmidt (1985) gave a detailed description of the manufacturing method for "hand-stretched flour tortillas" which is important as being representative of a modernized process actually being used to manufacture thousands of pounds of this product every day. The formula is basically as described above. High gluten flour is required for this product. A two-speed dough mixer is recommended, and it can be either the conventional horizontal mixer or the vertical type. The dough should be very stiff, so the batch should be smaller in size than that considered suitable for the mixer when conventional bread doughs are being mixed. The dough is divided into 25 gm pieces for tortillas of 6 inches diameter, 60 gm for the 10 and 12 inch diameter products, etc. Almost any kind of dough divider can be used. Rounding immediately follows the dividing. A rest period is then given so the dough can recover from the stress of dividing and rounding.

The dough ball is pressed into a circle 0.006 to 0.01 inch in thickness by passing it through two or more sets of sheeting rollers. In this particular process, the sheeted circle is somewhat irregular and must be finished by hand stretching the tortilla as it rests on (or travels across) a heated plate. Shaped tortillas then travel 20 seconds on mesh bands or baking plates through an oven heated to 500°F. There is little radiation or convection heating, most of the energy transfer occurring by conduction. The tortillas will be very near 212°F when they emerge from the oven, and will show some puffing. They must be cooled to prevent them from adhering to one another when they are stacked and to prevent condensation of moisture inside the container.

Filled tortillas, such as burritos, are important items of commerce, both for food service applications and in retail frozen packs. Roberts (1987) described an apparatus for making these automatically which is based, in large part, on a preceding patent. A tortilla having a filling deposited on it is conveyed through four stations or operations. In the first operation, a crease or groove is formed across the tortilla at right angles to the direction of travel by a roller which operates first in a forward direction and then in a backward direction. A filling is deposited in the groove by a stationary nozzle as the groove is being formed, then the leading or forward part of the tortilla is folded in a backward direction so as to form a fold which extends across the dough sheet near the front. Folding is done by conveyor-forming belts. In the second section, the sides of the tortilla are folded upward, inward, and then downward so as to form side folds partially inclosing the filling at the ends. In stations 3 and 4, the partially formed burrito is folded in a backward direction so that the dough sheet becomes completely wrapped around the filling.

BISCUITS

The type of biscuit discussed in this section is the soda biscuit or baking powder biscuit, a kind of chemically leavened bun or roll. Crackers and cookies, called biscuits in the U. K. and by some industry people in the U. S., will be discussed in the next chapter. For many decades, soda biscuits have been popular homemade breads in the southern part of the U. S. They have gained much wider acceptance in recent years due to their use as the basis for breakfast sandwiches in fast food restaurants. Variations on the theme have appeared in frozen form for retail sale, and these items are continuing to gain sales volume throughout the country. Many of the details about biscuit equipment given in the following paragraphs were taken from a paper by Poehlman (1985).

It is desirable to mix biscuit dough in relatively small batches because the chemical leavening system loses some effectiveness during long holding times and there is a tendency for the dough to develop a tougher structure than is desired. The dough should not be developed, which indicates that short mixing times at low speeds will be required. A vertical mixer with a hook will require about two minutes to cut in the shortening and one minute to properly blend in the water. Cool doughs are preferred, 55° to 60° being the suggested range, although slow acting leavening acids may allow successful processing of 75°F doughs.

If there is some question about complete hydration of the dry ingredients, a 10 to 15 min rest for the dough after mixing is acceptable. This not only leads to more uniform moisture distribution, but allows the dough to relax a bit if some development has occurred. In small operations—supermarket bakeries, restaurants, etc.—the dough can be cut into pieces of appropriate size for hand processing, and then sheeted out with a rolling pin. In hand sheeting operations, the dough may be given a couple of folds. Although this technique seems not to conform to the dictum that development should be avoided, folding does add to the fineness of the grain and tends also to control the rise, making the sides of the finished biscuit straighter. The crumb texture might be slightly tougher and breadlike, but not to the extent that the consumer finds the eating quality objectionable. To obtain the uniform thickness which leads to accurate control of piece weights, the rolling pin can be fitted with discs at both ends—the discs extending beyond the pin surface a distance equal to the desired thickness of the dough sheet—

or two metal or wooden rails of the desired height can be placed on each side of the dough and the sheet reduced until the pin rides along the rails. The latter method has the advantage that some shaping of the sides of the dough also occurs during the reduction process.

When the dough is sheeted by an extruder, the equipment should be of a type which imposes minimum punishment on the dough—roller feed types would seem to be a logical choice as opposed to auger feeders. Dough is extruded at 1 to 1.25 inches thickness. The width of the strip should be just slightly more than the cutters, so that scrap dough created at the cutter will be the smallest possible amount. Dusting flour applied to the belt and to the dough should be kept to a minimum. The dough strip passes through a compression roller which has the functions of sealing the sheet and insuring that it will feed uniformly through the main head rollers of the sheeter.

The head rollers would normally have a clearance of 0.625 inch. They would have scrapers to maintain clean surfaces so that sticking or marking of the dough surface will not occur. The sheeting rollers smooth the top and bottom of the dough ribbon to form a skin which will retain leavening gases in the oven, produce a uniform expansion, and lead to an attractive crust. The ribbon of dough, 0.625 inch thick, is transferred to the main conveyor. A small amount of shrinkage is obtained by reducing the main conveyor speed to slightly less than the speed of the head rollers. This reduces any tension which would cause misshapen dough pieces.

A rotary brush removes excess flour from the dough ribbon at this point. It is important to keep dusting flour at a minimum throughout the operation to avoid toughening the dough. The amount of rework is also minimized for the same reason. Reworked dough adds toughness because the gluten has been developed during the sheeting, rolling, and remixing process.

Round or hexagonal biscuits can be cut with rotary cutters. The round style, 3 inches in diameter and 2.5 oz in weight, is very common. Hexagonal biscuits are not standard, but have the advantage that scrap is greatly reduced with this style of cutter. Round cutters generate over 40% scrap, which has to be added back to the mixer. Hexagonal cutters of 2.75 inch size will generate about 11% scrap. The finished dough pieces are conveyed on to sheet pans moving in the opposite direction to the dough conveyor. Biscuits on the baking surface are separated from each other by about 0.5 inches, the distance in each case depending on dough expansion and the desired side appearance. Separation can be achieved without added equipment if the dough is cut with adequate distances between pieces and the speeds of the dough conveyor and sheet pans are adjusted correctly. It is customary to give the cut biscuits a few minutes rest; this allows the leavening system to generate some gas before the dough pieces are put into the oven. The sheet pans are transferred to a 450°F oven and baked for about 15 minutes. Biscuits can also be baked directly on the oven band.

BIBLIOGRAPHY

ANON. 1972. Automatic machine for crepes flambees. Crepmatic, La Garenne, France

ANON. 1973. The Gyrocrepe. Gideco, Paramus, NJ

ANON. 1982. Automatic Pancake Baking Machine Type PKB. Franz Haas Waffelmaschinen, Vienna, Austria.

ATWOOD, H. T. 1964. Mold for pie shells. U. S. Pat. 3.124,083.

BELSHAW, T. E. 1970. Developments in automated doughnut production. Bakers Digest *44*, No. 4, 50-51, 54-56, 66.

BELSHAW, T. E. 1976. Cutting and frying equipment. Proc. Am. Soc. Bakery Engineers *1976*, 112-120.

BRADEN, B. W., JR. 1976. Yeast-raised doughnuts. Proc. Am. Soc. Bakery Engineers *1976*, 127-132.

BURRIS, J. B. 1979. Fried pies. Proc. Am. Soc. Bakery Engineers *1979*, 111-118.

CANNON, A. S. 1987. Automated sweet goods production. Proc. Am. Soc. Bakery Engineers *1987*, 107-123.

CLARK, D. B. 1981. Corn tortilla and tortilla chip processing. Cereal Foods World *26*, 499.

FISCHER, L. G. 1976. Cake doughnuts. Proc. Am. Soc. Bakery Engineers *1976*, 121-127.

FREIHOFER, W. D. 1985. New trends in small cake production. Proc. Am. Soc. Bakery Engineers *1985*, 134-141.

GAGEANT, L. M. 1964. Automatic pie machine. U. S. Pat. 3,093,062.

GOODSELL, G. R. 1984. Cake doughnut production. Proc. Am. Soc. Bakery Engineers *1984*, 118-131.

HAVIGHORST, C. H. 1971. Mechanizes age-old process. Food Eng. *41*, No. 6, 62-63.

HAVIGHORST, C. H. 1976. Perky produces pies by the millions. Food Engineering *48*, No. 6, 62-63.

HOKES, J. J. 1977. Small cake production. Proc. Am. Soc. Bakery Engineers *1977*, 73-77.

MOYER, J. H. 1986. Doughnuts. Proc. Am. Soc. Bakery Engineers *1986*, 120-125.

PACYNIAK, B. 1986. Pudding pies may firm up soft snack cake/pie market. Prepared Foods *155*, No. 9, 200-201.

PLUTA, R. 1988. Personal communication. Colborne Manufacturing Co., Glenview, IL

POEHLMAN, R. W. 1979. Premium Danish production. Proc. Am. Soc. Bakery Engineers *1979*, 91-105.

POEHLMAN, R. W. 1985. Biscuits. Proc. Am. Soc. Bakery Engineers *1985*, 126-134.

ROBERTS, G. F. 1987. Apparatus for making filled food products. U. S. Pat. 4,691,627.

RUCKH, A. B. 1986. Cakes. Proc. Am. Soc. Bakery Engineers *1986*, 125-131.

RUHDORFER, A. 1974. Process for the production of frittaten. U. S. Pat. 3,830,946.

SCHMIDT, C. O. 1985. Tortilla production. Proc. Am. Soc. Bakery Engineers *1985*, 114-126.

SHAFFER, T. 1977. Automated yeast-raised production. Proc. Am. Soc. Bakery Engineers *1977*, 117-124.

STEVENSON, H. 1988. Personal communication. Colborne Manufacturing Co., Glenview, IL

VEY, J. E. 1986. Danish. Proc. Am. Soc. Bakery Engineers *1986*, 111-120.

ZONES, J. J., and ROBE, K. 1976. 1,000,000 pies a week. Food Processing *37*, No. 6, 60-62.

CHAPTER EIGHT

FORMING OF COOKIE AND CRACKER DOUGHS

INTRODUCTION

The manufacture of crackers and cookies in commercial plants requires certain kinds of equipment which have no direct application to other kinds of baked food production. These devices mostly fall into the category of forming or shaping machines and will be discussed in this chapter.

It is convenient to separate the discussion into cookie equipment and cracker equipment, since the machines used for these two types of products are quite different.

Nearly all large-scale cookie production nowadays is based on the use of band ovens. A discussion of this equipment is included in the chapter on ovens and baking.

FORMING COOKIES

There are several types of cookie forming equipment, each of which has limitations as to the physical properties of the doughs it can process satisfactorily. The three broad types of forming or shaping devices which are used to manufacture the great majority of commercial cookies are (1) extruders which push dough or batter through a constricting orifice, exemplified by deposit machines, bar presses, and wire-cut equipment; (2) rotary molders which shape the dough in die cavities cut into the surface of a metal cylinder; and (3) stamping machines or rotating cutters which cut shaped pieces from continuous sheets of dough. Each of these methods requires doughs of different rheological characteristics from those needed for the other methods although it is true there is some overlapping.

Stamping machines (reciprocating cutters) and rotary cutting machines must be fed continuous sheets of dough. An important requirement of these machines is that the scrap which is generated must be removed in one piece for efficient operation. Furthermore, the thickness of the sheet must be held within a narrow range so that the weight of the dough pieces will not vary significantly. If dough is to be formed into a sheet which will maintain its continuity so that it will not tear and will maintain uniform thickness, it must be cohesive enough to bear its own weight and have some elasticity. On the other hand, it must not shrink significantly after the piece has been cut—otherwise deformed pieces will result.

Doughs which run well on cutting machines are generally not suitable for forming into die cavities since they will not fill the die completely, they may not cut off cleanly, and they tend to shrink when baked, distorting the design. And, they seldom perform satisfactorily in extrusion equipment, since they tend to stretch and "string out" from the orifice making it impossible to control weight and shape.

Rotary molded doughs—those formed by pressing into die cavities—should form a solid lump when pressed together, but they should possess little or no elasticity and should exhibit minimal stickiness. Some of these doughs have the general appearance and texture of crumbly lumps of shortening and sugar, others are clay-like in texture and appearance. Nearly all of them are quite low in moisture as compared to other types of cookie formulas. They must, however, possess sufficient cohesiveness so that the individual dough pieces do not tear apart when they pass over transfer points, such as the transfer from the molder conveyor belt to the oven band. In spite of the seeming stringency of these restrictions on the physical characteristics of rotary molded doughs, a wide range of cookies can be made by this method. Some of the cookies sold in the largest volume, both plain and fancy designs, including most sandwich base cakes, are made by rotary molding. This process is not generally suitable for soft cookies.

Wire-cut, deposit, and rout press (bar press) cookies are formed by extruding doughs or batters through orifices. At some point before or after baking, the extruded strand is separated into individual pieces. For wire-cut goods, the dough is severed in a horizontal plane at the mouth of the orifice by a wire or knife. The cut surfaces form the top and bottom of the cookie and lead to a more or less flat and featureless top surface. Bar goods go on to the conveyor or oven band in a continuous string which is cut into appropriate lengths by guillotines or gangs of disc knives, generally before baking. By profiling the dies, the top surfaces of the strands can be given ridges or other longitudinal designs and their cross section can be shaped. The familiar coconut bar is an example.

By extruding two different doughs simultaneously, filled cookies such as fig bars can be made. In this system, the filling is extruded through a tube inserted in the center of the dough orifice. Deposit cookies are made from very soft doughs (fairly stiff batters) which are extruded directly onto the oven band. Dividing into pieces occurs as a result of mechanisms which increase the separation of the band and orifice so that the batter strand is stretched and breaks apart.

ROTARY MOLDING MACHINES

Design and Construction Details

A simple rotary molding machine consists of a hopper, a feed roll, a cylindrical die, a knife or scraper, a cloth web or apron, and a rubber covered compression roller (see Figure 8.1). There will also be a frame, motors, controls, etc. These machines may be permanently affixed to the oven band or they may be constructed as demountable attachments for cutting machine lines. They may also be mounted on casters so that the equipment can be moved out of the way when it is not in use. In any case, the frame must be very rigid so that the components maintain the same spatial relationships at all times.

The following discussion of the history of rotary molding processes is principally attributable to Weidenmiller (1966). The first known reference to rotary molding machines using rubber compression rollers can be found in *The Manufacture of Biscuits, Cakes, and Wafers* by Fritsch and Grosspierre, a 1932 publication. The earliest versions of this equipment were used for forming English style shortbreads and a limited range of other cookies. Modifications of these machines were made in the early 1930s by various manufacturers, and it was in this period of time that the rotary molded butter cookie became popular. The basic design of later equipment followed the same principles on which the earlier machines were based.

There have been different ideas of what actually causes the dough piece to be transferred from the molding roller to the conveyor belt. The most common early belief was that suction caused the transfer. Actually, the effective force is adhesion between the dough and the woven canvas belt. This adhesion is greater than that between the metal die and the dough. Therefore, the molded piece is transferred from the die to the belt, from which it is finally carried to the band of the oven.

The following general considerations should apply to the design of rotary molding machines.

(1) The equipment should contain the fewest number of parts to do the required work efficiently. This permits the equipment to be disassembled, maintained, and cleaned with minimum effort by unskilled laborers.

(2) All parts of the product zone must be readily accessible to sight and reach, or easily removable for cleaning and inspection.

(3) All interior corners must be formed with a minimum radius of 0.25 inch, with a greater radius if required to facilitate drainage.

(4) All permanent joints or connections must be butt welded.

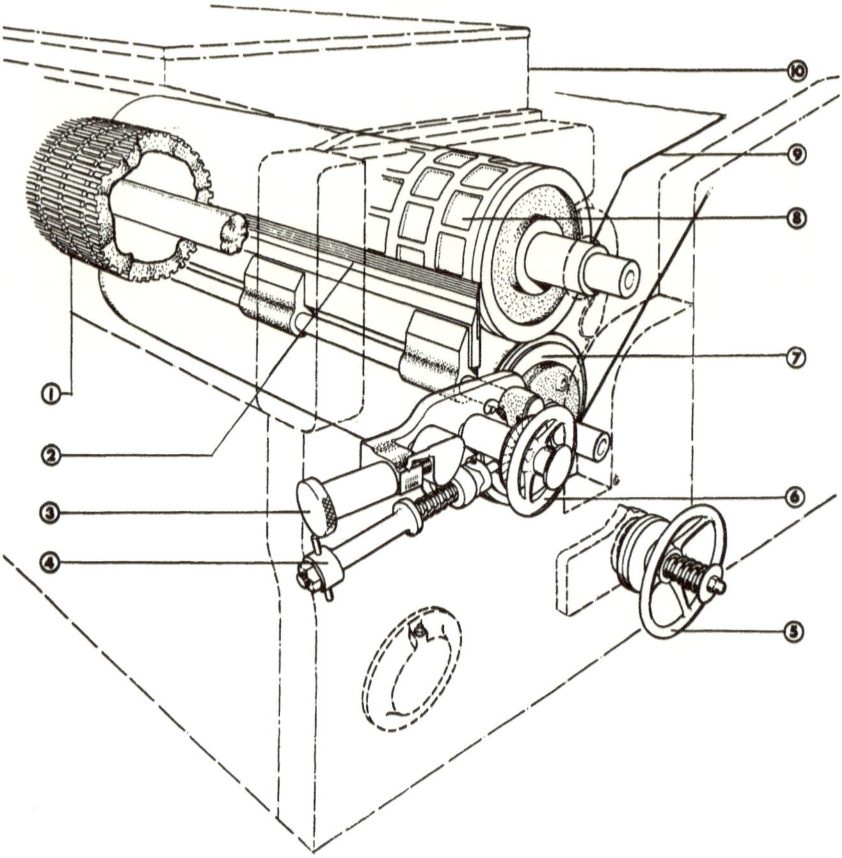

FIG. 8.1. OPERATING FEATURES OF A ROTARY MOLDER. (1) FEEDING ROLLER. (2) SCRAPER BLADE. (3) FORWARD AND BACKWARD ADJUSTMENT OF SCRAPER BLADE. (4) QUICK RELEASE FOR SCRAPER. (5) ADJUSTER FOR EXTRACTION ROLLER. (6) CONTROL FOR UPWARD AND DOWNWARD MOVEMENT OF SCRAPER. (7) EXTRACTION ROLLER. (8) MOLDING ROLLER. (9) EXTRACTION WEB. (10) DOUGH HOPPER.

SOURCE: VICARS

(5) All welding within the product zone must be continuous, smooth, even, and flush to the adjacent surfaces.

(6) All surfaces within the product zone must be smooth and free from pits and loose scale. They must also be nonabsorbent, nontoxic, odorless, and unaffected by the food product and cleaning compounds.

The hopper should be internally guarded with properly placed bars as a safety feature since the operator may have to distribute dough across the hopper with his hands. Delivery of dough may be done manually by shoveling from a trough at floor level or by continuous transfer methods. In some machines the hopper is set about 3 ft from the floor so that manual feeding is made easier, while in others the hopper extends upward several feet from floor level. The latter arrangement is suitable when feeding doughs from mixers located on the floor above. The hopper can be divided into two or more chambers by vertical partitions to permit running different doughs at the same time.

The exit slit at the bottom of the feed hopper is restricted by a feed roll and a die roll. The gap between the feed and die rolls should be about 0.25 inch. Adjustment of the separation can slightly alter dough weights. The feed roll has serrations which force the dough more effectively than do smooth rollers. It must be made so as to be adjustable in pressure to accommodate doughs of different flow characteristics. This can be achieved by screw-type adjusters in combination with springs, or by air actuated cylinders.

Die rolls are made of brass, bronze, or aluminum cylinders on steel shafts. Bronze is more resistant to wear and corrosion, while aluminum is cheaper and quite satisfactory for many applications. The rolls may have cast iron ends. Multiple die cavities in staggered array are cast or engraved into the surface of the cylinder. Cast impressions are said to have harder surfaces and to wear longer, but finer detail is obtainable by engraving. The die impressions on one roll may be of two or more different kinds and some manufacturers make sectional rolls enabling the baker to run different combinations of cookies at different times.

Extraction of the dough piece usually becomes more difficult as the design complexity increases. Cavity depth will vary according to the desired characteristics of the finished piece, but a typical depth for sandwich base cakes is 0.1 inch, while plain pieces such as butter cookies will ordinarily require depths of about 0.11 to 0.12 inch. Shortbreads and other rich cookies can be made considerably thicker, but the texture of most rotary molded cookies is adversely affected by increasing their thickness. Die cavities can be engraved to limits of about plus or minus 0.05 millimeters. It is important to use dies that are uniformly engraved with a predetermined angle that is suited to the style, size, and kind of biscuit to be formed. The speed at which the rotary machine will be operated can also be a factor in determining correct tool angle and points.

It is necessary that the rotary machine be made as a precision instrument having perfect alignment of the feed roll with the die roll, and that predetermined and uniform rubber roll pressure can be maintained by a system for equalizing and gauging rubber roll pressure. Light pressure on the rubber roll is desirable. If there is excessive sticking in the cavity, with failure to release properly, increasing the pressure will not solve the problem. Instead, the formula or processing conditions must be adjusted to yield a dough that performs satisfactorily. A sure sign that excessive pressure is being applied is the appearance of tails on the extracted pieces. The resiliency of the roll surface tends to change with age, and its performance varies accordingly. To minimize aging, the rolls should be kept clean. Occasional washing with a 5% soda solution has been recommended. The roll should be thoroughly dried with a clean cloth and wrapped for storage. A spare rubber roll should always be kept on hand, and it should be changed every three months.

The sharp edge of the knife or scraper bears on the molding roller and trims the bottoms of the dough blanks, in effect separating the dough in the mold cavity from the mass of dough in the hopper. Knives are set approximately 0.125 inch below the center, slightly lower for extremely wide dies. Fractures on the knife indicate too high a position. The angle of the blade edge at the roll surface must be such that the dough is sheared off smoothly.

The knife should be kept sharp and its edge must be straight. Any nicks or burrs will not only cause uneveness on the bottom of the dough piece (perhaps a minor consideration) but can scar the molding roller, perhaps ruining it. The blade must be strong and rigid so that vibration or other force does not push a part of the edge into the mold cavities with resultant damage to the knife or to docker pins in the cavity. A knife change should be considered with each molding roll change. If the dough is scraped out of the cavities to any significant extent, the blanks will be difficult to extract.

Conventionally designed machines have the rubber compression roller positioned under or slightly behind or ahead of the perpendicular center of the molding roller. This arrangement allows for a remolding of dough blanks that have been eased out of the die cavity by improper release at the point where the knife contacts the die roll.

Locating the rubber roller in front of the die roll and equipping it with air cylinders to adjust the contact pressure evenly across the die roll eliminates or at least minimizes the tendency to produce wedge-shaped dough blanks (i.e., the pieces have thinner leading edges that show indistinct design impressions or they have an irregular leading edge) and tails (thin protrusions of dough from the bottom edge of the extracted piece). The arrangement allows misformed dough blanks to be completely released from the cavity before they come into contact with the extraction belt, or causes them to be so badly formed that they are immediately recognized and signal the need for a correction in the dough or a machine adjustment.

Premature or delayed release from the die cavity is evidenced by a tailing-off of dough on the extraction belt, a double impression on the leading edge of the dough blank, or a thinner edge with indistinct design representation. Among the end results of these defects are weight variations, excessive scrap at the sandwiching machine, uneven appearance in the packages, excessive breakage when handling the finished package, and poor appearance of the individual cookies. These problems arise many times from inadequacies in the dough itself, although machine adjustment may allow marginal doughs to run satisfactorily. Individual variable speed drives on the molding, feed, and extraction rollers, combined with tachometers showing the speeds of feed and molding rollers, allow the supervisor to make quick and accurate speed settings to compensate for variations in the doughs.

The molding apron is an endless canvas belt of special weave. It can be made of linen. The tension is adjustable and the apron should be kept at the minimum tautness consistent with satisfactory running. Scrapers are provided to remove any excess dough which may cling to the web. There are special devices which correct off-track conditions by mechanical, pneumatic, or electrical means. In some newer machines, thermostatically controlled web conditioners apply steam to the belt to improve adhesion. Under ideal conditions, a good quality belt will have a useful life of about 1,000 hr. New belting should be conditioned by rubbing shortening and flour into its surface.

Various depositing actions may follow the molding operation. Registration can be excellent if the depositing action is coordinated with the rotation of the molding cylinder and occurs on the apron, before the dough pieces are transferred to another conveyor. Figure 8.2 shows a topping depositor placing a spot of filling on top of the unbaked dough piece coming from the molder.

A significant advance in machining capabilities was made when the coating of die cylinders with Teflon became possible. Teflon is the trademark for DuPont's fluorocarbon resins and products based on these resins. A very low coefficient of friction and good nonstick properties are among the outstanding characteristics of Teflon. It has poor resistance to abrasion, however. It is particularly useful in preventing the sticking of dough to metal parts. In pure form it may be applied by shrinking tubes of the molded plastic onto rollers, or by screwing or bolting to flat surfaces sheets of solid Teflon cut to the desired shape. Either of these techniques is expensive because of the large amount of this rather costly plastic which is required, but they do provide slicker surfaces and better resistance to wear than do some of the nonstick coatings applied as sprays or paints.

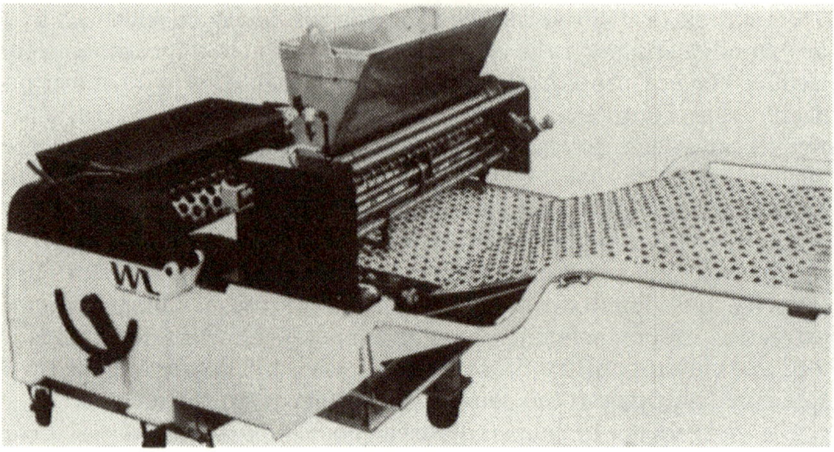

FIG. 8.2. PORTABLE ROTARY MOLDER WITH JELLY DEPOSITOR.
SOURCE: APV (WERNER LEHARA)

Most adhesives are inadequate to bind pieces of pure Teflon to other surfaces. The plastic itself will not adhere to metals. Special coatings, called Teflon TFE nonstick finishes, may be sprayed and baked on equipment, including irregularly shaped surfaces such as rotary dies, wires, etc. It is generally necssary to heat these coatings to about 700°F to cure them after they have been applied. Coating formulations contain additives which aid sprayability and promote good adhesion to the substrate.

Parts which have been coated with good results include sheeting rolls, feed rolls, die orifices for wire-cut machines, die rolls, wires, etc. It is not necessary or even desirable to coat all rollers and dies with Teflon. Many doughs will release very well from most dies without the necessity for applying the nonstick coating. In some cases, the release of doughs from Teflon-coated cavities is too rapid and interferes with the proper transfer of pieces. Teflon finishes do, however, permit the running of many soft, sticky cookie doughs which could not be handled by other techniques in rotary molding, wire-cut equipment, and cutting machines.

Teflon is generally applied only on engraved seamless bronze cylinders. For maximum adhesion of the coating, rolls should be satin finished before the application is made. Recommended depth of coating is 0.127 to 0.025 mm; heavier coatings encourage peeling. Cylinders may be recoated when the first application wears off. Repeated heat treatments tend to soften the metal surface, however, leading to faster wear.

Operation of Molders

It is good practice to start the machine with the hopper empty and the rollers turning at a moderate speed. The hopper can then be filled with an amount of dough that will form a fully compressed blanket on the feed roll and also completely fill all the die cavities. With the machine running slowly, determine if the die cavities are being firmly filled. If so, advance the rubber roll and keep just enough dough in the hopper to replace the amount being consumed by the die cavities.

It is also good practice to slightly moisten the extraction belt in order to cause adequate adhesion at the start of the run. If the dough is extremely dry, it may be necessary to provide a means of keeping the belt soft and tacky enough to ensure good adhesion throughout the run.

The rotary machine and the die roll must always be kept free of dry or crusted dough which may become lodged in or on the feed roll, hopper, knife, knife holder, or engraved areas of the die roll. Removal of the dough fragments should be done immediately after each run and before the dough fragments become hardened. It is important to include in care and maintenance programs specific procedures for cleaning and inspecting forming machines and dies. They will help to insure that the machines are ready to run when they are needed, will conserve labor and oven hours, and will reduce tonnage lost as scrap and misruns. Crusted dough fragments in the die cavities lead to scoring and grooving of the knife and subsequent damage to the roll.

Optimum rotary molding is being achieved when the transfer belt shows no signs of dough deposits or tails. Occurrence of tails is an indication that the dough forms are falling from the die cavity before they contact the extraction belt. If it is necessary to increase the rubber roller pressure, the amount of dough required to fill the cavity may be decreased. Since dough blanks approximately double in size during baking, the error in thickness introduced at this stage will be doubled at the packing machine. This is a major cause of loss of base cake height and scrap accumulating at the sandwiching machine. Die wear and uneveness can also cause these defects. To check these factors, the die cavity depth can be measured with a micrometer at three or four places across the cylinder and around the cylinder. The center of the roll and both ends should be checked. Variations from the original depth will be reflected in changes in count per pound, loose packages, and possible variations in color due to differential rates of bake-out.

It is necessary to use equalized rubber roll pressure so that once the proper amount of pressure is determined for each dough, the data can be included in formulation and procedure standards.

Die Design

If a new die is required for rotary molded cookies, a drawing is prepared showing the outline and features of the cookie. It is advisable, but not essential, to prepare this drawing in the actual size expected for the dough blank. The drawing is forwarded to the die manufacturer together with information on desired cookie size, etc. On the basis of the art work and other information, the engraver will make a sample cavity in a flat aluminum die. From this cavity, wax impressions will be taken to show the approximate dimensions and appearance of the dough piece. Finished weight of the cookie can be calculated within fairly narrow limits using the data obtained from the wax impression. It must be emphasized that the wax impression simulates the raw dough piece, not the finished cookie. Changes in design of the cavity can be made inexpensively and quickly at this stage.

If necessary, dough blanks can be formed in the flat aluminum die and baked to obtain an approximation of finished cookie appearance. The pressure used to hand fill the mold is always different from the pressures exerted in the molding machine, and this variation will introduce uncertainties into the evaluation. Transfer of the flat design into the curved format that will be engraved on the roller will also inevitably create differences in the dough blank and finished cookies. For these and other reasons, it is considered desirable to test run a curved die segment with the proposed dough formula before having the complete cylinder engraved.

To run a curved die segment, it is necessary to have on hand a cylinder provided with a removable segment. Normally, this space is occupied by a segment containing the same designs as the rest of the roll, and the die is routinely used for commercial runs. When it is desired to make a test, this segment is removed, and the test piece is put in its place. A run is then made using this cylinder and the proposed dough. Cookies formed by the other cavities in the roll will probably not be of a formulation meeting the ingredient declaration on the usual packages for cookies of this shape, so the test production may have to be sold through special channels. Furthermore, the other cookies deposited by the test cylinder will seldom resemble the test cookie closely enough to absorb heat in the same way, so they may appear underbaked or overbaked.

Jet-Cut Machine

The cookie former which is trademarked "Jet-cut" is described as an advance on wire-cut devices, but appears to be more of a modification of a rotary molder. These devices are based on a rotary molding cylinder, with the die cavity ending in a piston head. The cavities pick up dough from a hopper with the pistons in a retracted position, very much like a conventional rotary molder. When the cylinder moves to the discharge position, the piston pushes the dough blank out of the cavity. Release of the dough from the piston head is assisted by an air jet. Advantages claimed are: reduced weight variations (as compared to wire-cut machines), high speeds, precise positioning, greater band coverage, and applicability to a wider range of dough types.

EXTRUDERS

General Characteristics

Extruders vary widely in complexity, from simple equipment consisting of a hopper with feed rolls pressing dough through adjustable slits to very complicated devices which extrude deposit cookie batters through orifices moving in predetermined patterns. The most common type of machine consists of a hopper with one or more feed rolls which force the dough through a number of tubes usually called die cups. These dies may have orifices of different shapes—square, round, oval, scalloped, etc. In wire-cut machines, discs are sliced from the continuously extruded cylinder of dough and allowed to drop onto the oven band or transfer belt.

Deposit Machines

In deposit machines, the batter is extruded intermittently through shaped nozzles. Separation into cookie sized portions is achieved by lowering the oven band during the time the extrusion is stopped. As a result, the batter on the band pulls away from the batter still in the extrusion orifice. Anyone who has observed an operator fill ice cream cones from a soft serve freezer knows how the principle works. The band is then moved up toward the nozzle and extrusion begins again. These machines are relatively simple, but are quite demanding so far as dough characteristics are concerned. In more advanced machines, the nozzles can be moved to form various patterns such as curves, wavy fingers, swirls, and circles. A second depositor can be synchronized with the first one to put jelly or some other filling on top of the cookie.

Bar Presses

Bar presses are sometimes called rout presses, especially in the U. K. They employ a simpler method of forming than is used by deposit machines, since there is no need to separate the dough into cookie sized chunks at the extruder. A bar press extrudes continuous strings or strips of dough directly on to the oven band. Separation of these bands into individual bars can be made before or after baking by one of the usual cutting devices. The die plate may be inclined in the direction of the extrusion so that the ribbon is supported for a longer period of time, an arrangement which reduces breaking or thinning of the dough strand due to gravitational pull.

The die orifices are usually slots with a straight lower edge to give a flat bottom to the cookie and a grooved top edge to give a ribbed upper surface. Front to back dimensions can often be varied by moving one of the edges. The scope of possible variations in pattern is rather limited, but some character can be given to the top surace by modifying dough characteristics to give smooth or rough texture, strong differentiation in coloring between high spots and low areas, etc.

Simultaneous Extrusion of Two Different Components

By using concentric extrusion tubes, different doughs or other components can be combined in a single bar. This is a principle used in confectionery manufacture, extrusion of composite plastics, and many other technologies besides baking. In baked products, the usual combination has been an outer dough case with an inner jam-like filling, but there is no technical reason why three (or even more) components could not be co-extruded, or why types of components other than those mentioned could not be incorporated.

Fig bars—Fig bars and other fruit-filled bars are made by extruding the dough for the jacket and the jam for the filling through concentric orifices in a bar press machine. Two hoppers and two sets of forcing rollers (or screw extruders) are required to move these components through the dies. The fig paste is led through the dough hopper portion to its proper orifices by means of tubes or chutes. The dies are of conventional design, as previously described. The extruder must be adjusted to deliver a ribbon of dough that is moving at the same speed as the oven band on which it comes to rest to avoid stretching or wrinkling, and perhaps breaking of the ribbon.

Fig bars are baked in continuous strings which are cut into individual cookies when they come out of the oven. Some manufacturers (mostly British) have placed the cutters before the oven. The earliest examples of pre-oven cutters had a vertical knife which passed between two heavily oiled strips of felt before and after it contacted the dough strips. More recent versions have the guillotine separated from the dough by a thin sheet of rubber. Users of this system claimed the semi-sealed ends helped to retain moisture during baking, but it seems doubtful there is much of an effect. Those manufacturers who bake in an uncut strip say it reduces distortion of the ends of the bars, prevents flowing out of the hot filling, and retards loss of moisture from the filling.

Guillotines or rotary cutters can be used to sever the cooked strips. The former type is similar in principle to the other reciprocating vertical movement knives used in bakeries and which have been described in a previous section. Rotary fig bar cutters are designed to run two gangs of circular knives across the oven band. Cutter blades are heat treated and spaced along a shaft at the distance required to cut the desired length cookie. The shafts on which the knives are mounted are carried across the band by heavy roller chains running in guides (see Figure 8.3). The chain speed is fixed with respect to the oven speed. As the knives pass below the belt on their return to the cutting position, they can be carried through a stainless steel cleaning tank. The tank is provided with an overflow for maintaining the level, and it can be heated. These provisions keep the knives reasonably free of jam and syrup. Rollers to support the oven band as it passes through the machine and under the cutting knives are integral parts of the equipment. The cutter is driven from the main oven band drive by means of line shafting and bevel gear box.

Fig bars consist of about 60% fig "jam" as the filling and about 40% dough in the jacket. Formulas for jackets are similar to those for wire-cut cookies and mixing of the dough is conventional—a horizontal mixer works satisfactorily. Fig jam can be mixed on vertical mixers or horizontal mixers. Of course, many other types of fruit filling—apple, cherry, blueberry, etc.—can be made in exactly the same way as fig bars. Why not a mince pie filling?

Dual textured cookies—There has been much interest in recent years in creating dual textured cookies, often described as cookies with a crisp outer crust and a chewy interior. Several methods have been suggested for obtaining this texture differentiation in the finished cookie. One method is to extrude a continuous cylinder having around its circumference a dough high in crispiness-inducing ingredients and

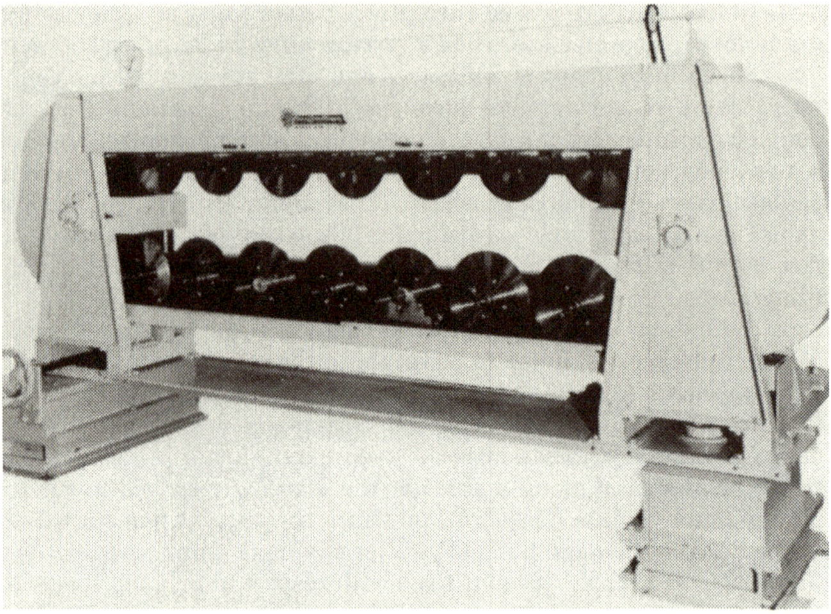

FIG. 8.3. ROTARY KNIVES FOR CUTTING FIG BARS AND THE LIKE.
SOURCE: THOMAS L. GREEN CO.

in its center a dough containing a large amount of ingredients contributing to chewiness. The similarity to fig bar extrusion is obvious.

Before baking, however, the continuous tube is severed into pieces by a type of reciprocating die or roller which presses the outer dough through the inner dough and seals it to itself. The purpose is to completely inclose the inner dough in the outer dough. If the dough consistencies and dimensions are correctly chosen, this seal can be made by a guillotine blade with a flat bottom 0.25 inch or so in thickness, perhaps followed up by a sharp blade which cuts through the flattened areas. Some rounding of the piece occurs as the doughs flow and expand during baking so that the square edges are smoothed out, and under certain circumstances approximately circular cookies will result. Reciprocating cutters which make a curved outline are also known to have been tried in attempts to get a more nearly circular piece. Other cutters pinch the dough tube from the sides to give a more three dimensional effect.

Wire-cut Machines

These devices represent an advance in complexity over the bar press and deposit machines in that they include a device which cuts off pieces of the extruded dough as it is emerging from the die orifice. The cut-off device consists of a wire or blade which is quickly drawn through the dough by a reciprocating "harp" (see Figure 8.4). A considerable body of information has been accumulated on the construction and operation of these machines, and it will be reviewed in the following section.

Hoppers—Dough is often fed to the hopper manually, but it may be fed by gravity from a mixing room on the floor above the extruder or from a trough which is elevated from the floor of the extruder room so its contents can fall into the hopper. Mechanical systems for transferring doughs or batters to the hoppers are also fairly common. Since the weight of the dough resting on the feed rollers affects their performance, greater accuracy and uniformity can be obtained by having an auxiliary hopper which meters the dough to the wire-cut hopper at a rate which keeps the dough in the latter constantly at the same level. Soft batters can be fed continuously by pumps and this would seem to be the method of choice when dough properties make it feasible.

Vertical separator plates can be inserted in the hopper to allow feeding of two or more colors or flavors of dough at the same time. Some authorities have recommended using short partitions between each die cup to lessen the difference in feed rates between the outside and inside cups. Keeping the feed rate steady and maintaining hopper contents at approximately the same height at all times are the initial steps in assuring a constant pressure and rate of extrusion.

Hoppers have been jacketed and warm air or warm water circulated to improve uniformity of extrusion rates. The ends of the hopper have been curved so as to reduce tendency of the dough to stagnate in this area.

Feed rollers—The characteristics of the rollers which press dough through the die cups are important factors in maintaining uniform size and weight in the finished cookies. Feed rollers can either move continuously or intermittently, depending on the result intended. In some extruders, separation of the rolls is adjustable up to two inches or more, while in others these rolls are permanently set at what is thought to be the optimum separation. It is desirable to run the rollers at a speed which does not heat the dough through excess friction.

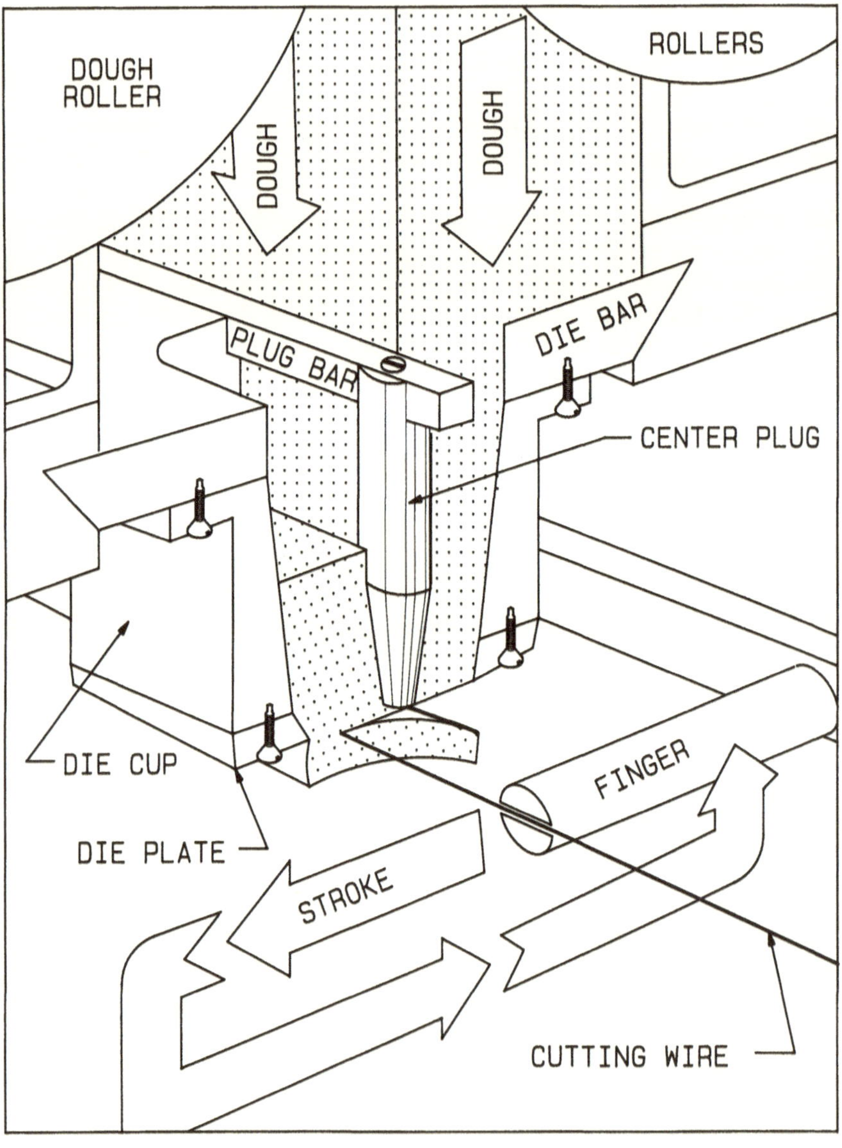

FIG. 8.4. OPERATION OF THE WIRE-CUT MACHINE.
SOURCE: WEIDENMILLER CO.

Feed rolls can be either smooth or grooved. If there are grooves, they can run parallel to the axle or have a slight spiral orientation. Grooved rolls give a more uniform pressure at lower speeds. There is some risk of catching the roll knives or scrapers in grooves which run straight across the roll, parallel with the axle. This problem can be minimized by having grooveless bands running around the circumferences of the roll for the knives to ride on. Of course, there will be less pressure on the dough near the bands. If the grooves are spiraled, the knives will be supported at some point along their length at all times, and the force on the dough will be uniform across the hopper. If one of the feed rolls spirals in one direction and the other roll spirals in the opposite direction, the dough will tend to circulate in the hopper, preventing hardening or toughening at the ends.

Two large feed rolls are generally used, although it is possible to use a three-roll arrangement, with one large roll on one side and two small rolls on the other side. The point of minimum separation of the feed rolls should be adjustable to compensate for different dough flow characteristics. A review by Wittenberg (1965) showed that companies were using separations ranging from one-sixteenth to one inch. The average separation used today is probably between one-fourth and one-half inch.

The speed of the feed roll can be varied, but usually the gearing is interlocked so that a change in the speed of one roller results in a corresponding change in the speed of the other. It is possible to obtain a machine in which the feed rolls have separately adjustable drives, but the advantage of this arrangement is thought to be relatively minor and, since the cost is substantial, separate drives are seldom ordered. Nonetheless, it may solve some difficult machining problems. The average roll diameter is about 8 inches, but they can be obtained in various other sizes. It is generally accepted that larger rolls disturb the dough less, while at the same time they exert a more even pressure than smaller rolls.

The amount of time following mixing and before forming (floor time, lay time) affects the response of cookie doughs to forming procedures. With longer floor time, flour in the dough tends to hydrate more, there is some drying at the surface with crusting possible, some of the chemical leaveners can begin to react causing changes in the dough density, and the temperature can change with effects on shortening crystallization and other physical attributes. All of the changes can affect the size and shape of the dough as it feeds through the wire-cut machine and give noticeable differences in weight, size, and appearance of the finished product. It is important to process a mixed batch quickly, and to establish a uniform schedule and adhere to it.

One method of improving the uniformity of extrusion across the width of the hopper has been to insert a filler block. This device is a bar of machined metal placed up against the feed rolls and having a series of tubes running directly from the nip of the feed rolls to the die cups. This has the effect of giving each cup its dough supply directly from the feed rolls.

Die cups—There are several designs of die cups. The common beveled channel cup tapers from the hopper nearly to the orifice and then has a straight section for a short distance. These cups are used for wafers and other cookies which require a thin edge. Other cups have a constriction around the inside of the exit opening. Such dies cause the dough to roll outward at the perimeter as they emerge from the cup. They are used when cookies with relatively thick edges are desired. For cookies which are to have holes in the center and also for cookies which tend to rise too high in the center during baking, cups are made with a plug running down the center. Depending on the size of the plug and the consistency of the dough, the hole which is put in the center of the extruded piece will fill up either partially or completely during baking.

It will frequently be observed that the dough pieces coming out of the end cups are lighter than those coming out of the center cups. In order to compensate for this effect and obtain equal weights across the band, it is may be necessary to use different size cups in a set. This arrangement does not allow for compensation of flow rate differences when several doughs having different consistencies are used with the same set of cups. Flow can also be adjusted by using a screw inserted through the wall of the cup and into the dough stream. By changing the distance the screw projects into the channel, the rate of extrusion can be changed in small increments. Adjustments can easily be made to meet the requirements of whatever dough is being run. It is also possible to retard dough flow by using cups containing crossed wires in the channel. Wires are seldom used as a means of controlling flow, however, because they are not readily adjustable to meet the demands of varying conditions and different formulas.

It is sometimes necessary to make the cup's orifice oval in cross-section so that a round cookie can be formed. This is intended to compensate for the distorting effect of the cutting wire which is forced through the dough at each stroke of the harp. Here again, there is no easy way to compensate for the different degrees of distortion which occur when doughs of different consistencies are being processed. The die designer will generally base his calculation of the width to length ratio of elliptical dies (i.e., the minor and major axes) on a reversal of the ratio of length to width of a cookie baked from a round die of approximately the same size using the same dough and the same baking conditions. Most manufacturers of wire-cut dies are prepared to make these calculations if the necessary data can be supplied.

Most die cups made for wire-cut machines will be 1 to 1.5 inches in diameter, although the previously mentioned survey showed that companies were using from 0.25 to 2.75 inches in diameter. The cookies obtained ranged from 1 to 3.5 inches in diameter. It is generally observed that the larger deposits tend to spread less, percentagewise, than do the small deposits.

Several of the die cups will be held on a bar that fits snugly into a channel at the bottom of the hopper. Usually, a handle is fastened to one end of the bar so it can be readily slid into and out of the channel. It also facilitates carrying the bar. As a result of this convenient design, changing from one variety to another is a matter of a few minutes, at least so far as the die is concerned. It also makes stripping the machine for cleaning very convenient. The outlet ends of the die cups should be parallel to the path of wire movement and they should be accurately machined and polished. Nicks and barbs resulting from mishandling damage can cause wire breaks.

It is possible to have die arrays consisting of two rows, usually staggered, in each set. This arrangement is especially applicable for the smaller cookies, since one pass of the harp can be used to cut double the usual amount of pieces with each stroke. A harp having a separate wire for each cup will be necessary.

Wires and harps—The wire used for cutting pieces from the extruding cylinder of dough is held in a harp. A harp consists of a frame holding a wire at each end and at intervals along its length. Wires from 0.013 to 0.031 inch in diameter have been used successfully. Wire 0.024 inch in diameter has been recommended for most common types of dough while 0.029 inch diameter wire is thought to be more suitable for certain dry or sticky doughs. Heavier wires are used when the dough contains particulate ingredients, such as chocolate chips, or when a rough topped cookie is desired. Thinner wires lead to smoother tops and they also are more suitable for soft doughs. Wires may be circular in cross-section or they may be saw-toothed or flat, like a narrow blade.

Teflon-coated wires have been suggested to reduce sticking of the dough pieces to the wire, a problem which creates much scrap when it occurs. Wires may also be vibrated continuously in order to increase cutting effectiveness and to obtain better release of the dough pieces. Oscillating knives are driven in a reciprocating motion by a power unit placed at the side of the wire cut machine. A stationary back-up plate or board, which the wire can press against as it completes its stroke, can be used to complete the cut in soft doughs. This design also reduces the tendency of thin pieces to flip over when they start to drop at the end of the stroke.

The mechanism controlling the travel of the harp is arranged so that the wire passes close to the orifice on the cutoff stroke while on the return stroke the wire is lowered slightly so it does not interfere with the dough which is constantly extruding down into its path. Cuts may be made either in the same direction as the band is moving or in the opposite direction. If back-up plates are used, cutting can be done only against the flow.

Since wires wear out during usage and will eventually break, it is good practice to have on hand a replacement harp with wire in place, ready to use in these emergencies. An even better practice is to replace the wire about every 1 or 2 shifts in order to forestall the breakage that occurs with wear. The cost of these precautions, both for labor and materials, is nominal. Burrs or rough spots on the edges of the dies cause rapid wire failure.

Although wire-cut machines are very frequently used with doughs containing particles such as nuts, chocolate chips, and raisins, there are many difficulties encountered in such applications. For instance, the wire tends to carry these relatively large particles through the dough, breaking up the structure and making it uneven. Furthermore, such doughs have a strong tendency to cling to the wire. These tendencies can be reduced by using stiff doughs and by having the particulate materials as soft as practicable. Reducing the size of the particles may improve cutting although an increase in the size of chocolate chips may actually improve cutting if the chocolate can be made soft enough (as by slight warming) to yield readily to the wire. In the latter case, the extension of the chip into the unsevered mass of dough tends to hold the particle in place while the cut is being made.

Post-baking Treatment

Wire-cut cookies are very soft and porous as they leave the oven, and they have a tendency to stick to the band unless the baking surface is well greased either by a separate fat application or by fat leaking from the cookies. It is desirable that the band should extend beyond the oven exit for not less than 50 ft (Gadams 1984) so the cookies can dry out for easier removal by the stripping knife. Cookies should pass across this knife in one continuous motion without having to be pushed by the following cookie. The knife should be less in width than the cookie diameter and sloped downwards at an angle of about 25° to the horizontal plane of flow of the cookies.

Wire-cut cookies are particularly suitable for various decorating and flavoring operations. A wide range of toppers, jelly and icing applicators, folders, and enrobers are available for attaching to the production line (see Figure 8.5) or on separate lines for use with cookies which have been collected off another line. Most of this equipment is of a non-specialized nature suitable for other types of baked products as well as cookies, and it will be discussed in a subsequent chapter. A popular type of soft wire-cut cookie is based on the application of a central jelly spot by a piston type depositor followed by a folding of the still warm cookie by a simple belt and roller mechanism. Almost every conceivable type of icing has been applied to wire-cut cookies, using many versions of the waterfall or curtain icers or roller icers. Sometimes the icing is followed by sprinkling on coconut shreds, or nut pieces, chocolate vermicelli, or candy bits.

SANDWICHING MACHINES

These devices assemble two base cakes and a layer of creme filling to form a sandwich cookie, the prototypical example of which is the Oreo cookie known to every child. The manufacture of sandwich cookies has gone through several phases. From hand spreading, the process advanced to the old stencil machine in which the base cakes were pressed against a stencil plate which was basically nothing more than a metal plate with circular holes cut in it. Filling was forced into the stencil holes, and the cakes with filling sticking to them were knocked away from the plate.

Modern devices work automatically at a rapid rate producing uniform products with very little scrap. Base cakes are received at the sandwiching machines from cooling conveyors leading from the oven or from chutes which are hand fed from storage boxes. Converging devices can be used to bring together multiples of the desired number of rows emerging at the end of the cooling conveyor. Vibrating conveyors jog the rows of cookies into a stacked-on-edge position. Half of these rows have the top, with its embossed design, leading while the other rows contain cookies with the design trailing. Cakes are fed into the magazines in the proper position for sandwiching and are removed from the magazines, one at a time, by means of double pins on double chains.

As the bottom cakes travel through the machine with their embossed side downward, they receive deposits of creme extruded through a rotating sleeve with shaped orifices. Extrusion pressure is supplied by

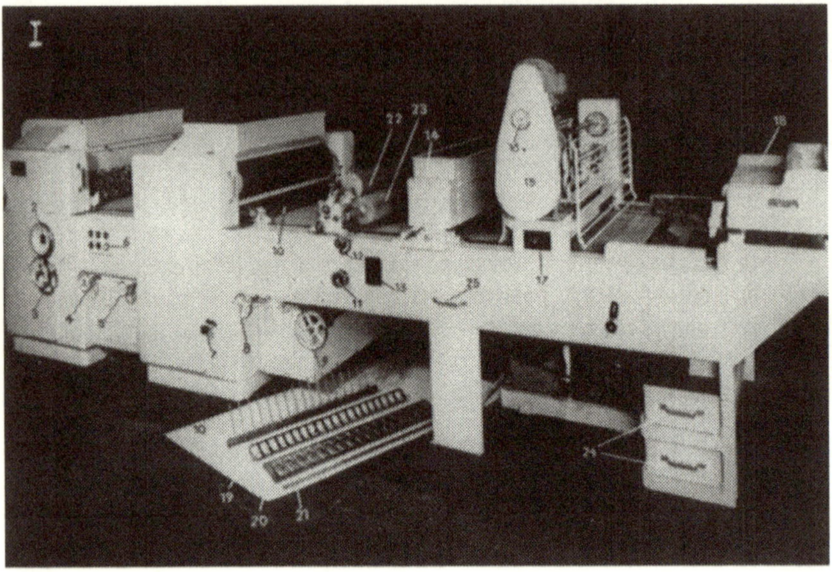

FIG. 8.5. Line incorporating molding and wire-cut equipment with decorating and treating devices. First head for molded and roll cut cookies; second head for bar press and wire-cut cookies; followed by egg washer, dough decorating device, guillotine cutter, and sugar sprinkler. (1) Adjuster for knife position. (2) Safety coupling. (3) Adjuster for rubber roller. (4) Speed controller. (5) Band adjustment. (6) Motor switch. (7) Coupler for the cutter. (8) Adjustment for wire cutter--ripple roller. (9) Tension adjustment for cloth band. (10) Wire-cut frame. (11) Coupling for coating device. (12) Vertical adjustment of cloth band. (13) Switch for dough decorator. (14) Dough decorator. (15) Guillotine. (16) Cutter speed adjustment. (17) Cutting machine switch. (18) Sprinkler for sugar or salt. (19) Equalizing chamber for dough pressure. (20) Wire-cutter frame. (21) Bar cookie frame. (22) Transfer roller for coating. (23) Coating brush. (24) Dust collector. (25) Scrap container.

Source: Hecrona

the pump which moves the creme from hopper to depositor. The creme deposit is cut off by adjustable stationary wires (see Figure 8.6). Size of deposits can be adjusted by changing settings on the variable speed motors driving the auger paddles and the pump. An air valve halts the creme delivery in case the sandwich machine stops for any reason.

As the bottom base cake with its creme deposit reaches the second set of magazines, the top cakes are dropped onto the creme. Then, the sandwich is gently pressed together to assure adherence of the components and establish uniform thickness of the finished cookie.

FIG. 8.6. SANDWICHING MACHINE. SHOWING CREME BEING DEPOSITED ON THE BASE CAKES.

SOURCE: PETERS MACHINERY

A typical sandwiching machine will take round base cakes from 1.5 to 2.62 inches in diameter or rectangular, square, and finger-shaped base cakes up to 3.25 inches long and as narrow as 1.16 inch. Speeds are variable up to 1,600 sandwiches per minute on a two row machine. Operation of these machines can be made automatic except for filling the creme hopper. Peanut butter and cracker sandwiches are usually made on a somewhat different machine.

The rows of sandwiches are conveyed to the packaging area in stacked or flat array, depending upon the type of container to be used. Several machines can feed one packing conveyor. Sandwiching machines can be combined with tray loaders and overwrap equipment to give a completely automatic packing operation.

TROLLEY COOKIE EQUIPMENT

The manufacturing process for trolley cookies has some unique points which should repay study by the baker looking for ideas leading to new products. The unusual aspects of trolley cookies lie in the finishing operation; the base cake which forms the heart of this confection is normally just a wire-cut cookie of standard formulation.

Trolley cookies are not a predominant factor in today's market. There has been a gradual decline in the number of firms offering these cookies because the production method is slow, difficult to control, and costly in labor. The cookies do not have enough appeal over more economically manufactured goods to justify the elaborate processing method. As a result, older equipment is gradually being abandoned and it has not been replaced by equivalent capacity in new plants.

The distinguishing feature of trolley goods plants is that the base cakes are suspended on pins attached to a movable framework (the trolley), so that one or more coats of sticky or slow-drying coating can be applied to all surfaces of the base cake. One plant describes the trolley as being four stories high and holding tens of thousands of cookies at any one time.

It is possible to completely cover the piece with marshmallow, and then enrobe it (after a drying period) with a kind of water-icing. This process may take more than a day to complete, especially in humid weather. Air-conditioned rooms and special drying facilities have speeded up the procedure to a total of 5 to 6 hr in modern installations. A view of the trolley is shown in an article (Anon. 1973) listed in the bibliography of this chapter.

The base cakes are usually wire-cut cookies although cutting machine bases are known to have been used. Many flavors of base cake have been used, e.g., vanilla, chocolate, and coconut. Pieces with fairly open texture, such as exhibited by the usual wire-cut cookie, seem to contribute better eating quality to the finished cookie. Base cakes must, however, have sufficient structural strength and rigidity to hang on the trolley hooks during the dipping and transfer operations. Dipping the hooks or pins into corn syrup before they are pressed into the cookies helps to hold the base cakes and reduces scrap. Excessive tenderness at this stage leads to serious difficulties.

The base cakes are dipped 2 or 3 times. The first dip can be a jam, jelly, or marshmallow, the middle layer can be a chocolate flavored coating consisting mostly of sugar, corn syrup, cocoa, and shortening, and the outside layer is generally a thin coating of a glaze or water icing. Trolley marshmallow is generally a tougher, denser material than normal deposited marshmallow.

Other types of confectionery based on the trolley cookie concept are described in the Yoon (1986) patent, which sets forth a method for stabilizing the moisture content of the cookie and jelly and adding a "nutritive" coating.

MANUFACTURING SUGAR WAFERS

The Basic Process

The forming of sugar wafer base cakes is inseparably connected with the operation of the wafer oven. Therefore, the wafer oven will be considered in this chapter, not in the chapter on ovens.

The type of sugar wafer commonly found in U. S. retail outlets is a cookie with a top and bottom layer of the characteristically crisp wafer sheet inclosing either one layer of creme or alternating layers of creme and wafer. The cookies are generally rectangular in form. There are, however, many other forms which are possible. Furthermore, there are many combinations of fillings and coatings which, in the most complex forms, resemble candy more than cookies. Ice cream cones and cups are made in the same general way. The equipment for hollow wafer sticks, fan wafers, shells, etc., appears considerably different, although the principles are the same. Rolled sugar cones are produced on a different type of equipment. The first part of this section will be devoted to explaining the method used for producing the most common variety of sugar wafer cookie.

Figure 8.7 is a typical layout for a complete sugar wafer plant. The sequence of processing steps in a factory of this type is as follows:

(1) Ingredients are mixed in whisks and the batter pumped to a supply tank.

(2) Batter is dispensed by an accurate measuring piston pump onto wafer plates at the entrance to the oven. The dispensing device travels with the book (a pair of plates) while the latter is open and then withdraws and returns to its initial position as the plates close.

(3) After it closes on the batter, the book continues to travel through the oven absorbing heat and cooking the batter.

(4) The baked wafer sheets are removed from their plates by take-off units and conveyed on a steel band to collection containers from which they are transferred to the wafer-builder equipment.

(5) At the filling applicator, creme from the filling mixers is spread on the wafer sheet, which is now at room temperature. Alternate layers of wafer and creme are built up until the required thickness of cookie is obtained.

(6) Completed sandwiches are passed through a pressing unit which causes the components to stick together.

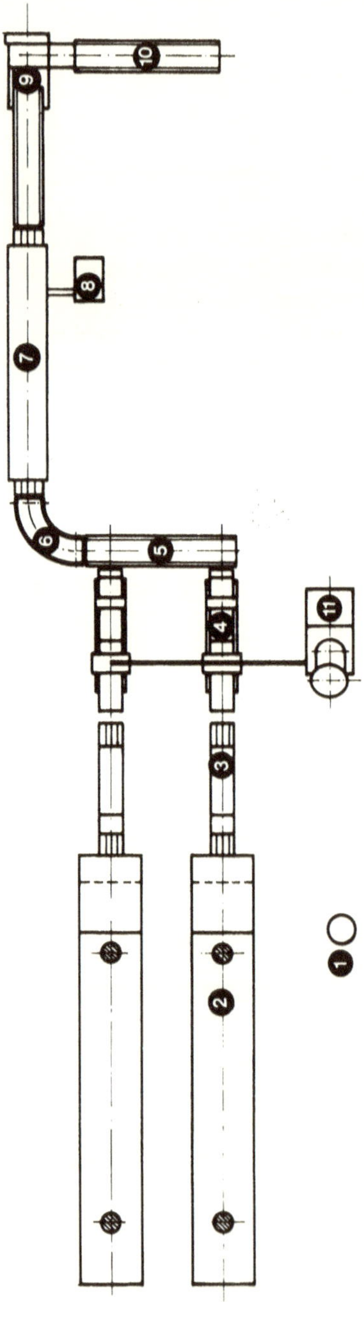

FIG. 8.7. Layout for a complete sugar wafer line: (1) Batter mixer. (2) Wafer ovens. (3) Cooler. (4) Creme spreader. (5) Marshalling conveyor. (6) Conveyor. (7) Filled wafer cooling cabinet. (8) Refrigeration unit. (9) Automatic tandem cutting machine. (10) Packing table. (11) Creme mixer.

Source: Franz Haas Waffelmaschinen

(7) From the pressing unit, the sheets pass through a cooler which sets up the filling.

(8) Next, the sheets are collated into a stack of the required thickness and fed to the cutting machine to be sliced into individual cookies.

(9) The cookies are packaged.

The Oven

The ovens made by different manufacturers can vary substantially in their structural details. The oven described below represents a hypothetical model of relatively simple design incorporating all of the important operating functions.

In the traveling book type of oven, sets of two plates each are mounted on wheels that support them on rails fixed on each side. A chain conveyor system pulls the books continuously through an upper course and then through a lower course in the baking chamber. The bottom plate is fixed solidly to the carrier chains and the top plate is hinged to the bottom plate. Books open automatically to accept their charge of batter, then close and lock together. The closed books, containing their deposit of batter, are carried through the heated chamber. When it reaches the front of the oven, the book opens, the baked sheet is ejected, and the book moves on to receive its next charge of batter (see Figure 8.8).

Wafer ovens are made in standard models containing 12,18, 24, or 30 plates. Models with 36 plates, and perhaps even larger units, are available from some manufacturers. The plates are usually 290 by 470 mm, or about 11.5 by 18.5 inches, in length and width, but there appears to be a trend toward larger plates. Plates can be either mounted on frames or self supporting. The main advantage of self supporting plates is that they contact the flames directly. Also, they can be adjusted more easily. A special gray cast iron has been recommended for the construction material. Hard chrome plating can be applied to extend cleaning cycles and promote release of the wafer from the baking surface. Oven books are cured by the application of vegetable or animal fat to heated plates, similar to the treatment used for other baking surfaces.

Production rates vary depending on several factors, including thickness of the sheet, type of batter, extent of baking (how much moisture is removed), etc., but one oven manufacturer estimates a yield of about 320 lb from a 24 plate gas-fired oven in five hours (including 35 min preheating). This poundage refers to the wafer sheets only, not the finished cookies.

A modification of the traveling plate oven utilizes two sets of books

FIG. 8.8. SUGAR WAFER OVEN SHOWING OPEN PLATES READY TO RECEIVE BATTER.
SOURCE: FRANZ HAAS WAFFELMASCHINEN

conveyed side by side through a common baking chamber to increase output per unit of floor space and per dollar of capital expenditure.

Another variation, the continuous band oven is designed in such a fashion that the batter is deposited on the surface of a reeded drum similar in construction to the drive drums of band ovens. A reeded steel band forms the top plate, heat being applied to the inside of the drum and to the outside of the band. Batter is injected between the drum and the band, and a continuous wafer sheet is withdrawn from the outlet. This is not a common piece of apparatus.

Ovens for making ice cream cones are designed in a circular format and look quite different from wafer ovens. The basic principles are the same.

Wafer ovens can be heated by gas or electricity. Cost considerations will often favor the use of gas-fired equipment. Gas consumption has been estimated at 1780 Btu per pound of output, including the gas used for preheating. The same source estimates electrical usage at about 1 kwh per 1.1 lb of wafers, including preheating.

When gas is used, the entire oven chamber is heated, electrical baking is done by means of elements built into each plate. Electricity should be supplied through an automatically controlled voltage regulator with infinitely variable output. A set of bus bars within the oven frame is connected to the voltage regulator. Current is transferred to the moving plates through a carbon brush arrangement, and it is cut off automatically when the machine is stopped and cannot be switched on until the oven has been restarted.

The effectiveness of the equipment which removes the wafer sheets from the books is a factor in the amount of scrap and other aspects of subsequent operations. One device fits directly onto the end of the oven and is completely automatic in operation. To synchronize the ejection movements, it is driven from a special star wheel which engages the wafer-plate carriage rollers. An oscillating take-away carriage which bears an air-blast device on its forward end follows the wafer plate closely as the book emerges from the oven and moves upward on its track. Controlled jets of air from the nozzles on the blowing device loosen the wafer sheets as the plates start to open, so that the wafer can slide down to receiving rods on the take-away carriage. Many of the beads of scrap dough formed on the side and ends of the plates are blasted off by the air jets at this time. A lost-motion mechanism operating here allows sufficient stopping time for the carriage to receive the wafer sheet, then the carriage starts its return trip to be in position for receiving the next wafer.

Driving brushes move the wafer sheet forward through a scrap removal brush and scraper into the collector box. Two connected boxes are provided. Either one can be positioned to receive a full load of wafer sheets. When full, the box is moved over for emptying, leaving the other box in the loading position.

Other Equipment

For optimum precision in post-baking handling, wafer sheets should be "conditioned," to relieve them from stresses which result from nonuniform distribution of moisture and from temperature differentials. Absorption of moisture increases the size of wafer sheets. Since the sheets as they leave the oven differ throughout in baked structure and in moisture content, they may warp, causing serious difficulties in applying the filling and in stacking the layers. In modern plants, conditioning is achieved by slowly conveying the sheets through a chamber where the relative humidity and temperature are controlled (Haas *et al.* 1985). From the conditioning cabinet, the sheets are conveyed to the wafer-building equipment.

An automatic wafer-building unit consists of a scrap trimmer, a creme depositer, and auxiliary equipment.

Creme depositers and spreaders for wafer sheets usually consist of hoppers with drum extrusion feeders. Filling is removed from the feed rolls by applying heat to the rollers, or by a doctor blade, and is laid on the wafers in the form of a continuous sheet as the wafers are being carried on a moving belt (see Figure 8.9). These units may be set up in tandem to make the single or multiple spreads required. The thickness and width of the creme deposit is determined by the adjustment of roller speeds, belt speeds, the distance rollers are separated and the distance the creme drops from the hopper onto the wafer sheet. An ideal extrusion of filling leaves the hopper in a continuous sheet which is uniform in thickness and covers the wafer sheet completely to within one-eighth to one-sixteenth of an inch from the outside edges.

When depositing very soft cremes and other semifluid fillings, it is helpful to lay the creme on the wafer sheets as several separate ribbons and then spread them across the sheet with a spreading bar. This type of application may be less than satisfactory, however, since it tends to cause considerable scrap.

If heated rollers are employed to release fillings, the creme may be made excessively soft, causing difficulties with the sandwiching operation. The heated roller can sometimes be used satisfactorily with cremes of low density or those made with fats of rather low melting point. In general, however, use of a knife to insure separation of filling from the roller is preferred even in these cases. When a low density creme is being used, it may be necessary to open the gauge rollers and run them slower to avoid deaeration.

According to Haas et al. (1986), in the industrial production of filled wafer sheets, coated wafer sheets in a lower position are inserted succesively into two non-moving helical conveyors which are rotatable in opposite directions about parallel axes, then the helical conveyors are moved so that they raise the coated wafer sheet until it joins the underside of a cover sheet or a coated wafer sheet which is in an upper position. After this contact is made, the helical conveyors are stopped, and the next coated wafer sheet is inserted into the lower position and the process is repeated until a block of the desired thickness has been formed.

Either a wire cutting device or a rotary saw is used to divide the large slab of wafer and creme composites into finished product size. One commonly used method is to run the continuous sugar wafer sheet through wires to cut in the direction of product flow and then make the right angle cut with an oscillating band saw. The wire size is usually 0.019 inch. The rotary saw produces more waste since the saw blade is itself 0.035 to 0.050 inch thick.

A simple but effective arrangement consists of two gangs of saws set at right angles to each other so that an operator could lay a pile of sugar wafer slabs in front of the first gang of saws and push the slabs through

FIG. 8.9. CONTACT CREAMER FOR WAFER SHEETS.
SOURCE: VICARS

to make the first cut. The second cut is made by pushing the pile through the second gang of saws and then onto a packing table.

Completely automatic equipment is available for feeding and cutting the slabs in both directions.

CRACKER EQUIPMENT

The traditional process for preparing soda crackers or saltines involves the cutting of square or rectangular shapes from a continuous sheet of dough. The dough will have been laminated by special machinery to help insure the flaky, layered structure that is so characteristic of these products. Most cutters are the reciprocating, or up-and-down, kind, but perfectly satisfactory crackers can be made with roller dies, although two consecutive rotary cutters are generally necessary.

Doughs used in this method of forming must be capable of holding together in a continuous sheet during conveying and cutting. When scrap is generated, as is nearly always the case, it is removed from between the biscuits in a continuous web and this imposes additional restrictions on dough properties (see Figure 8.10). Tough doughs with a considerable amount of elasticity can be handled very well on cutting machines. High absorption, high flour content doughs which are completely unsuited for wire-cut machines or rotary molders are particularly suitable for cutting. Such doughs are capable of considerable expansion in the oven and will bake into products of relatively low density compared to other types of crackers and cookies. On the other hand, hard sweets and semihard sweets, which are also traditionally made on cutting machines, undergo virtually no expansion during baking. The range of products adaptable to cutting machine processing is, then, rather broad although the method is not as flexible (so far as dough characteristics are concerned) as wire-cut machining.

The nature of equipment used for saltine cracker manufacture and the consumers' perception of the product as a commodity make these items unsuitable for small shops and for most medium-sized bakeries. Saltines and similar crackers can only be made economically on high speed lines operated, preferably in around the clock operations. There may be some opportunity for profitably making specialty crackers—dietetic, "natural," super premium, etc.—on relatively small lines, but the possibility of commercial success seems rather remote.

Laminating Devices

Many of the doughs processed through cutting machines first undergo multiple sheeting and folding steps. For years, the standard equipment for laminating doughs was the hand-fed high-speed dough brake. A given quantity of dough, usually 40 lb or more, was formed into a thick strip on the brake and then folded into two, three, or four layers. If desired, interleaving materials such as shortening, could be applied to the strip by hand and the dough folded to enclose them. Additional materials were not applied to saltine doughs, the layer effect observed in the finished product being obtained only from the orientation established in the dough plus some effect from the dusting flour.

The folded assemblage was sent through the brake again in a direction perpendicular to the initial sheeting. Folding and sheeting were repeated one or more times, with or without cooling and rest periods between the steps. Although vast quantities of saltines and other laminated products were made in this way, it was always recognized that brakes severely punished the dough, making it difficult to obtain recognizable laminations and causing irregular shapes in the finished products. They do have the advantage of being versatile, however, since

FIG. 8.10. STAMPING MACHINE WITH SCRAP RETURN.
SOURCE: VICARS

they can be used with different kinds of layering materials and the laminating procedure can be easily varied.

Their versatility makes reversible sheeters practical for medium-sized shops which produce several different kinds of laminated products—Danish, puff pastry, snack crackers, etc. Medium duty equipment with automatic programming of the reduction steps can be obtained. A large conveyor surface permits folding and turning of the dough piece without removing it from the machine. Two operators can process 1,000 lb of dough per hour, giving each piece 3 or 4 three-folds. By proper adjustment of piece dimensions and the folding procedure, pieces can be brought off as long thick strips suitable for feeding to the reduction rolls on a forming line.

In the usual cracker bakery, combination machines for sheeting bulk dough and then lapping and cross-rolling it to form the structure characteristic of saltines are widely used. These have been discussed in detail in preceding chapters.

Cutters

Stamping machines—Vertically reciprocating cutters or stamping machines are used mostly for crackers, although embossed cookies are also made in this manner. The following discussion will deal mostly with the former class of products. Dies for crackers consist of two dough-contacting components supported by a frame and attachments. The docking pins which form holes in the dough sheet to prevent undue puffing in the oven are attached to springs and slide in and out of holes in brass dies. They remain in a raised position until pressure is exerted on the frame and the bottom is held in a fixed position, i.e., when the cutter is at the bottom of its stroke. At that point, the docking pins are pushed through the dough. They retract before the cutter starts its upward motion so that the dough is held down while the docking pins move upward through it.

The other part of the dough cutting apparatus consists of serrated strips or knives, often of brass. The serrations, or saw-toothed effect, can be either triangular or approximately rectangular. Their number and width must be adjusted so that the crackers break cleanly along the desired lines before they reach the stacking station, but they must hold together in required multiples during packing and distribution. The arrangement of serrations which accomplishes these goals may vary for different doughs. Needless to say, if there are two or more packages requiring that the sheet of crackers be broken in different combinations, separate cutters should be used for each.

Cutting edges of the knives tend to wear unevenly. The usual pattern is for the sides to wear more than the front or back edges, and for the corners to wear most of all. A new cutter will show a drop of about 0.1. to 0.2 mm from center to edge, depending on width and other factors. When wear is excessive, as shown by uneven cutting, the knives can be planed or ground to the original specification, and then sharpened and renotched. Knife edges which are too sharp will sever the dough layers instead of sealing them together, as desired.

To compensate for lengthwise shrinkage in the band of cracker dough, the knives are arranged to cut a rectangle. The cracker before baking is about 6% longer than it is wide. For a two-inch cracker, the cutter is frequently made 2.125 inches between centers of the knives from front to back. The side-to-side dimension can be held satisfactorily if the proper kind of dough is being machined, and so the cut can be made equal to the desired baked product dimension. Similarly, cutters for round crackers must be made elliptical, with the long axis in the direction of band travel. Cracker length can also be adjusted by changing the speed of the apron leading on to the oven band, this having the effect of stretching or contracting the piece depending on the relative speeds of the two bands.

Since the usual shape for crackers is rectangular or square, the cutter will not yield any scrap, except for a small strip at each side of the dough web. If the width of the dough strip is properly adjusted, there should be virtually no cutting scrap. Hexagons, as for oyster crackers, also allow scrapless cutting except for a slight amount of scrap along the edge. Fancy interlocking designs such as fish, animals, and geometric shapes yielding little or no scrap can be made on stamping dies, but require rather sophisticated art work. When scrap is generated, it is drawn off in a continuous web above the dough sheet and, often, dropped back on the dough at an early stage of the lamination process.

The number of docker pins in cracker cutters differs according to the ideas of different manufacturers, but a common pattern is 13 pins arranged in alternating rows of 2 and 3. Sixteen pins in 4 rows of 4 pins are also used. Number, size, and pattern of docking pins affects appearance and texture of the cracker. Increasing the number of docker pins will, in general, reduce the baking time.

The usual widths of the cutting surface of dies to be used in conjunction with band ovens are 32 or 40 inches. The modern tendency is for wider dies to be used in conjunction with wider oven bands. Room is allowed for an inch of trim on each side of the cutter. Cutters can be made 1, 2, 3, or 4 rows long. The higher multiples may allow greater production without causing the excessive wear and damage that may result from increasing the strokes to a very fast rate.

Rotary cutters—Rotary cutting machine dies are either cast or fabricated cylinders, often of relatively small diameter. Cylinders fabricated from sheet metal are generally useful only for simple outline cutting, as for square or rectangular pieces, but fairly elaborate designs can be cut from cast cylinders. Dockers are applied to cutters for hard sweet and semihard sweet cookies as well as to large base cake dies. The entire machine will consist of a cutting cylinder geared to have a peripheral speed equivalent to the linear speed of the apron traveling under it, the canvas web or apron, and the supporting frame with motor and controls.

The main disadvantages of rotary cutting are the tendency to pick up and retain dough pieces in the cutter and the inflexibility of the method with respect to workable doughs and shapes. Advantages are less wear and less capital expense than with reciprocating cutters. Since there is less wear, the labor and down-time needed for maintenance are reduced and accuracy retention is improved so that more exact finished shapes are obtained. Rotary cutters are also much less noisy than stamping machines, which may be a consideration.

An improved type of unit for embossing and rotary cutting consists of two synchronized cutter cylinders (see Figure 8.11). The first roller impresses the biscuit design on the dough sheet while the second carries the ridges which cut the cookie outline as well as the docking pins, if the latter are used. The embossing roller also presses the dough to the correct thickness, thus improving weight and dimensional control, and makes the dough piece adhere to the conveyor belt so that it is not as likely to be picked up in the cutter. Scrap is lifted off in the normal way, leaving the dough pieces to be transferred to the oven band.

Cutting pads—Resiliency and firmness of the supporting surface under the cutting machine conveyor are important factors affecting cracker design and the release of doughs from the cutter. For precison high quality cutting, a sheet of hardwood having its cross-grain presented to the die is a very good choice for the cutting pad. Maple, and occasionally teakwood, are used. The position or orientation of the wood must be changed frequently to prevent the cutting of deep grooves in the material. Linoleum is a satisfactory material. Such pads will accentuate any differences due to uneveness in cutting edges, however, and therefore more resilient supports are far more common.

One or more thicknesses of canvas will give a moderately firm pad which yields enough to compensate for minor variances in cutter height. The canvas should not be obtained by cutting up worn out stamper aprons. If it is necessary to use rubber pads, several thicknesses of cloth, or even worse, newspapers and other temporary stuffing, the operator can conclude that the cutter is in poor condition or poorly aligned or designed. Uneveness in any kind of pad is an indication of the necessity for replacement.

The more the pad yields to cutter pressure, the more the dough is pushed up into the center of the dies. The dough piece will tend to cling more to the edges of the cutter and to the docker pins. Border designs and brand names or symbols will be impressed too deeply. Release is interfered with and the dough sheet is distorted into folds and ridges. The ridges can sometimes be brought down by a free-running roller, but tensions put into the dough by the stretching and folding actions still remain and can lead to misshapen crackers, irregular breaking, and checking.

Use of tape to build up areas on the pad where cutting is indadequate is a strictly temporary tactic which should not be depended upon for any long-time adjustments. The cutter should be corrected instead, or the pad replaced if it is at fault.

FIG. 8.11. CUTTING MACHINE WITH SEPARATE ROLLS FOR CUTTING THE OUTLINE AND DOCKING.

SOURCE: VICARS

Cutting aprons—Aprons—the conveyor belts carrying the dough sheet under the cutting machine—are generally made of heavy canvas but, of course, various combinations of textiles and plastic can be employed. Some operators prefer to run them wet, others run them dry. Water or shortening can be used on the apron. Heavy applications of dusting flour should be avoided. A seamless belt would be preferred, but is rarely used. Both the apron and pad should be kept as free as possible of dough accumulations. In the stamping machine, the apron should be drawn fairly tight to reduce distortion of the dough pieces.

Trouble-shooting Crackers

Some defects in cracker appearance can be related with good assurance to certain mistakes in processing (Somers 1984). For example:

(1) Fishmouthing and poor separation at the cutoff knife can result

from insufficient pressure on the cutting roll. If the cutting edges are not true, the edges of the cracker will spring open during baking.

(2) Setback or pyramid type separation of the cracker sheets can be due to the cutting machine running faster than the oven band or to intermittent motion as the dough sheet slips on the oven band.

(3) Streaks appearing on the baked cracker may result from improper mixing, particulary nonuniform incorporation of soda.

(4) Flat or dead areas on cracker surfaces can also result from inadequate mixing.

(5) Uneven or patchy blistering may be due to either improper setting of burners or positioning of the cracker toward one side of the oven band.

BIBLIOGRAPHY

AICHELE, W. J. 1981. Cookie and cracker processing. Cereal Foods World *26*, 161-165.

ANON. 1973. Distributors help Johnson Biscuit up sales 8% to 10% per year. Bakery Production and Marketing *8*, No. 4, 106-107.

BARTA, B. B. 1988. Personal communication. Franz Haas Machinery of America, Richmond, VA

GADAMS, F. 1984. Wire cut cookie production. Biscuit Bakers Institute 59th Conference, Biscuit and Cracker Manufacturers' Association.

GIORGETTI, P., ROMANI, F., and STRINO, E. 1986. Confectionary product and process for producing same. U. S. Pat. 4,569,848.

GREENE, H. L. 1982. Equipment manufacturer helped biscuit industry get its act together. Snack Food *17*, No. 2, 26-27, 40.

HAAS, F., SR., HAAS, F., JR., and HAAS, J. 1985. Apparatus for conditioning wafers. U. S. Pat. 4,524,682.

HAAS, F., SR., HAAS, F., JR., and HAAS, J. 1986. Process and apparatus for producing filled wafer blocks. U. S. Pat. 4,567,049.

MC KEE, H. B. 1985. Method for feeding cookie preforms. U. S. Pat. 4,562,084.

MORETH, N. W. 1967. Modern sugar wafer production—a technological breakthrough. Manufacuring Confectioner *47*, No. 11, 16-22.

MORETH, N. W. 1982. Increased cookie production through laydown control. Bakers Digest *56*, No. 2, 8-10, 12, 14.

REGET, G. 1966. Wire-cut cookie manufacture. Proc. Am. Soc. Bakery Engineers *1966*, 263-271.

SMITH, A. 1970. Soft cookies. Proc. Am. Soc. Bakery Engineers *1970*, 114-123.

SOMERS, J. F. 1984. Trouble shooting crackers. Biscuit Bakers Institute 59th Technical Conference, Biscuit and Cracker Manufacturers Association.

THORNTON, I., DE WITT, K. W., and ROBERTSON, S. A. 1985. Method of preparing a biscuit or cookie product. U. S. Pat. 4,517,209.

WEIDENMILLER, E. A. 1966. Personal communication. Morton Grove, IL

WEIDENMILLER, T. 1988. Personal communication. Weidenmiller Co., Elk Grove Village, IL

WITTENBERG, H. L. 1964. Rotary cookie production. Biscuit Maker Plant Baker *16*, 26-30.

WITTENBERG, H. L. 1965. Wire-cut cookies. Biscuit Bakers' Inst. Training Conf. *1965*.

YOON, Y. 1986. Method for manufacturing a jelly confectionery coated with chocolate. U. S. Pat. 4,563,363.

CHAPTER NINE

OVENS AND BAKING

INTRODUCTION

In most bakeries, ovens are the most conspicuous and characteristic pieces of equipment. With their loaders, unloaders, coolers, depanners, and conveyors, they dominate the layout and determine—in large part—the arrangement and location of the other pieces of machinery. Baking is also the operation which limits output in most plants. For this reason, selecting the oven, maintaining it properly, and operating it at the maximum rate consistent with good product quality are key elements in the successful management of a bakery.

The oven has an important influence on product quality. It cannot compensate for errors committed earlier in the processing sequence, but a well-adjusted oven of the proper design can bring out the potential of a well-processed dough piece. The principles of oven design and operation which govern its effectiveness in optimizing product quality are not completely accessible to scientific analysis. The mechanical details of oven construction are, of course, important in that they are related to labor requirements, efficiency of fuel utilization, frequency of product damage, and sanitation. But of more fundamental importance and not as well understood are the effects of heat transfer mechanisms on product quality. All ovens transfer heat by conduction, convection, and radiation, but the differences in the percentage of heat transferred by each method during each stage of baking determine the variation in baking results in different ovens.

There are many ways of classifying ovens, and each of the ways has some value for a specific purpose, but the two major classifications used in this chapter are "Retailer Ovens" and "Wholesaler Ovens." This scheme was adopted mostly for the convenience of the reader, since there is necessarily considerable overlap between the two groups in technical details of heating devices, controls, enclosure construction, etc. It is expected that the individual reader, however, will have his or her attention concentrated on one of the two groups and the classification method chosen will save time and make it easier to find pertinent information.

Fryers and their appurtenances are discussed in Chapter Eleven.

GENERAL CONSIDERATIONS

History of Oven Development

Although it is not within the scope of this volume to give extensive discussions of the history of equipment development, a brief description of some older types of ovens should contribute to a better understanding of the advantages of modern tray and band ovens.

The most primitive type of oven we know about was a chamber of mud or brick in which a fire was built. When the oven had been thoroughly heated, the ashes were scraped out and dough pieces inserted into the chamber. The same general pattern applied to the largest and the smallest ovens. All of these ovens were very inconvenient to use and very wasteful of fuel. A great advance was achieved when it was discovered the baking compartment could be isolated from the fire chamber. Then, the heating could be carried on continuously during baking and while the dough pieces were being transferred into and out of the oven.

The peel oven, in which a fixed hearth is surrounded by a fire box, was the predominant type used in commercial bakeries until fairly recent times. Peel ovens are generally of massive construction, with very thick brick walls serving as heat sinks and minimizing temperature fluctuations. Sand was sometimes placed between brick oven walls for the same purpose. These ovens are still in use in some parts of the world. The "deck" oven, nowadays used for smaller scale roasting and baking in some food service operations, is similar to a peel oven except that it can be heated by electricity or gas and convection assisted by fans. They depend on insulation rather than heat sinks to maintain uniform temperatures.

The hearth in a peel oven is wide and deep, but the height is normally rather small. This creates timing difficulties in charging and removing dough pieces. The baker uses a peel, or long wooden spatula, to insert dough pieces into the baking chamber and to remove the loaves. Since the back part of the oven is loaded first and unloaded last, there may be several minutes difference in baking time between the first and last loaves. The difference is accentuated by the fact that the last doughs in are the first doughs out and are near the front door, which is probably the coolest spot in the oven. They not only have the shortest baking time but the lowest baking temperature. Somewhat greater uniformity was achieved through the use of a portable metal hearth which was completely loaded with dough outside the oven before it was pushed into the baking chamber. The problem of uneven temperatures remained, however, since doughs near the side walls and back wall received considerably more heat than the remainder. This situation was alleviated only by inventing the moving hearth.

The first type of moving hearth was a circular metal plate which rotated in a horizontal plane around a pivot or axle placed approximately in the center of the oven. Since each dough piece was carried in a path around the oven, temperature differences were more-or-less equalized. Baking times could also be kept constant because different parts of the hearth could be loaded and unloaded in the proper sequence. The motion of the hearth and its load created some air movement in addition to that normally resulting from temperature differentials, and thus increased the contribution of convection to the total heat transfer. As a result, baking times decreased. One of the chief disadvantages of these ovens is the large amount of floor space they require. The hearth being circular, rectangular pans cannot be fitted efficiently on it. Also, the corners of the square space in which the oven must fit are wasted. Rotary ovens are rarely, if ever used in commercial installations today, but some laboratory ovens for test baking use this principle.

The reel oven was a considerable improvement in that it not only alllowed front-to-back movement of the dough pieces but also moved them vertically, providing an additional leveling out of differences in temperature distribution. The only variations it does not even out are the side-to-side differences, which can be considerable. As an additional advantage, the reel oven increased the amount of hearth area which could be put in a given floor space. Convection was greatly increased by the fanning action of the moving shelves. Some manufacturers designed ovens to give a steam trap through which the pans moved. Because of its many advantages, the reel oven became very popular and many of them are in use today. The chief disadvantages are the difficulty in automating the loading and unloading operations and the inability to provide zone heating.

As more efficient heat sources were made available, it became obvious that transferring the heat to the product was the limiting factor, especially in smaller and medium-sized ovens. This led to the development of forced convection ovens, in which currents of air created by fans greatly increase the transfer of heat from the electric elements, or other heat sources, to the product. These ovens also tended to minimize heat variations through the baking chamber. Many ovens for home, retail baker, institutional caterer, and in-store use are based on the forced convection principle.

Large wholesale bakers began to demand levels of output which could not be met by any existing type of commercial oven. In response, oven designers and manufacturers began to design tunnel ovens in which the dough pieces were conveyed continuously from the input to the discharge areas. In some cases, trays or shelves were used to convey pans containing the loaves or rolls, while in other cases the hearth was a continuous band. The former type was most useful for bread, rolls, sweet doughs, etc., while band ovens were extremely convenient for cookies, crackers, and other small pieces. Because of their efficiencies in loading and unloading, and economies in construction band ovens have gradually taken over some of the areas thought to be necessarily restricted to tray-type ovens. Even loaf bread has been baked on a band, though probably not in this country. The major differences in draw plate, rotary hearth, and traveling tray ovens can be seen in Figure 9.1.

A troubling problem in the baking of loaf bread is the necessity for storing and withdrawing straps of pans between the times they are being used in the proofing and baking cycle. Various automatic devices were invented for moving and controlling empty pans in a continuous loop. The latest advance, starting perhaps 15 or 20 years ago, is an integrated proofing and baking system which carries loaf bread, or the like, from the panning station through the proofing room and oven to the depanning station on a completely automated conveyor. Essentially, the pan is never idle. In these systems, the oven is a rectangular chamber in which the conveyor loops around the interior to provide the needed baking time. Although lacking somewhat in flexibility, these systems are high in efficiency and have many other advantages.

Heat Transfer Mechanisms

Heat may be generated within a mass (such as a piece of dough) by radiation, friction (work), or chemical reactions, but in baking we mainly have to deal with the heat transfer mechanisms of convection, conduction, and radiation. The relative effectiveness of each of these means of transfer varies with oven design as well as with the conformation of the dough pieces, the size, shape, and construction materials of their container or pan, and the distribution of the dough pieces on the hearth.

Radiation—Transfer of energy by infrared radiation is a significant factor in most ovens. These radiations are not in themselves heat, but are converted into heat through absorption by and interaction with absorbing molecules. The effectiveness of electromagnetic radiations in transferring heat depends upon their wavelength. Light, which is an electromagnetic radiation, is an extremely poor means of transferring

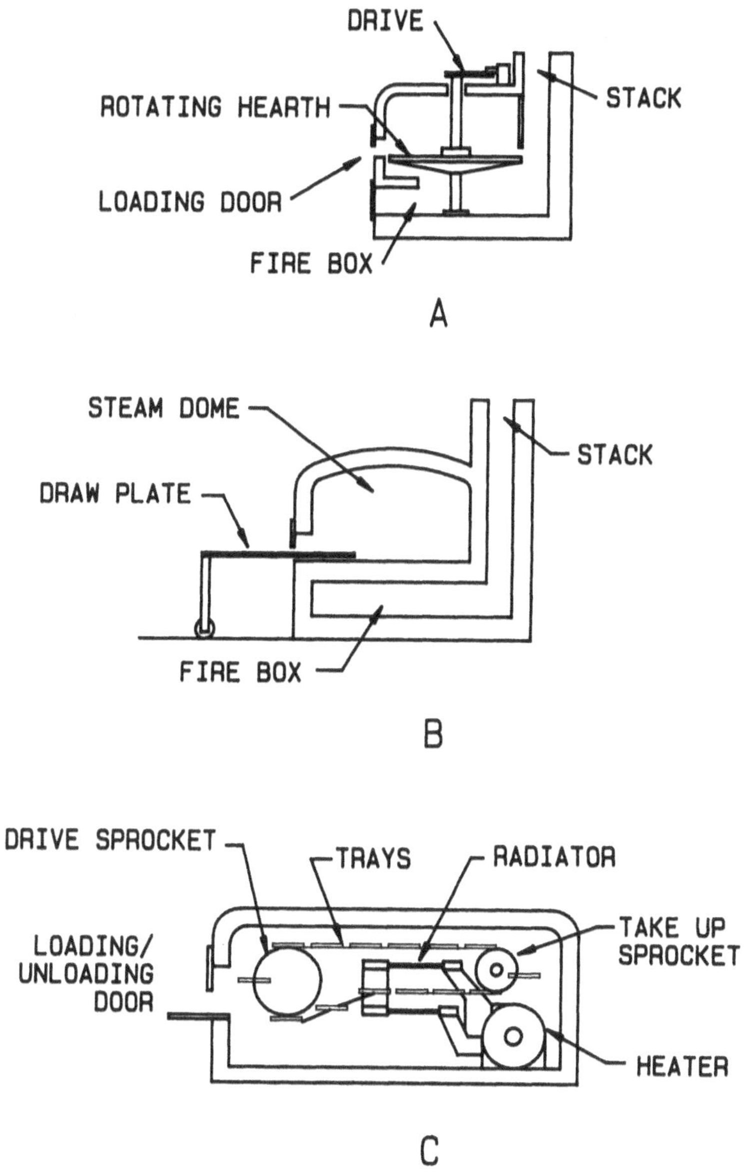

FIG. 9.1. Three types of oven. Top, rotating hearth. Center, Draw plate. Bottom, single lap tray oven.

heat. Invisible infrared radiation is quite effective. The quantity of heat transmited between two bodies by radiation is directly proportional to the differences of the fourth powers of the absolute temperatures and inversely proportional to the square of the distance between the bodies. This indicates that increasing the temperature of the radiating material (flame, oven walls, etc.) is a powerful tool for increasing the heating effect. It also tells us that, as the bakery product is moved further from the radiating object, the heating effect rapidly diminishes.

Radiation has two characteristics different from the other means of heat transfer which affect its action on dough pieces: (1) it is subject to shadowing, or blocking by intervening layers that are opaque to the radiation, and (2) it is very responsive to changes in absorptive capacity of the dough, e.g., to changes in coloration in the case of infrared radiation or to changes in water content in the case of radio waves of certain frequencies.

Changes in color affect the progress of baking principally through increases in the absorption of infrared rays. The darkening which accompanies baking is accompanied by an increase in the dough piece's absorption of visible wavelengths. An increase in absorptive capacity for infrared rays, though not apparent visually, is an almost invariable concomitant of the visible change. As a result of the increase in heat absorption in the darkening areas, there is a tendency for the color changes to accelerate after the first browning appears. This means that ovens relying on radiant energy for a relatively large proportion of their heat transfer will tend to accentuate color differences. Such a tendency may be either good or bad, depending on the characteristics desired in the finished product.

Radiant energy comes from the burner flames and all hot metal parts in the oven. It is not necessary that the oven part be red hot, or otherwise visibly heated, for it to radiate energy. This radiant energy travels in a straight line and much of it never reaches the product because it is intercepted by some substance that is not transparent to the radiation. Of course, the metal or other substance that receives the energy can become hot enough to re-radiate a significant amount. Shadowing, or blocking out of radiation by some intervening substance, can occur from pan walls, parts of the oven, or parts of the dough piece itself. The shadowing due to parts of the dough is most apparent with radiation near the visible range (i.e., infrared) and is much less with radio frequency waves. Radiation from below the band is completely intercepted by the band and partly reradiated as infrared energy (normally, the band will never get hot enough to radiate part of the energy as light) and partly converted into conducted heat. The sides of the pan shadow the dough piece until it rises above them, at which time radiation becomes much more effective in promoting color development in the crust.

If a dough piece is shaped approximately like a segment of a sphere, with a relatively smooth surface, the reception of radiation during its trip through the oven will be about equal over all parts of the surface. Absorption will also be approximately equal as long as the color (indicative that changes have taken place which also affect infrared absorption) remains the same over the entire surface. This also applies to flat-surfaced doughs covering most of the band. Products with irregular surfaces, such as wire-cut cookies with holes in the middle, twist bread, fancy pastry shapes, and pies with raised rims will have parts of their surface in a better position than other parts to absorb radiant energy. If these surfaces are sufficiently abosrptive, heating will occur at an accelerated rate in the prominences. The shadowed areas will receive less energy and will tend to heat slower and brown slower. It must be understood, however, that all of these effects of radiation can be offset by complementary inputs of conducted or convected heat.

The radio frequency waves generated in electronic ovens are efficient transmitters of energy, but they have characteristics very different from those of infrared radiation. They will be discussed separately in a subsequent section.

Convection—Convection is the transfer of heat from one part to another within a volume of gas or liquid by the gross physical mixing of one part of the fluid with another. In the oven chamber, molecules of air, water vapor, or combustion gases heated by whatever means, circulate throughout the oven, constantly mixing with other gases and transferring heat by conduction when they contact solid surfaces. Within the dough piece, convection occurs as the result of the movement of water vapor and other gases. Furthermore, some translocation of liquid water, melted shortening, and other liquids cause a transfer of heat from one region of the dough to another.

If the overall results of convection can be generalized, it would be as a smoothing or evening effect of heat distribution. The gases within the oven mix readily as long as there are no mechanical barriers. Their mixing tends to make more uniform the temperature throughout the chamber. Within the dough piece, the principal effect of convection is also a smoothing or blurring of temperature differentials. The hotter parts of the dough give off more water vapor than the cool sections, and this loss of hot vapor from one area and its gain in others tends to minimize differences.

When different temperatures are required in different zones of the oven, it is necessary to isolate these zones in some manner to retard convective heat transfer between them. Normally, the zones are separated by walls or solid partitions which extend from the top of the inside of the chamber almost to the tops of the dough pieces and from one side to another.

Convection directly affects exposed areas only, of course. The bottoms of the dough pieces, protected as they are by the band or pan, do not participate directly in this type of heat exchange. It is doubtful that much convective transfer occurs at the bottoms even when mesh bands are used. All exposed parts probably participate about equally in convective heat exchange. No doubt some protective effect of protuberances could be shown, but the extreme turbulence of the gases above the band and the relatively low relief of most dough pieces suggest that variations in reception of convective energy by the different parts of the dough cannot be very great. On the other hand, products at the edge of the band may receive substantially more heat because of the more rapid flow of hot gases and the higher temperatures in this area.

Especially in a small oven heated by electricity, or by gases circulating outside a closed baking chamber, air currents exist but they are relatively minor and contribute little to heat transfer. This situation can be improved by placing a fan so that a rapid current of air passes over the heating elements and into the baking chamber. Convection ovens based on this principle can significantly increase the speed and uniformity of heat transfer.

Convected heat becomes conducted heat when it raises the temperature of a solid body—pan, dough, hearth, etc. Thus, its indirect effects on the reactions in a dough held in an oven can be substantial.

Conduction—Conduction is the transmittal of heat from one part to another part of the same body, or from one body to another which is in physical contact with it, there being no appreciable displacement of the particles contained in the bodies. The differential equation for heat transfer by conduction is called Fourier's law:

$$q = \frac{dQ}{d\theta} = -kA \frac{dt}{dx}$$

where q is the rate of heat flow in Btu per hr, Q is the quantity of heat transferred in Btu, theta is the time in hours, A is the area (sq ft) normal to the direction in which the heat flows, and dt/dx is the temperature gradient, that is, the rate of change of temperature (°F) with the distance (ft) in the direction. The factor k is called the thermal conductivity and is dependent upon the material through which the heat is flowing and upon temperature. The negative sign indicates that the net flow of heat is in the direction of lower temperatures. The thermal conductivity of water at 32°F is 0.343 and at 100°F is 0.363 Btu per sq ft per °F per ft.

When baking in a band oven, conduction of heat to the dough occurs only through the band. The band receives its store of energy from radiation, from convection, and from heat conducted through the supports on which it rides. Because of the localized nature of conductive transfer, steep gradients of temperature are set up within the dough piece, the hottest areas being the ones in contact with the pan and particularly where the pan is in contact with the band. Unwanted differences in the rates of heat-catalyzed reactions can easily occur unless these gradients are carefully controlled. Conduction from one part of the dough to another is a force tending to reduce temperature differentials.

To summarize the effects of the three types of heat transfer that occur during conventional baking, it can be said that conduction and radiation tend to cause localized temperature differentials, conduction acting to raise the temperature of the bottoms and radiation acting to raise the temperature of exposed surfaces (and especially darkened areas and protuberances), while convection tends to even out temperature gradients.

Electric Heating—It is possible to heat bread or other dough or batter products which conduct electricity by placing electrodes on each side of the piece and passing a current through it. This is an old idea; possibly Volta thought of it when he was working with his electrified frogs. Yonezawa (1986) described an apparatus consisting of an endless belt conveyor having an insulated bottom belt and a pair of insulated side belts having slots for receiving electrically conductive plates. The plates were constructed so as to form a series of moving compartments on the belt. Dough pieces are placed in these compartments in contact with the conductive plates, and a three phase, 200 volt alternating electric current is passed through the piece to bake the dough as it moves along with the conveyor. Baking with electric current, using the dough as part of the circuit, has been mostly a laboratory exercise up to this point. On the other hand, using radio frequency waves to bake certain dough and batter products has been applied on a large commercial scale for about 30 years.

Electronic heating—Electronic ovens generate heat within the product as the result of vibrations set up in some of the product's molecules when they absorb electromagnetic radiation of high frequency. The water molecule is particularly effective in this regard. The radiation in question is created by electronic circuits resembling radio transmitters in a very general way. There are two principal types of electronic ovens, the dielectric heating variety using frequencies in the range of 30 to 40 MHz and the microwave or radar types using frequencies of 915, 2,450, 5,800, or 22,125 MHz. The most common present day ovens employ the 2,450 MHz frequency, a few others use 915 MHz (896 MHz in Europe). The lower frequencies are better for thawing and have been suggested for baking because they penetrate better.

The basic form of a dielectric heater electrode system consists of two plates, one of which may be grounded while the other receives input of high frequency (30 to 40 MHz) electromagnetic energy. The material to be heated is placed between the two plates. When such a system is applied to an oven, electrodes are generally placed above and below the band, allowing the band to be grounded without making it carry a large radio frequency current. A horizontal field electrode system can be combined with a plastic web for heating products outside the hot oven chamber. There are said to be several of these systems installed in U. K. biscuit plants. Figure 9.2 illustrates the wave distribution as related to four different electrode and band arrangements.

Microwave systems are the basis of the radar ovens used in homes and restaurants for the rapid heating of miscellaneous foodstuffs. Microwave baking can be very efficient because the oven cavity is not held at a high temperature and therefore does not lose large amounts of energy to the environment as do conventional ovens. Since electricity rather than gas must be used as the energy source, the overall economics may not be as favorable as the efficiency, however.

Two main difficulties with microwave heating for general baking purposes are that it does not contribute to crust formation and it tends to be absorbed by the outer layers, giving insufficient heating to the interior. Consequently, most dough and batter products cooked entirely by microwave tend to have nontypical appearance and texture. Since the compounds normally formed during browning of the crust are absent, the taste may also be affected. These difficulties are reduced if the product is thin, as pizza crusts, and reduced even more if the product normally has a light colored crust, as in the case with soda crackers (for some markets) and some brown 'n serve items.

If the product is not cooked directly on the oven band, an additional problem is finding a pan which will not interfere with the microwave radiation. Metal loaf pans, as for bread, cakes, etc., would generally be considered out of the question, but some authors have suggested ways in which they might be used. Most plastics are not suitable because they cannot withstand the temperatures which are reached without deforming, deteriorating, or giving off noxious gases. Some plastics

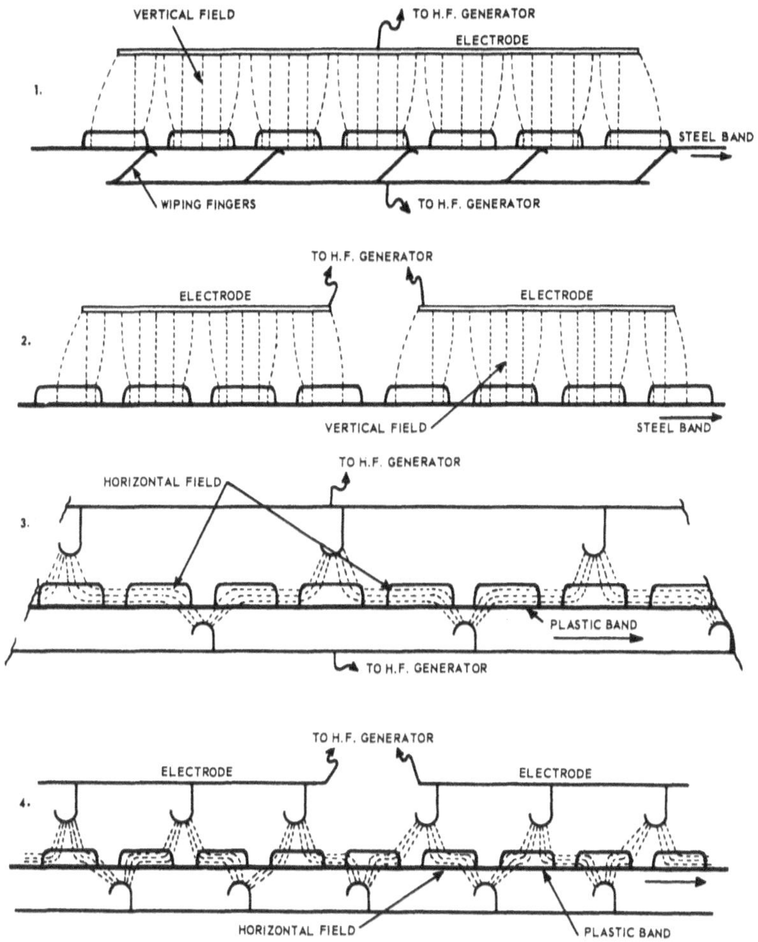

FIG. 9.2. HOW THE WAVE ENERGY IS DISTRIBUTED IN DIELECTRIC HEATING OF DOUGH PIECES ON (1 & 2) STEEL BANDS AND (3 & 4) PLASTIC BELTS, WITH (1 & 3) CONTINUOUS ELECTRODES AND (2 & 4) SPLIT ELECTRODES.

SOURCE: HOLLAND (1962)

have been developed specifically for containing products which are to be heated in microwave ovens, but some (or all) of these are too expensive for baked products or have other disadvantages.

Because electronic heating will not give normal crust coloration, and for other reasons, there has never been a large scale installation for bread baking. It can be useful for rapidly thawing frozen prebaked bread and rolls and unbaked dough pieces. Its most important commercial application has been in the final stage of baking soda crackers where it is used to reduce moisture content without causing additional browning. The installations in cracker and cookie plants have been mostly the dielectric heating type because their economics appear to be considerably better than radar ovens at high heat transfer rates. Some commercial microwave baking units have apparently been used for pizza crusts and similar products.

Many inventors have proposed combining conventional baking, or irradiating with infrared waves, with microwave heating to give products with a browned, firm crust and uniform textured interior. Rosenberg and Bogl (1987) reviewed many of the recent publications on combination baking procedures. Conventional and microwave heating can be done simultaneously, as has been demonstrated by a number of workers.

Use of electronic heating during the early stages of the baking process seems to offer few advantages over more conventional heating techniques. In the first, or development stage, of baking, the dough piece can be rapidly heated on the band, unless it is large and foamy like an angel food cake. In the second stage, when most of the spring occurs, the application of electronic heat can lead to an excessively vigorous evolution of gas, with some distortion of structure. It is in the final stage, where a delicate balance between coloration and moisture reduction must be achieved, that electronic heating is most useful. Moisture can be reduced with little or no browning because the localized dehydration with rapid temperature rise which occurs during conductive, convective, or infrared heating does not occur. Regions with the most moisture are heated the most, and the heat input drops as the moisture content approaches zero.

This relationship of the moisture content of the piece to energy input is a very important feature of electronic ovens. The wetter the dough, the more heat is generated in it. Because of this relationship, the ovens are self-compensating for variations in moisture content. When the final moisture content of the baked product is low (5% or less), as is the case with cookies and crackers, the moisture content in different parts of the product tends to be evened out by this effect. Not only can excess or uneven browning be alleviated, but checking can also be reduced. If the microwave heating is continued after the moisture content has been brought to a very low level, however, some browning may be induced in the interior of the product. This is sometimes desired in baking certain kinds of cookies and crackers, and might be advantageous in toasting croutons and the like.

As stated previously, microwave heating can be applied at the same time as conventional heating. Pei (1982) describes a combination system in which simultaneous microwave and convection baking is applied to bread, with good results. According to several published reports, however, installation of electronic heating in existing ovens, though possible, has many disadvantages. Heat tends to break down insulation and the accumulation of deposits on the conductors and electrodes causes current leakage. Furthermore, any uneveness in the band, which must form part of the electronic system, causes nonuniformity in the current flow. The solution to these problems has been to place the electronic heating unit just beyond the terminus of the band oven, using a plastic conveyor or belt.

Although numerous investigations into the use of microwave radiation in cooking bakery products can be found in the literature, there seem to be possibilities which have not yet been investigated. Microwave heating in the final stages of cooking two-crust pies might avoid some of the excessive browning that is observed when it is necessary to completely cook raw fruit in the filling. Also, a better cook of custard-type fillings might be obtained by an early or late application of microwaves.

ENERGY SYSTEMS IN CONVENTIONAL OVENS

The energy systems in conventional ovens are generally considered to be comprised of the fuels, the devices which turn those fuels into heat, the control apparatus, and safety mechanisms. Some of these topics will be discussed in the sections dealing with specific oven types, but the general topics of fuels and control systems will be covered below. The effects and control methods of steam introduction into the baking chamber will also be considered in this section.

Fuels

The fuel crises which have occurred in recent years, sometimes taking the form of a shortage in natural gas, a shortage in some kinds of fuel oil, high prices for electricity, etc., have made it difficult to predict which fuel will be the most economical over the long term. Some bakers have opted for ovens which can be quickly converted from gas to fuel oil and vice versa. It is probable, however, that most of the large ovens in the United States burn natural gas. Oil of different grades, manufactured gas, propane-butane, and electricity are other energy sources being used become of some special situation or requirement.

Electricity has many advantages as an energy source, some of them being cleanliness, low maintenance requirements, and ease of control found in ovens using it, but the cost of electricity is prohibitive in all but a few locations. There may be some ovens burning coal or coke still being used, though the author knows of none in this country. They have so many disadvantages as compared to natural gas that they can be regarded as being of no current interest to the bakery technologist. Average theoretical Btu availability per unit for the various fuels has been reported as: electricity 3,142 per KWH, natural gas 1,000 per cu ft, manufactured gas 550 per cu ft, light fuel oil 140,000 per gal, and coal 13,500 per lb.

The average one pound loaf of bread will require from 150 to 200 Btu for baking to completion. Additional heat will be required for the pans, and some will be lost from the oven to its surroundings, so a considerably larger amount of fuel will be required than is indicated by the preceding figures. Heat taken up by the pans will obviously vary depending on their weight and composition, as well as their initial temperature, but an average pan requirement of 40 Btu per lb of dough is often used in calculations of oven capacities.

Efficiency of the oven in utilizing fuel depends upon the details of construction, such as type and thickness of insulation, manner of transferring the heat, etc. Skarin (1964) estimated that 400 Btu of gas were consumed per pound of bread in direct-fired ovens equipped with air-agitating systems. With gas burning at 85% efficiency, 340 Btu are delivered into the oven (from the original 400), and of these about 50 Btu are lost rhough the walls or by other routes. Indirect-fired ovens require an additional 20% fuel because of the loss of efficiency in supplying heat to an enclosed system. For oil, about 80% of the theoretical heat content can be recovered in an efficient combustion system.

Control Systems

On-and-off and modulating control systems can be used for controlling the input of heat to the oven. On-and-off controls consist of a thermocouple which activates a thermostat switch; they can maintain the temperature within a range of about 10°F. The more expensive modulating-thermal-control systems can adjust the amount of gas flowing to the burner as required to maintain the temperature within plus or minus 2°F.

There are three types of controllers used for ovens and driers: pneumatic, electric, and electronic. Pneumatic controllers receive a mechanical incoming signal and use a relatively complicated air-pressure balancing system to produce a pneumatic force that operates the heater valves. The sensing element can be a liquid-filled bulb connected by a capillary tube to the control instrument. A system of levers multiplies the slight motion of the liquid to an extent sufficient to activate the balancing section. An electric controller also receives a mechanical incoming signal, and uses a simple electric resistance slide wire balace to generate the electrical impulse that operates the gas valves. The electronic controller receives an incoming electric-voltage signal from a thermocouple and processes this through an electronic circuit to modulate the electric current that operates the valves. Most, if not all, modern equipment is based on electronic control.

Steam

Steam is introduced into ovens primarily to modify crust characteristics. It tends to reduce the rate at which the dough surface dehydrates. Consequently, the crust remains elastic for a longer time, preferably during the entire period of dough expansion, and ragged breaks are avoided. The delay in crust stiffening may allow the development of greater volume. The crust becomes smoother leading to a glossier appearance. Browning reactions are also modified so as to give better crust colors, in most cases.

In order to achieve the desired improvements, it is necessary that moisture condense on the dough surface. Injection of high pressure steam is worthless for the purpose. Authorities agree that saturated steam of 2 to 5 psig should be injected into the chamber with orifice velocities of 200 to 500 ft per min. Violent turbulence is not desired, since it would tend to reduce the opportunities for moisture vapor to condense.

As the crust temperature approaches 212°F, steam has little effect, since condensation will not occur then. The major effect occurs in the first 1 to 2 min in the oven, as the temperature of the crust is rising from about 90°F to near the boiling point of water.

According to Dersch (1960), the steam in a typical oven comes from the following sources: 2.7 cu ft per lb of dough from vaporization of water in the dough, 0.9 cu ft per lb of dough from products of combustion, and 9.25 cu ft per lb of dough from the steam injectors. It is obvious that far more water vapor originates from steam injection than from the other two sources.

RETAILER OVENS

Retailer ovens, so far as that description is used in the present chapter, are nearly all batch type ovens such as deck ovens, reel ovens, rack ovens, and rotating hearth ovens. There are a few models of small continuous band ovens suitable for baking at individual outlets, primarily at pizzerias.

Deck Ovens

Deck ovens are often found in small retail bakery operations, and at one time they were practically the only type of oven used in pizzerias.

They are not particularly energy efficient and occupy a lot of space per unit of product baked. Bakers find them difficult to charge and empty, and the first row in, last row out feature of the larger models is not always convenient although this is not a factor in the smallest models where only one row (or even one pan) is charged. Since there is no transport mechanism, these ovens tend to be cheaper and considerably easier to maintain than reel or rotary ovens. Design of the baking chamber of deck ovens owes a great deal to the old coal fired ovens though most modern deck ovens are heated by electricity or gas. The relatively small height (baking chambers range from about 7 to about 16 inches) is not suitable for some products, but is, of course, no problem with pizzas, rolls, most hearth breads, etc. Space utilization can be improved by stacking these ovens, but the lowest and highest decks become even more awkward to load under these circumstances. When the advantages and disadvantages of these ovens are examined, it becomes clear they are not suitable for any kind of a quantity production operation. They still have a place in the pizzeria and in other restaurants, and in small specialty bake shops.

Reel Ovens

A reel oven consists of an approximately cubic compartment six or seven feet high. This insulated enclosure has a door in front almost the width of the oven but usually less than a foot high. Inside the chamber is a ferris wheel type of mechanism which moves 4 to 8 shelves in a circle centered on the sides of the oven, so that each shelf is brought by the door during every rotation. An indicator on the front of the oven indicates which shelf is passing the door at a given time, so that the operator can stop the rotation and add pans to, or remove them from, specific shelves. Doors either open outward on a bottom hinge or slide upward in a counterbalanced arrangement. Ovens specialized in size and other construction features for cookies, rolls, pies, etc., are available. The drawing reproduced here as Figure 9.3 shows the most important features of a typical reel oven.

These ovens are usually heated by gas or electricity. In some cases, oil is used. With gas or electricity, the heating means are placed on the floor of the oven. Gas ovens usually have a baffle placed above the burner to change part of the flame's energy to radiant heat. Indirect fired ovens, generally oil heated, burn the fuel in a separate chamber located beneath the baking chamber, and the hot vapors are conducted through radiators in the oven floor before exiting up the back and out the flue.

Steam can be applied in these ovens. Perforated steam sprays effec-

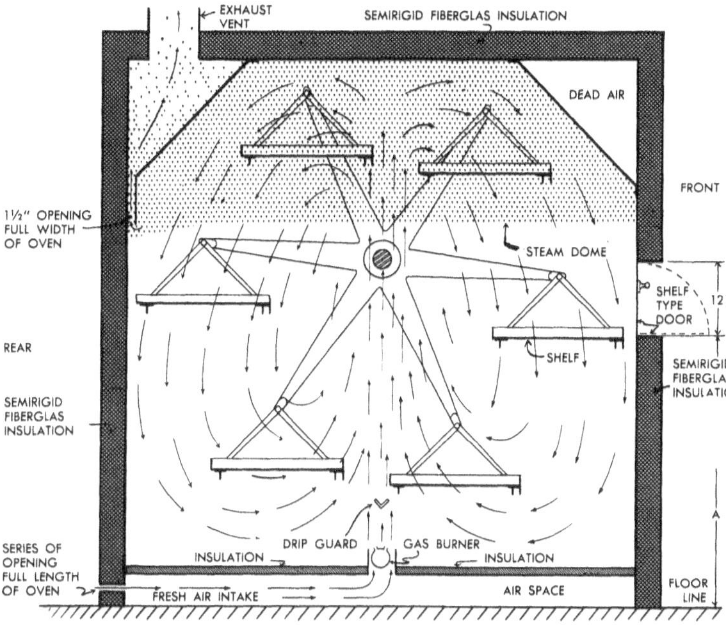

FIG. 9.3. MECHANISM AND CONSTRUCTION OF A REEL OVEN.
SOURCE: DESPATCH OVEN CO.

tive in the entire chamber or in a partitioned zone are common accessories.

There is a timer attached to the front of the oven. That is generally all it is attached to. It indicates baking time and signals completion but does not generally control heating time. Thermostats responsive to 3°F changes are the usual temperature control measure. This does not mean that products will encounter temperatures differing by only three degrees. There are side-to-side variations in temperature and top to bottom variations. The vertical differences are canceled out for all practical purposes by the rotation of the reel, but the horizontal differences are not, and bakers often find there is a significant difference in the baking response of doughs placed on the left side and the right side of the shelf. Some fluctuations occur when the door is opened, allowing hot air to exit and cold air to enter in a draft which can affect sensitive products. In spite of these negative features, the reel oven has been found to be satisfactory for baking many kinds of products. It is probably the best choice for retail bakers making several hundred pounds to a few thousand pounds of several varieties of product each day.

There must be at least ten U. S. and Canadian manufacturers producing reel ovens, and considerably more in Europe and Asia. A wide range of sizes, features, and quality is available. Not all of these ovens are used for bakery products, of course. They have a wide range of applications in restaurants, institutions, small food factories, and the like for roasting meats, preparing trays of food, etc. The baker would do well to investigate the reputation of the supplier as well as the design of the equipment before contracting for one of these ovens. Also, it is important to get a size sufficiently large to take care of all forseeable needs—it is not possible to expand the capacity of these ovens by any practical means.

Rack Ovens

Rack ovens consist of a chamber into which are rolled, or otherwise transported, racks of shelves containing pans of dough products. The racks may remain stationary or be rotated while the baking is taking place. Some kind of forced convection arrangement is nearly always present in these ovens. Figure 9.4 is a diagram of a retail bakery based on a rack oven.

The following description of a commercially available oven, taken from the manufacturer's literature (Anon. 1986), gives a good explanation of the operation of a typical rack oven.

From initial loading, fermented product can stay on the same rack through proofing, baking, and cooling without handling at each stage. A two rack oven can be loaded easily in under 30 secs, or an eight rack completely unloaded and reloaded in less than three minutes. The floor space is a fraction of that required for an equal capacity tunnel oven. A two rack requires one-third the area of an equivalent reel oven or 45% of that required by a double deck oven of the same output.

The high efficiency heat exchanger and insulation coupled with cross flow forced air circulation reduce running cost. A two rack oven producing mixed types of product averages one gallon of oil used per hour. The soft turbulent air heating system creates a faster heat transfer to the products, permitting a reduction in baking time or baking temperature. The capacity of the burner and heat exchanger is sufficient to permit continuous baking without having to wait for oven recovery between bakes. When the oven door is opened, the rotating turntable brings each rack to the loading position and stops automatically. When a rack has been loaded or unloaded, a simple operation of the indexing button brings the next rack into position. On completion of the loading, the oven door is closed and the automatic control system starts the

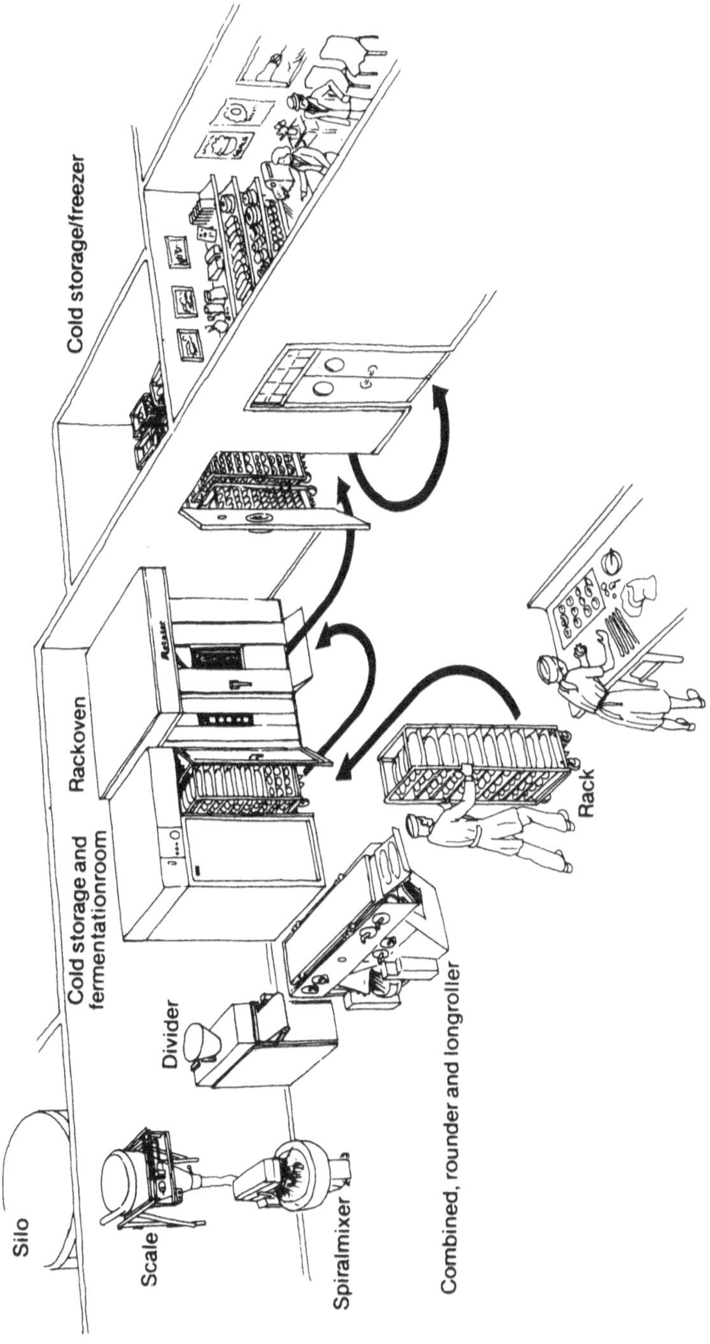

FIG. 9.4. RETAIL BAKERY USING RACK PROOFING AND RACK OVEN.
SOURCE: SWEDISH BAKING OVEN FACTORY LTD.

turntable rotation, air recirculation and temperature control. Baking is then carried out by a horizontal forced air flow between the baking pans or trays, giving excellent bottom heat condition in addition to an evenly fired top coloring. The operator can manually switch off separately the burner, rotation, and air flow, if necessary. Air circulation is automatically turned off when the oven door is opened, and, as the fan drops in speed, air is drawn from the opening door into the heat exchanger, producing little or no spill out of hot air into the bakery. The heat exchanger and residual heat in the oven have sufficient capacity to retain the oven temperature during the loading and unloading period during continuous production.

The oven is fitted with a high humidity, self generating, steaming system suitable for the production of crusty bread. A damper controls the evacuation of steam from the baking chamber.

Maintenance is minimized by having only one bearing in a warm zone. All others are on top in ambient air temperature and lubricated from one point.

Other Types of Ovens

Forced convection can be applied to almost any kind of oven. The units most often called forced convection ovens are shelf ovens which are supplied with high capacity fans moving strong currents of hot air through the chamber. Several manufacturers, in the U. S. and abroad, supply this equipment in a broad range of sizes and features (Lohrer 1987). Figure 9.5 shows the fan and associated details of a small convection oven.

Some of the specifications for a 28 tray forced convection oven, nonrotary carriage, are given here (Pamart 1987). The 28 trays are transported on a carriage. The oven is furnished with two of these carriages so that one can be loaded or unloaded while the other is baking. The surface offered by the 28 trays is about 80 sq ft. A window in the door of the oven allows constant inspection of the interior.

Two reversible centrifugal turbines driven by 2 hp motors provide the convected air. An automatic programmer/timer can be used to reverse the operating direction of the turbines so that the uniformity of heat distribution can be increased even more. When the door is opened, heat and air are turned off.

Oven temperature can be built up at the rate of about 20°F per minute. Either gas or electric heating can be specified. Accessory options include oven humidifier, iodine vapor lamp, low voltage control, and stainless steel carriage.

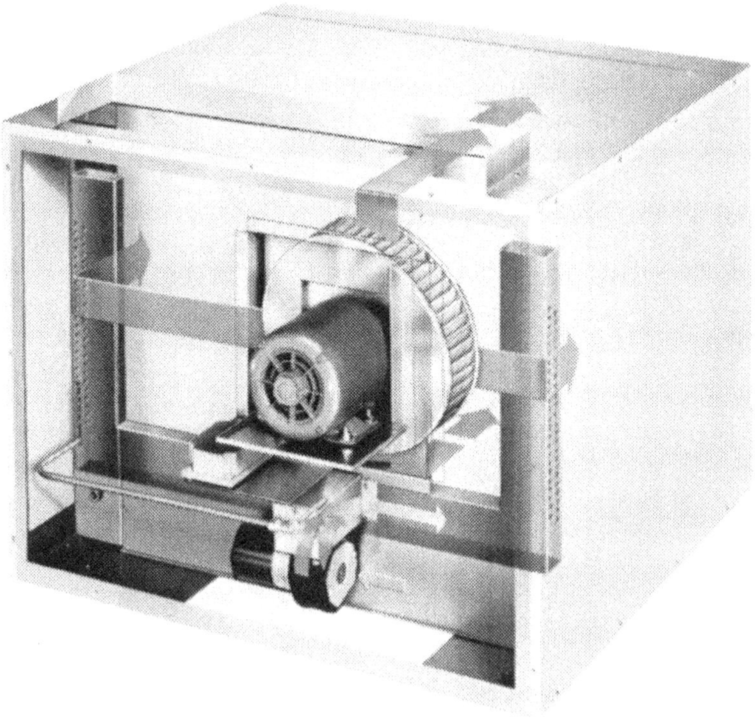

FIG. 9.5. THE AIR MOVING SYSTEM IN A FORCED CONVECTION OVEN, RETAIL SIZE.
SOURCE: HOBART CO.

WHOLESALER OVENS

According to Lugar (1962), manufacturers designing and building good baking ovens must take into consideration the following basic essentials: (1) Application of heat—in what manner is the heat to be applied; (2) Method of control of the required heat; (3) Best manner of insulating the heat for best efficiency; (4) Flexibility of the oven to bake a variety of products; (5) Cost of construction; (6) Overall ease of operation and maintenance of the baking unit; and (7) Ability to bake the most desired products in the most efficient manner. As vice president of Thomas L. Green Co., Mr. Lugar was at the time writing mostly about biscuit ovens, but his words were applicable (and are still applicable) to the design of ovens for all kinds of bakery products. Some of these general topics will be addressed in the following paragraphs before specific types of ovens are discussed.

Heating the Oven

The two main types of heating systems for modern conventional large-scale ovens are ribbon burners in the baking chamber itself and combustion chambers located outside the oven proper. Ribbon burners are placed approximately 2 ft apart, while the recirculating ovens have a heating unit about every 50 ft. There has been a great deal of controversy about the relative desirability of these two systems, but it is well known that both types are being used satisfactorily. Open flame or surface combustion (ceramic) elements have been used, but the former type are more common. Figure 9.6 illustrates some of the possible arrangements of burners and and convection means.

Several burner designs have been developed to obtain maximum safety and control of the heat sources within the oven. The open inspirator system uses gas at 1 to 5 lb presure and mixes it with room air drawn into the inspirator by the Venturi effect. Good combustion is obtained at or near the maximum output, but at lower settings adequate air for complete combustion may not be entrained. The question arises as to the fate of solid particles drawn in with the room air. Smaller particles will undoubtedly pass through the jets and be burned without causing any problem. Large particles which might clog the burner can be entrapped by filters fitted over the intakes to the inspirator. Even so, some accumulation of deposits inside the burner occurs, and these systems are designed for easy and fool-proof cleaning of the inspirator and other critical parts.

The zone proportional mixer system uses air pressurized to approximately 1 psig to draw gas at substantially atmospheric pressure into a proportional mixing unit. This gas is depressurized to zero gauge by a special valving arrangement. The combustible mixture is distributed to several burners, usually to a complete zone. Control is effected by changing the the volume of air going into the proportional mixer. Any change which is made affects all of the burners in the zone, making this system rather inflexible. The air is filtered before being pressurized.

A modification of the zone proportional system mixes the air and gas in proper ratio at each burner. It is necessary to lead a zero pressure gas line and low pressure air line to each burner and to have a separate proportional mixer and gas valve for each burner. This system provides excellent flexibiliity although it may be questioned as to how often the high level of control achievable by it is actually necessary.

Gas and air for all the burners in an oven are premixed in a special device in the premixed gas system. The disadvantage of this method is the presence of relatively large quantities of an explosive mixture

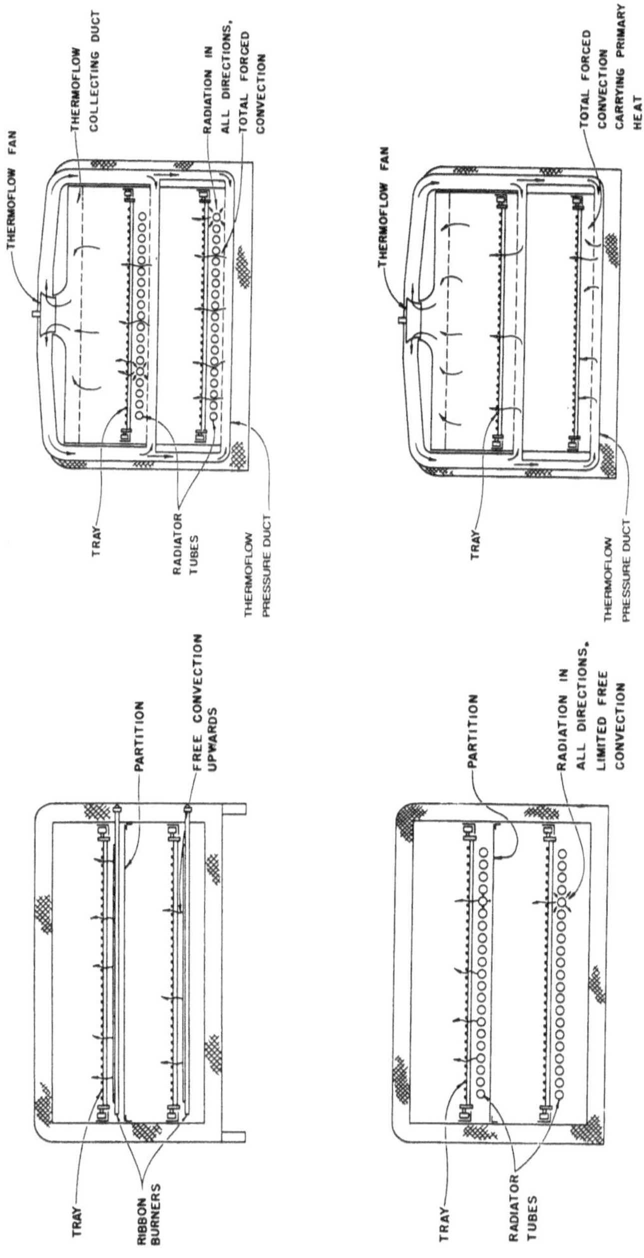

FIG. 9.6. SOURCES OF RADIATION AND CONVECTION IN FOUR TYPES OF TRAVELING HEARTH OVENS.

SOURCE: J. A. DERSCH

throughout the distribution system. Each burner must be fitted with a fire check to prevent backfiring. The system is flexible because each burner can be adjusted separately, but because of the potential dangers is not much used. Some ribbon burners can be adjusted so as to give different flame height in three sections across the band.

In the exterior-combustion chamber type of oven, several arrangements are possible. Generally, there is one combustion chamber per zone, which restricts the number of adjustments which can be made. On the other hand, relatively close control of the temperature of the circulating gases can be achieved. The type of burner is used in the exterior combustion chamber is more-or-less immaterial so far as effect on the dough piece is concerned.

Steam can be used to transfer heat from the combustion chamber to the baking tunnel. In one steam system, metal tubes sealed at both ends and partially filled with water or some other suitable liquid extend from the fire box into the oven. The flames vaporize the liquid, and the turbulent gases under high pressure carry heat throughout the length of the tube. Since heat is transferred to the product mainly by radiation, very high temperatures must be reached by the tube before satisfactory baking conditions are attained. This system was successfully applied in commercial operations but is no longer used to any great extent in bakery ovens.

Zone Control

Although sequential application of different temperatures might be applied in convection ovens without moving the dough piece, and is even done sometimes in home baking, it is difficult to implement in ovens where the product remains stationary. If the dough is carried through a tunnel which can be partitioned off into chambers, however, it is easy to apply different temperatures (as well as different top and bottom heating conditions) at different points in the baking process. For example, traveling hearth ovens have the advantage that baffles can be placed in the baking chamber to restrict heat circulation. If the heating elements can be adjusted separately in each of these zones, different temperatures can be applied to the product at different times in its baking cycle. This gives the baker more flexibility in adjusting conditions to the optimum for a given product. Generally, the burners can also be regulated separately for the top and bottom of each zone and the exhaust control dampers are separate for each zone. The use of zones is particularly helpful in baking crackers and cookies, and the following paragraphs will use these products as examples of the methods and benefits of zone control.

The number of zones in an oven will vary according to the specifications established by the engineer, who will take into consideration the needs of all the kinds of products the user expects to bake in the oven. Although many different brands of cookies may be baked, the number of types requiring different baking conditions will rarely exceed eight, and is usually less. In setting up baking conditions, it is often constructive to consider the oven as being divided into three sections, even though more than three zones are present.

The recommended temperature pattern for most wire-cut cookies is a fairly low temperature in zone 1, where the spread and rise of the dough takes place, a considerable increase in temperature in the intermediate zone(s) to set and bake the cookie, and a slightly lower temperature in the final zones to give the desired color and moisture content. Relatively low bottom heat, as compared to top heat, is used in the first zone to promote good spread and a finer, more uniform cell structure.

In baking saltines, the first zone (or section) of the oven is adjusted for high bottom heat and relatively low top heat. This arrangement leads to proper bottom development and starts the gas evolution needed for adequate spring, while at the same time maintaining a soft uncrusted top which will allow moisture and gases to escape. Shelliness and premature development of blisters result from excess top heat in the first zone. In the middle section, the top heat is brought up and the bottom heat is usually reduced somewhat. The top of the cracker is set, and the final shape of the biscuit develops, although moisture is still being driven off. In the third section, moisture is reduced to the point where dehydration can be satisfactorily completed during cooling, and coloration is begun.

Rotary molded goods and some rich wire-cut doughs require low heat in the first zone, slightly increased heat in the middle section, and the correct intensity of heat for bake-out and coloration in the final zone. This provides the gentle, slow development necessary for these doughs, and prevents excessive or premature rise. Higher bottom heat in the early zones tends to set the bottom early and may facilitate the removal of sticky doughs from the band.

Traveling Hearth Ovens

In tunnel or traveling hearth ovens, the baking hearth is made of steel segments which move through the baking chamber on conveying chains. Loading and discharging are at opposite ends of the tunnel, which is frequently a considerable advantage when positioning the auxiliary equipment. Straps, individual pans, or even unpanned dough pieces can be placed on the hearth. Because the baking surface is not divided into relatively small compartments, as in the traveling tray oven, it is more flexible in the size of pans it can accept and is often more efficient in capacity, especially where several sizes of straps are utilized. If pans are not at the loading area at the precise time a tray becomes available, the whole tray remains empty, but a traveling hearth oven can accept pans at any place on the hearth so that it is not necessary to leave empty spaces which are full multiples of the tray width. Thus, additional efficiency in hearth utilization may be obtained if timing happens to be somewhat irregular. See Figure 9.7 for an example of a traveling hearth oven.

This type of oven seems to have lost some of its popularity in recent years. It is generally regarded as being more expensive and more difficult to maintain than a band oven of equal capacity, and traveling tray ovens are somewhat more flexible and no more costly. Nonetheless, traveling hearth ovens do have a place in modern bakery practice.

Band ovens carry the continuous concept one step further and make the hearth one uninterruped strip of metal. The band may be solid, perforated, or woven. Doughs can be deposited directly on the band, as in the case of some sheet cakes and most crackers and cookies, or pans can be used.

Because band and traveling tray ovens are the most common baking units in commercial operations, they will be discussed in considerable detail in the following sections.

Traveling Tray Ovens

General characteristics—Because provision can be made for long horizontal runs, traveling tray ovens are more efficient than reel ovens in space utilization. Furthermore, they do not require high ceilings. Each tray holds several pans or straps and is permanently fixed to a conveying chain. Trays are pulled by the chain from front to back of the oven, then moved to a lower track so baking can continue on the return trip to the front, where they are unloaded. In the early designs, problems were encountered in obtaining a smooth, vibrationless transfer as the trays were moved from the upper to the lower track. The jarring and vibration sometimes observed were not particularly harmful to bread, but could cause damage to delicate or fluid products such as pies and sheet cakes. Various methods of stabilization were invented, among them the shoe system and the transfer arm. Figure 9.8 shows some of the tray stabilization devices.

Small size traveling tray ovens may be operated as batch or multicycle equipment. After the oven is loaded, the pans are permitted to make two or more round trips before they are unloaded. In longer

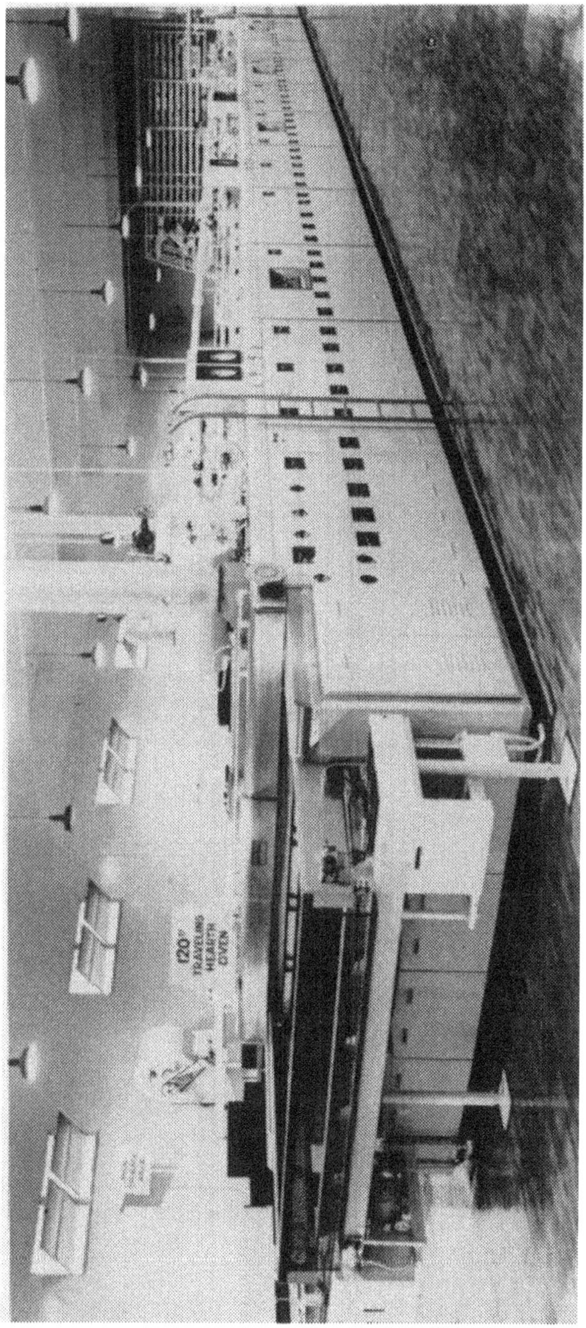

FIG. 9.7. A DIRECT GAS-FIRED TRAVELING HEARTH OVEN WITH AUTOMATIC SIDE LOADER AND AUTOMATIC UNLOADER. SOURCE: APV (BAKER PERKINS)

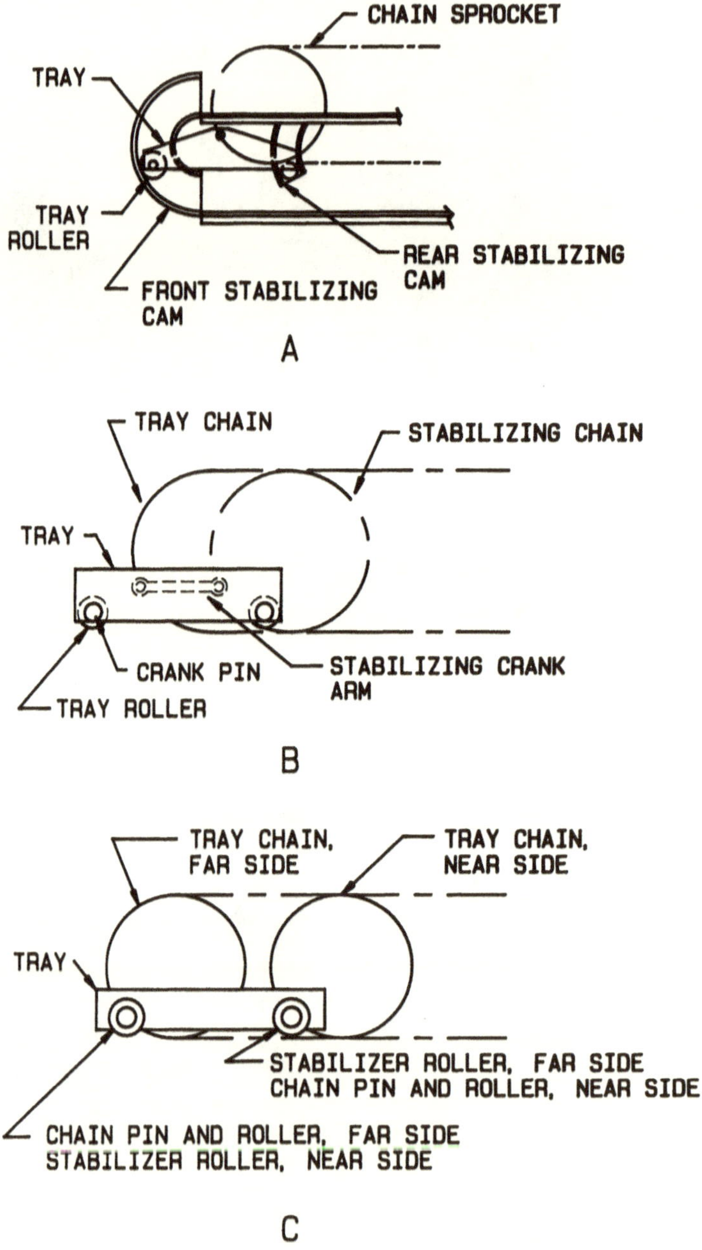

FIG. 9.8. Three methods of controlling the random movement of trays in traveling tray ovens.

ovens, however, the travel time required for one complete cycle is sufficient to complete the baking so the oven can be operated continuously, allowing automatic loaders and unloaders to be used.

By making the trays go through the oven two times in one baking cycle, approximately twice as much capacity can be obtained in the same floor space. Of course, the oven has to be much taller to contain the extra trays and machinery. An example of a double lap oven is shown in Figure 9.9.

Design features of traveling tray ovens—Construction details will be given for single-lap, direct gas-fired tray ovens for commercial pan bread.

The shell of the oven consists of a steel frame, resting on a level concrete foundation. This frame supports the steel lining sheets which form a rectangular baking chamber. Expansion joints allow the lining to expand and contract as the temperature changes without affecting the outer shell. A horizontal baffle divides the baking chamber into upper and lower compartments. The oven can also be divided into "zones" along its length, to allow application of a sequence of temperatures.

The top, bottom, and sides of the oven are insulated with about two inches of rock wool, and then plastered with fireproof cement to meet sanitary standards. The insulation is covered and concealed by the exterior finish sheets which are usually enameled white. Generally, the exterior sheeting is designed so as to be readily removable for inspection and servicing of the mechanical components located between the inner and outer walls.

Rows of direct gas-fired burners extend into the oven cavity below and above the tray conveyor. The burners are arranged in groups to form control zones for regulating the temperatures in different parts of the oven. Every zone has a separate air and gas supply, a modulating temperature controller, and a group of burners. To permit adjustment of the heat balance across the oven, some or all of the burners are constructed so as to allow variation in flame intensity along their length. Each burner has an inspirator for fuel mixing where gas at about 0 psig and air are combined. Ignition can be made continuous.

Indirect fired ovens have a more complex heating system. They will have one or more heating units located outside the chamber area. A heating unit will consist of a burner, combustion tunnel, heater body, radiator tubes, delivery and return ducts, circulating blower, exhaust stock with damper, controls, and safety devices.

Safety devices are of three types: (1) controls regulating the mechanical functions of the loader, unloader, and tray conveyor, (2) controls

FIG. 9.9. Double lap, indirect oil fired tray oven with tilt tray unloader and front feed loader.
Source: APV (Baker Perkins)

regulating the heat in the baking chamber, and (3) devices to protect against damage to the conveyor system, loader, and unloader and to prevent overheating of the oven and flues.

The conveyor drive is protected by three safety devices. If a pan or some other object jams the conveyor, it will either break a shear pin to disengage the main drive sprocket or activate an overload clutch which shuts off the motor. In the event these do not operate, the excessive current will cause the overload-protected magnetic starter to interrupt the electricity flowing to the motor.

There is ordinarily a safety bar mounted across the top of the oven loading door, and it will cause the conveyor to stop if it is contacted by a pan or lid overhanging the edge of the tray.

Pressure relief panels are held in place by thin strips or special bolts which release the panel when a sudden increase in pressure occurs. This arrangement helps to prevent serious damage to the oven if an explosion occurs.

Maintenance of tray ovens—Oven manufacturers can provide detailed oven maintenance service manuals for their machines and it is advisable to make several copies and distribute them to persons responsible for oven inspection, lubrication, cleaning, and repair. This manual should be studied by the plant engineer and every maintenance mechanic as an indispensable part of their job. Using the manufacturer's recommendations as guidelines, lubrication and inspection programs should be instituted.

The lubrication schedule should be based on a chart prepared by the plant engineer or supplied by the manufacturer. Every item to be lubricated, the frequency of lubrication, and the type of lubricants should be listed. A check-off system should be inaugurated so that the mechanic who does the lubrication can mark down the date each task is performed in a permanent record.

A chart showing the needed inspection of mechanical and electrical components should be made up by the plant engineer and used by the mechanic or electrician making the inspections and repairs. The manufacturer's suggested check list may be used or a special list more applicable to local requirements may be developed.

An adequate stock of repair parts should be kept on hand to reduce downtime, overtime labor, and inconvenience. Where disassembly and reassembly of a unit becomes time consuming, as in the case of clutch and brake assemblies for motors, it may be more economical to stock at least one complete replacement assembly. Further details on maintenance for tray ovens can be found in the article by Anderson (1970).

Band Ovens

General characteristics—The characteristic feature of band ovens is, of course, the continuous steel belt which forms the baking surface and which turns around two large metal drums, one at each end of the oven. There are many advantages to this form of construction. For example, the baking chamber can be very long (300 ft is a fairly common length) leading to fast transit times and high production rates. There is no need for elaborate conveying and transfer mechanisms with all of the maintenance and control problems inherent in such designs. In fact, the band can be extended beyond the oven entrance sufficiently to accommodate certain forming operations so that no dough piece transfers are necessary, and extended past the exit end to allow a modest amount of cooling and some additional drying so that the product does not have to be removed until it is well set, thus providing support for the piece during the whole period of its maximum fragility. Because there is no need for elaborate transfer mechanisms, the baking chamber itself can be restricted in volume to almost the minimum theoretical space required to hold the product and the hearth, so that heat control is simplified and heat loss is reduced.

Although bands can be used as hearths for baking most kinds of bakery products, and as supporting surfaces for pans, their greatest utility has been for cookies and crackers. They seem to be nearly ideal for these products and much of the information in the literature deals with the use of bands for biscuit baking. Consequently, most of the following discussion will deal with the experiences obtained and data published on this type of use.

According to Fischer (1977), the wholesale baker expects from a band oven: (1) Uniformly baked product; (2) Constant baking rate; (3) Flexibility; (4) High baking rate; (5) Efficient removal of gases; (6) Automatic operation; (7) Minimum heat loss into the bakery; (8) Maximum use of fuel; (9) Stable baking surface; (10) Safety for people and machinery; (11) Mechanical and electrical dependability; and (12) Sanitation compatibility.

The following description of a particular model is generally applicable to band ovens for cookie and cracker baking. In these ovens, raw dough pieces are dropped onto a flexible steel strip at the entrance and baked cookies are taken off this strip at the other end.

The original models consisted of substantial tunnels of brick through which moved a motor-driven traveling conveyor. Dough pieces were placed on top of the conveyor, as in later designs. Heat was supplied by banks of steam tubes above and below the baking conveyor, the steam being produced in furnaces located outside the ovens. These steam tube ovens were very satisfactory and were used for a long time.

Modern band ovens are outstanding triumphs of engineering and technology. With very little maintenance, an oven 300 ft long and 39 inches wide will routinely bake 2,000 to 3,000 lb of cookie dough per hr continuously for long periods. On certain doughs, much higher outputs can be obtained. Assuming proper adjustment, the millions of units put out in 24 hr of operation will be uniform enough to satisfy the most discriminating of consumers. The feeding and removal of dough pieces is routinely automated so as to require minimal attention by the operators.

In a survey of the types of continous band cookie ovens being used in commercial bakeries (Light 1965), the following information was elicited: (1) The ovens ranged from 100 to 400 ft in length, and the bands were from 26 to 60 inches in width; (2) Direct-fired, indirect-fired, and infrared means of heat generation were being used; (3) Energy sources were natural gas, manufactured gas, propane, oil, and electricity; (4)

Ovens had been manufactured by Baker Perkins, Thomas L. Green, J. W. Greer, Spooner, Vicars, Werner, etc.; (5) There were single-ribbon, double-ribbon, high intensity, and double-flame burners, with premix, inspirator, and aspirator means for combining air and fuel. Sometimes blowers and heat tubes were used, as well as various types of baffling, dampering, and exhaust systems; (6) The baking surfaces were solid or perforated continuous steel bands, and several types of wire mesh varying from very open (for some types of pretzels) through continental types, with smaller wire but still with open mesh, to heavy cord weave bands utilizing both straight-and crimp-wire designs; and (7) Basic adjustments which could be made included variable band speed, number of burners in top and bottom of each zone, pressure valves to regulate top and bottom burner flames separately in each zone, and dampers to remove or retain moisture in specific areas, vertical baffles between zones, and controllable exhaust systems. Of course, some changes in design and improvements in materials have taken place since this survey was taken, but many of these same ovens still exist and are working today.

Baking chamber construction—The oven chamber consists of a frame supporting the necessary rollers, guides, burners, and the like, together with insulation on top, sides, and bottom. Vents with dampers, clean-out doors, and observation ports are included as necessary.

Fiberglass wool is a common insulating material. When this is used, the oven is cased in enameled metal such as 10 gauge steel with a white vitreous enamel. A representative arrangement would be 10 inches of 3 lb density fiberglass on top, 8 inches on the sides, and 6 inches on the bottom. Sheet material can be used both as insulation and a structural component. In this case, the outer covering of metal is not necessary.

The baking chamber is usually manufactured in modular units, for example, 10 ft lengths. Ovens up to 400 ft are in operation, while ovens as short as 60 ft are used for special items.

Bands—Baking surfaces can be either solid steel bands, perforated steel bands, or wire mesh bands. The earlier mesh bands had a fairly open weave, but the more recent versions have had a finely textured, closely woven structure. Perforated bands and wire mesh allow steam to escape from the bottoms of dough pieces and help to prevent gas pockets and distortions. They also tend to reduce spread. The unbaked dough or batter is either deposited directly on the oven band, or formed on auxiliary equipment and then transferred to the band by conveyors. An oven with wire mesh band is shown in Figure 9.10.

According to a survey compiled by Suarez (1963), solid steel bands

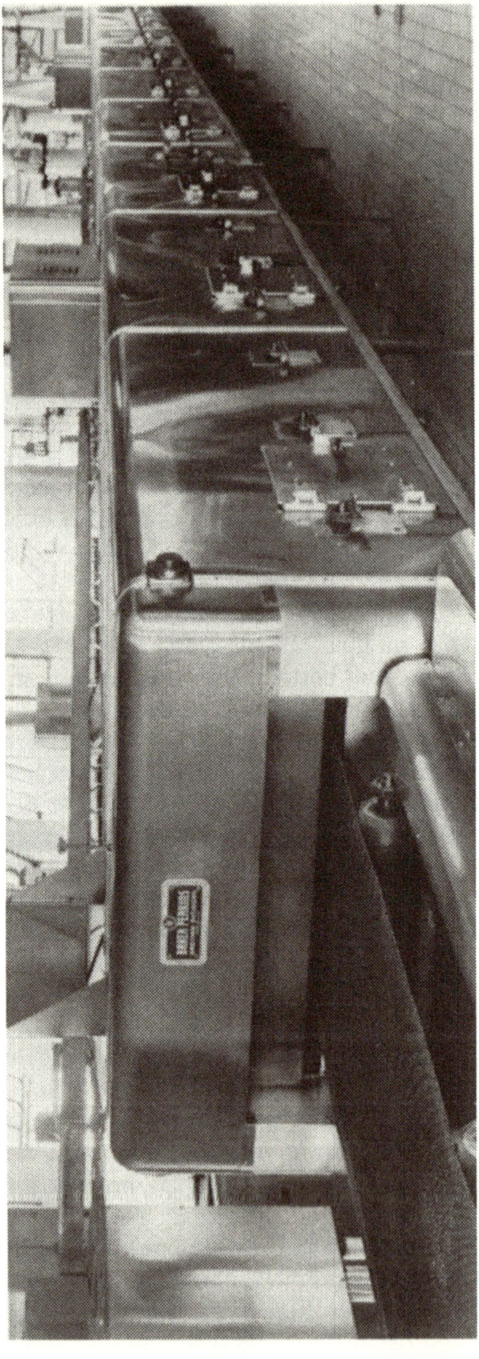

FIG. 9.10. BAND OVEN WITH WIRE MESH BELT.
SOURCE: APV (BAKER PERKINS)

then being used ranged in thickness from 0.032 to 0.092 inches, but most were 0.062 inches. Thinner bands allow use of smaller drive drums and assure rapid transmittance of heat. Wire mesh bands reported in this survey had an average thickness of 0.22 inch.

Most steel bands are low carbon alloy steels which have been cold rolled according to given specifications. The cold working process generally increases the yield strength and tensile strength and reduces ductility. The hardened steel is tempered to restore the ductility. The necessity for a straight, flat band is obvious. When properly made and maintained, a split band provides satisfactory service but patched bands are highly undesirable. Uniform thickness is also essential. The expansion and contraction of the band caused by temperature changes is compensated for by pneumatic cylinder adjustments at the axle of one of the drums around which the band turns, or by heavy counterweights on one of the drums.

On occasion, cookies will stick to the band. The presence of particles of sugar or milk solids in contact with the band will inevitably lead to sticking. Certain doughs are much more troublesome than others. Lean doughs which are relatively high in water are more prone to sticking. Sticking can also be accentuated or alleviated by band characteristics.

Dirty bands with built-up carbon deposits and bands with numerous pits and scratches are conducive to sticking. The band should be slightly oily from previous runs of rich formulas. If it is not, it may be necessary to grease it lightly or dust it with flour. Use of liquid shortening in the dough helps to eliminate sticking, though it may cause other difficulties.

New bands are usually conditioned by applying sufficient shortening to thoroughly saturate the surface. Sometimes beeswax or other special materials are used. The shortening is usually rubbed on at the inlet and the excess rubbed off at the exit end of an oven held at between 300° and 350°F. Temperature is then increased in 100°F steps until the highest operating temperature is reached. Lard, lard plus lecithin, corn oil, and coconut oil are some of the band conditioning materials which have been used.

Bands are powered by variable speed motors. Relatively low horsepower units are required. The time of baking is adjusted by varying the speed of the band.

Band guides—In long band ovens, even a slight amount of sideways movement will soon accumulate to the point where the band will be damaged and the oven jammed unless corrective measures are taken. This sideways movement is prevented by band guides which may be of several kinds. The simplest, and least satisfactory, consists of angle irons welded at strategic places along the band length. They will force a wandering band back into place, but they can also cause serious damage to the band in the process. Friction quickly wears out these simple guides.

Spring-mounted vertical spools fixed at numerous places along the band are reasonably satisfactory. They do cause edge wear, however. Canting the rollers supporting the band will guide it without edge wear, but this system requires close attention by the operator. Overcompensation or failure to correct the faulty condition soon enough can ruin the band and damage the oven. A combined system utilizing spools as a sensing device to warn the operator when to adjust the support rollers is probably the best compromise solution. A patented system may be even better; the spools themselves cant the rollers when the spools are driven off center by a wandering band. In Figure 9.11, the guides can be seen alongside the band just in front of the opening into the oven.

CONVEYORIZED PROOFING AND BAKING SYSTEMS

The basic principle of these systems is that the pans containing the dough pieces circulate continuously through the panning, proofing, baking, and depanning operations without the need for human intervention for handling or timing, for the most part. In some cases, the trays or racks may be subject to momentary delays for one reason or another, as while waiting for a space in the oven or proofer or while being transferred from one conveyor track to another. In the continuous system, however, the pans move from the panner through the proofer and the oven without stopping and while attached to the same "endless" conveyor. In conventional processing, the pans would be handled individually (of course, as straps in some cases) and would be stationary at some points in their progress through the stations. The newer method greatly reduces the complexity of the equipment needed to handle pans. To fully achieve the benefits of this concept, it was necessary to design new types of proofers and ovens.

The following discussion of plants based on the new types of proofing and baking systems is based in large part on a paper by Wells (1983) which dealt with both traveling tray and rack equipment as well as continuous conveyor methods. Additional material has been obtained from an article by Grogan (1980).

There are four sets of equipment to be considered in the traveling tray and rack systems. These are the conveyor grouping system prior to the proofer, the proofer, the conveyor system prior to the oven, and the

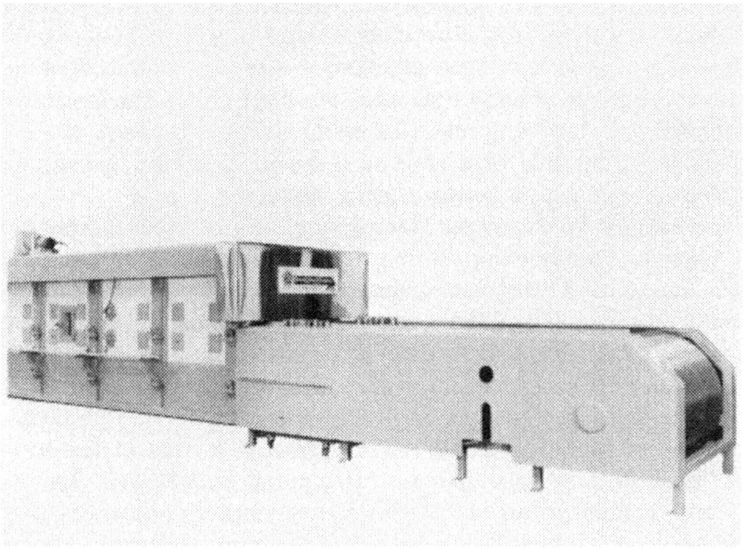

FIG. 9.11. AUTOMATIC BAND OVEN WITH BAND GUIDES VISIBLE AT ENTRANCE TO OVEN.
SOURCE: THOMAS L. GREEN & CO.

oven. A typical arrangement for a bread system of this type has a grouping conveyor for the proofer that is designed to separate a continuous flow of pans into groups of a predetermined size. There are two basic types of groupers, the full-complement style and the random-load style.

The full-complement grouper separates pans into groups which have approximately the same length as the size of the tray being fed, and allows space between pans in the group. If a full group is not available for loading when a tray or shelf of the proofer is in the loading position, then that tray or shelf will be allowed to circulate empty.

The two-pan stop, two-speed full-complement grouper is said to be the most reliable type of this equipment for handling bread and roll pans in a high speed operation. This grouper is equally reliable for feeding both proofers and ovens. Conveyor grouping speeds greater than 100 ft per min are not recommended since greater speeds could cause dough pieces to shift in the pans while grouping into a proofer or to collapse the dough when grouping pans into the oven.

It is essential to the traveling tray and rack equipment that sufficient conveyors be available to allow for the accumulation of pans. Provisions must be made to ensure that pans do not back up and cause unnecessary waiting of the production equipment. Back-up conveyors are usually located before the grouper. The required amount of back-up conveyor space is at least two and one-half times the length of the proofer tray or shelf. Any special feeding systems between the grouper and the proofer, such as a step or a feeder preceder operation, will decrease the amount of accumulating conveyor.

The second style of grouper, the random-load grouper, allows any size of pan group from one pan up to a full complement to be loaded into a proofer or cooler. This type of grouper should only be used as a proofer or cooler grouper. If used with the proofer, the pans must be regrouped into a full complement before the group is loaded into the oven. Random loading into the oven can upset the lateral heat distribution.

The two-pan stop grouper is probably the most reliable of the random-load groupers. The only critical setting is that of the back pan stop. This stop must be set to come up against the bottom of the last pan of a fully-formed group at a point approximately one inch from the leading side of the pan. From 5 to 15 ft of accumulator conveyor should be placed ahead of these groupers. Longer stretches will require an interlock to temporarily stop all conveyors between the grouper and the make-up equipment any time there is a back up of more than 12 ft of pans.

Load and discharge conveyors are mounted just outside the front end of the proofer. Each of these conveyors has its own drive and the space between them is equal to the shelf spacing of the racks. Racks are made out of aluminum structural frames, with each rack having stabilization rollers mounted on its sides. The removable grids, or shelves for the racks, are stainless steel. The rack elevator section at the front of the proofer consists of two vertical pawl assemblies, one set on each side of the machine. These movable pawls are moved by the main proofer drive and pick up the rack in a repeated step-by-step action until the rack has moved to the top of the elevator.

The lowering mechanism for the racks (called a "Lowerator" by the manufacturer) at the rear of the proofer has a single chain circuit on both sides. Each chain is equipped with attachment lugs and swivel links to move the racks from the upper to the lower horizontal runs. These chains are driven continuously from the main proofer drive. Rack loaders and unloaders operate only while the rack is temporarily halted in its upward travel through the elevator. The loaders are driven by separate motors energized through an electrical cam switch so that they function only while the rack is at rest.

The air conditioning unit is located on the outside of the proofing enclosure and has provisions for adjusting the humidity as well as the temperature of the proof room. The walls are made of panels two inches thick faced on the inside and outside with painted aluminum or stainless steel sheets and having foamed polyurethane cores.

Racks are periodically cleaned by a washing device consisting of spray headers which are placed at the rear of the machine about midway in the Lowerator section. Racks travel between the sprays during their descent.

After pans leave the proofer, they are taken by a conveyor to the oven. The three arrangements which have been used all transfer groups of pans in steps from one unit machine to another without the use of pan stops. The conveyors are designed to start and stop smoothly while holding the pans in separate groups spaced out between the unit machines. The system is said to smoothly and gently handle sensitive proofed doughs. In each of these conveyor systems, the number of pans loaded on each proofer shelf will be the same as the number of pans loaded on each oven tray. The first type is the single-line conveyor, consisting of a single row of conveyors which take a group of pans laterally from a proofer and then in stages to the adjoining piece of equipment. The second type is the direct transfer system, which transfers a tray or shelf load of pans directly as separate groups or rows in the lengthwise direction from the proofer to the oven. The third type is a combination of the other two systems. In it, the pans are first transferred lengthwise out of the proofer for two or three pan groups, then they are advanced laterally in a single row to the oven. It allows slower conveyor speeds along with adaptability to seeding, slitting, and pan lidding operations.

Regardless of the system being used, there must be a sufficient number of storage spaces to prevent holding up proofer discharge because of delay in loading trays into the oven. The minimum amount of accumulating time required to prevent halting of the proofer under normal conditions is equal to the sum of the fastest cycle time of the oven and the cycle time required by the oven loader to clear the loading area. Bread usually requires five stages or conveyor sections and a high speed roll operation usually requires six stages for accumulation. Stages consist of the proofer discharge conveyor, three or four intermediate steps, and the oven load conveyor.

A conveyor carries the pans through the oven, which may be of the traveling tray design (taking in and discharging at the same end) or the tunnel type (entrance at one end, exit at the other). Tray type ovens may consist of a baking chamber, structural framework, insulation, and outside finish sheets. The baking chamber is a box-type structure formed by steel lining sheets joined together by a bellows type expansion joint which allows complete internal expansion without movement of the oven as it is heated. The structural steel framework encloses the baking chamber and supports such mechanical components as main conveyor shafts, tracks, chain, trays, gas and air headers. The baking chamber is insulated on sides, top, and bottom and covered with a mastic-coated fiber glass insulation board.

Typically, the oven would be heated by a direct gas fired system which is divided into zones. Each zone would have independent air and gas supplies, a modulating temperature controller, and burners that have three lateral flame adjustments. A tubular air convection system functions throughout the baking chamber, except in the steaming zone, to provide air flow which helps control the top and sidewall bake. Dampers are present to change the amount of air going above and below the products.

Tunnel ovens for rack systems have three exhausters, one at the front door, another near the center of the oven, and a third at the rear door. The tray oven requires only two exhausters, one at the front and one at the rear. Loaders and unloaders are designed to be used with a continuously driven oven and are integral parts of the ovens.

Conveyorized proofer and oven systems represent an advance in automation beyond the systems described previously. For full utilization of the potential of these plants, it was necessary to redesign both the proofer and the oven. There are perhaps three major U. S. manufacturers offering these factories today. They seem to have found their greatest acceptance in high volume bread and roll plants, although they are no doubt adaptable to any bakery producing long runs of any item.

The conveyor systems also have unique features, which vary from manufacturer to manufacturer and are covered by some basic patents. One circuit of this "endless" conveyor goes through the proofer and one through the oven. The two basic types of conveyorized proofer and oven systems in use today are (1) the style in which the pans or straps are center mounted on the chain and (2) the style in which the pans are side mounted on the chain drive. The side mounted style is constructed of universal joints that are linked together with steel bands. Each universal joint supports four ballbearing wheels—two on each axis. Other conveyors which have been reported are rod type belts which will convey pans or peel boards, and another which uses triple strands of large link chain. The latter type appears to have been made and used primarily in Europe.

Connected to the drive chain are grid support struts that run free on an idler track constructed of angle iron. The carriage grids that fit into the support struts firmly hold the pans, which overhang the grids, and they are attached without fasteners onto the endless floor mounted conveyor chain. Grids are made of steel for the oven and stainless steel for the proofer. Advantages claimed for the side mounted chain, as compared to the center mounted, are a shorter turning radius, closer pan spacing, and a smaller module to house the conveyor.

Details follow for a side mounted version specifically designed for high volume roll production. The proofing module receives pans from a conveyor system interlocked with two make-up units. Pans enter the proofing module at a low level on the right side of the box, spiral upward to the top of the first loop, then over to the second loop where they spiral downward to a low exit point. The double spiral grid conveyor is supported by floor-mounted aluminum uprights and cross braces and runs on a steel track. Twin air conditioning units are located on the floor between the double spirals; they control the proofing temperature and humidity. All air-conditioning duct work is stainless steel and contained within the enclosure. The outer walls are prepainted aluminum with a foamed polyurethane core two inches thick.

Pans enter the oven module at a high level on one end of the box, and travel downward through a series of race tracks in a spiral to a low exit point. These units are fabricated of steel.

The oven is direct gas fired and uses ribbon burners that are located beneath the conveyor loops on each side. An air agitation system located at each end of the oven module distributes hot air to the ends of the oven. Hot air from near the top of the oven is directed through duct work and past a damper that vents a portion out the stack and recirculates the balance into the turns. This process is said to result in good heat distribution throughout the oven. A section of the oven can be partly walled off as a steam zone. The enclosure has an aluminized steel interior and a sheet steel exterior painted white, with an eight inch thickness of mineral wool between the two walls.

A synchronizing feeder is used by both the proofer and oven modules during loading. Pans are positioned width-wise on a conveyor leading to the feeder. This continuously moving synchronizer positions the pans so they can be loaded onto the holding grid. Proper spacing is accomplished through the use of springloaded synchronizing lugs. These traveling lugs hold the pans in position until the grid can move upward on the track and pick up a pan. Since the pans overhang the grid, a pair of stripper belts which overhang the grid can remove the pan at the end of the proofing or baking process.

In the proofer, the chain is hydraulically driven by multiple small load-sharing motors in a twin caterpillar chain drive arrangement. If excess pressure should be exerted on the chain, back pressure on the remotely located hydraulic power unit will automatically stop the drive motors and chain. Although either hydraulic or electric drive systems can be used, the former assures slow, easy starts and quick, smooth stops if the forward movement of the chain should be hampered. The chain in the oven is driven in the same manner, except the power unit is smaller and only one hydraulic motor, externally located, is present.

This author (Wells) says, early indications are the longevity of tray and rack equipment is greater than that of the continuous conveyor systems. Other factors to be considered in making a choice between these alternatives are the production rate (which appears to favor the continuous systems), floor space (greater density of product can be achieved in the tray and rack equipment), and flexibility (variety of products and pan size can be accommodated better by the tray and rack equipment). Maintenance and production people can more easily understand maintain continuous conveyor equipment because it has fewer mechanisms and a smaller number of moving parts. The open construction of continuous conveyor systems makes sanitation easier and faster.

Proofers for the rack or tray system use about half as much floor space as the continuous type, for equal production. Conveyor ovens probably take up less floor area than the others, but they occupy more cubic footage (Wells 1983).

BIBLIOGRAPHY

ANDERSON, R. C. 1970. Oven maintenance. Bakers Digest *44*, No. 8, 58-62, 76.

ANON. 1986. Rotorack Ovens. Andrew Denholm Ltd., Midlothian, Scotland.

DERSCH, J. A. 1958. Principles of heat distribution in baking ovens. Bakers Digest *32*, No. 5, 37-40.

DERSCH, J. A. 1960. Bakery ovens. *In* Bakery Technology and Engineering, S. A. Matz (Editor). AVI Publishing Co., NYC

DIXON, F. R., and CLEMENTS, R. G. 1963. Impovement in heat treatment of baking ovens. British Patent 924,071.

FISCHER, H. A. 1977. Efficient oven design and imperative for optimum fuel usage. Candy & Snack Industry *142*, No. 5, 17-18, 20-21.

GRISSINGER, G. R. 1973. Automated proofing and baking advances. Proc. Am. Soc. Bakery Engineers *1973*, 128-137.

GROGAN, P. E. 1980. Conveyorized proofing and baking systems. Proc. Am. Soc. Bakery Engineers *1980*, 113-118.

HOLLAND, J. H. 1962. High frequency baking. Biscuit Maker and Plant Baker, Aug. 1962, 698, 700, 702, 706, 708-709, 711-712.

HOLLAND, J. H. 1966. High frequency baking. Presentation at B&CMA Annual Meeting, Feb. 8, 1966.

JOHNSON, L. A., and HOOVER, W. J. 1977. Energy use in baking bread. Bakers Digest *51*, No. 5, 58-60, 62-64.

KAISER, V. A. 1974. Modeling and simulation of a multi-zone band oven. Food Technol. *28*, No. 12, 50-53.

KELLEY, H. N. 1969. A Brief Encyclopedia of Commercial Ovens. Middleby Marshall Oven Co., Morton Grove, IL
KOCH, A. 1983. Oven energy efficiency. Proc. Am. Soc. Bakery Engineers *1983*, 90-97.
LIGHT, H. J., JR. 1965. Baking principles by zone. Biscuit Bakers Institute 40th Annual Training Conf., Montreal.
LOHRER, W. F. 1987. Personal communication. Hobart Corp, Troy, OH
LUGAR, T. R. 1962. Economic Biscuit and Cracker Baking. Thomas L. Green & Co., Indianapolis, IN
MORETH, N. W. 1981. Optimal energy utilization in ovens. Bakers Digest *51*, No. 5, 26-31.
PAMART, P. 1987. Personal communication. Arpin, Gennevilliers, France
PEI, D. C. T. 1982. Microwave baking—new developments. Bakers Digest *56*, No. 1, 8, 10, 32.
PIRRIE, P. G. 1934. Use of steam in bakers' ovens. Am. Soc. Bakery Engineers Bull. *98*.
ROSENBERG, U., and BOGL, W. 1987. Microwave thawing, drying, and baking in the food industry. Food Technology *41*, No. 6, 85-91.
SEROTA, R. 1973. Heating with radio waves. Automation *20*, No. 9, 102-106.
SIEVERS, R. S. 1978. New baking methods. Proc. Am. Soc. Bakery Engineers *1978*, 98-105.
SKARIN, R. 1964. Selecting an oven for maximum performance. Proc. Am. Soc. Bakery Engineers *1964*, 88-93.
SMITH, D. P. 1984. Thermal treatment of food products. U. S. Patent 4,479,776.
SMITH, W. H. 1966A. Steam, humidity, and damper control in bakery ovens. Biscuit Maker Plant Baker *17*, 730-732.
SMITH, W. H. 1966B. What happens in the baking oven? Biscuit Maker Plant Baker *17*, 652-656.
STANDING, C. N. 1974. Individual heat transfer modes in band oven biscuit baking. J. Food Sci. *39*, 267-271.
SUAREZ, P. E. 1963. Oven bands—their use and maintenance. Biscuit Cracker Baker *52*, No. 9, 74, 76.
UNKLESBAY, K. B., and UNKLESBAY, N. F. 1985. Effect of dough temperature and infrared radiation on crust color of pizzas. J. Foodservice Systems *3*, 243-249.
VARILEK, P., and WALKER, C. E. 1983. Baking and ovens—history of heat technology. Bakers Digest *57*, No. 5, 52-54, 56-57, 59; No. 6, 24, 26-27.
WELLS, R. A. 1983. Proofing and baking systems. Proc. Am. Soc. Bakery Engineers *1983*, 119-124.
WHITESIDE, R. L. 1982. Energy use in the baking industry. Bakers Digest *56*, No. 4, 30-34.
YONEZAWA, M. 1986. Apparatus for making bread. U. S. Patent 4,592,273.

CHAPTER TEN

PANS, PAN HANDLING EQUIPMENT, AND SLICERS

INTRODUCTION

This chapter is a heterogeneous mixture of subjects. The equipment discussed includes pans and the handling devices they encounter outside the oven, as well as devices affecting the product, such as depanners, slicers, and coolers. It does not include sprayers, enrobers, icing applicators, and other machines which apply something to the dough piece or baked product; these will be considered in chapter twelve. It does not include packaging materials and equipment; these will be covered in the fourth volume of this series.

PANS

There are so many different sizes and shapes of pans, not to mention strapping features, materials, covers, etc., that just a listing of the variables would exceed the space which can be allotted for this chapter. The permutations must run into the tens of thousands. This section will deal with general principles; a few specific examples will be given to clarify important points. Disposable foil, paper, or plastic baking pans will not be covered.

Baking pans can be considered as interacting with the oven and the dough during the baking process. It is generally recognized by both practical bakers and technologists that their conformation, construction, and surface characteristics (color, reflectivity, and adhesive qualities, for example) affect not only the finished product but also the efficiency of production.

The cost of pans for a bread bakery represents a considerable percentage of the total capital investment. About 6 to 8 times as many pans as are in the oven at any one time will be required. This allows two sets in the proof box, 3 to 5 sets in the cooling, greasing, and panning operations, and one set in the oven. When the need for this many sets for each size and type of bread is considered, it is obvious that substantial sums of money are involved. For this reason, careful evaluations of pan and pan set requirements must be made before a purchase order is signed. Thought should also be given to the possibility that a different production line, if that is in the works, may not be compatible with current or on-order pans.

Materials

The materials which have been used in baking pans include black iron (blued steel), tinplate, steel coated with aluminum alloy, aluminum, and type 3 stainless steel (Anon. 1973). Stainless steel has many advantages, but is expensive as a material and to fabricate; it also has relatively poor heat transfer. Black iron is cheap, strong, and has good heat response, but has corrosion and resistance problems. Tinplate has good fabricating properties and strength, is relatively cheap, and has adequate heat absorption, but has problems with high temperatures, corrosion, and wear. Aluminum itself is easy to fabricate, has excellent corrosion resistance, and good heat transfer properties, but is expensive and not very strong; also, it cannot be economically welded. Steel coated with aluminum alloy combines most of the advantages of the other materials and has few disadvantages.

About 20 years ago, practically the only sheet metal in wide use for pans was tinplate (either hot-dipped or electrolytic plated steel), but this material has lost ground to aluminum coated steel, or "aluminized steel". Tinplate for baking pans is low carbon steel coated with about 90 millionths of an inch of commercially pure tin. One difficulty with tin is that it has a melting point of about 450°F, well below some of the temperatures reached in bakeries today. Therefore, the tin coating may reach a temperature at which it melts and runs together to form drops and streaks on the pan surface. This leaves uncoated, or very thinly coated areas, which are highly susceptible to corrosion and rust, so that the pans are permanently damaged.

The perforations often made in the bottom of bread pans to allow escape of air and moisture often corrode progressively in tinplate pans. Eventually, the round holes develop an elongated space from the center toward the ends, causing imperfect loaves and reduced pan life. The rusted and pitted areas which ultimately form in all tinplate pans will cause discolorations on loaves.

Aluminum will not melt at any temperature encountered during bakery processing and aluminum coated steel has the metallurgical, physical, and electrochemical properties necessary to make it useful as the structural material for baking pans (Inzerillo 1979). These properties are described as:

(1) Workability—it can be satisfactorily folded, drawn, and formed into the many sizes and shapes of pans used for bakery products.

(2) Strength—aluminum coated steel of the same thickness as tinplated steel can withstand the impacts encountered in automated pan handling over extended periods without greater changes (than tinplate) in the original configuration of the pan. One manufacturer claims

their aluminized steel pans have a service life of 3,000 production cycles under normal conditions.

(3) Corrosion resistance—the aluminum coating is more than ten times the thickness of the tinplate used for the same purpose. Coating of this thickness is essentially free of voids and micro-pores, making it highly resistant to the penetration of moisture.

(4) Temperature resistance—aluminum coating is heat stable at 1,200°F, far above any temperatures required for baking, so that melting is never a problem.

(5) Heat absorption—the material can be processed to impart a permanent heat absorptive coating.

Aluminum contributes so many important advantages to baking utensils that it is unfortunate its disadvantages have not been overcome by pan manufacturers. Its structural weakness and unsuitability for welding are serious defects, but could perhaps be overcome by innovative design, less damaging handling procedures, and protective shrouding. Such improvements could enable the baker to take advantage of aluminum's perfect resistance to corrosion, excellent heat transfer properties, adaptability to deep drawing and forming into complex shapes, and relatively low density. Great quantities of baked foods are prepared in disposable foil pans for frozen distribution and other applications, but they require special handling techniques.

Surface Treatment

There was formerly a requirement for "burning in" new pans. Bakers knew when they received a shipment of pans it was necessary to modify their surfaces by a heat treatment before the pans would perform satisfactorily. Burning in was accomplished by heating for about 3 or 4 hours in an oven held at 400° to 420°F. This treatment produces a darker pan color which leads to a better coloration of the baked loaf, and it also produces an alloying or firmer bonding between the tin coating and the steel sheet. Manufacturers now will deliver pans having already oxidized tin surfaces or etched aluminum surfaces which do not have to be subjected to the burning-in treatment.

It is very common to apply silicone coatings to pans, especially tinplate bread pans, to improve the release of loaves. This reduces cripples and often enables increased speeds at the depanning station without the use of pan oil. The pans usually remain cleaner, and there are some other advantages. Many bakers use pan oil (but smaller amounts) with silicone treated pans.

Coatings are usually applied by firms which specialize in this service. Some professional applicators offer a program which includes monitoring pan condition and advising on re-coating as needed.

Sizes and Shapes

Bread—The reader does not have to be told that bread is baked in many different sizes and shapes of pans. Rectangular, square, round, and lidded are some of the common variations. And of course, pans for hearth breads are cleverly designed to give the impression the loaf was baked without any container.

For common white loaf bread baked in open top pans, 5.8 to 6 cubic inches of pan volume for each ounce of dough is considered satisfactory. Pullman breads will require six cubic inches per ounce for a closely grained loaf and seven cubic inches per ounce of dough for loaves of more expanded texture.

It is necessary for the top of the pan to be larger than the bottom. This flare facilitates detachment of the bread from the sides when the loaf is depanned and makes it possible to stack the pans.

Selecting the proper angle of flare is an important decision, especially when specifying deep containers such as bread loaf pans, because of the effect of flare on the ease of nesting the straps. Straps of pans can occupy a large amount of space in a factory, especially if several different loaf sizes are being made, some of which are infrequently scheduled. Furthermore, nesting and de-nesting are essential operations in some conveying and transport systems. With large flare, release problems are reduced and pan life is increased because dented pan walls will not come in contact with one another. These pans do have less stability when stacked, however. With decreasing flare, there is less clearance between nested pan surfaces.

For bread pans 2.75 or 3 inches deep, the bottom length should be 0.5 inch smaller than the top inside length. The bottom width should be 0.625 inch smaller than the top inside width. These are general rules; obviously, there are conditions which may make different specifications desirable.

There should also be a flare in pans for the ostensibly square pullman loaf—the loaf baked in a covered pan. If the top and bottom dimensions of the pan are the same, difficulties in dumping the bread will be encountered. Pans four inches or more deep, should be flared 0.125 inch per side and end, or a total of 0.25 inch. This provides ample clearance for release of the loaves and compensates for the usual shrinkage during cooling.

Construction and design of baking containers for hearth breads has proven to be a difficult task for pan designers. Of course, the term "hearth bread" implies that the loaf is baked on the oven surface, as these breads originally were, but most modern systems use some sort of pan, tray, screen, or sheet to hold the dough piece during its movement through the oven. For many years, pumpernickel and other rye breads were baked on perforated screens. Such containers were not very suitable for high speed production, so the perforated basket hearth pan was developed. These consist of an inner container with a rounded bottom having only a small flat area. These utensils will not run satisfactorily on conveyors, so an outer pan or cover is put around them. The bottom of the outer pan must be largely open so that the inner container is not shielded from heat. To offset the structural weakness caused by the open areas, the outer pan must be reinforced by adding a wire to the outer rim. This supports and strengthens the flanged edge. Panning bars are installed on the bottom to allow the indexing finger from the molder to slide under the pan without catching in a hole. Some of the outer pan constructions which have been used include full open bottom, flanged bottom platform, and corrugated sidewalls. In addition, other features include extended strapping, various sizes of holes in the inner as well as the outer pan, contoured wires to support the insert, strapping under the wire, welded pan rim corners, stitch welded pan rims, complete perforations in the insert, and many combinations of solid areas in the bottoms, sides, or ends (Inzerillo 1979).

Other types of hearth bread pans are combined in sets that are modifications of a regular pan bread set where the strapping provides some of the strength that is contributed by the outer container of the pans described above. The pan bottom is flat and there is a 0.625 inch radius along the bottom sidewall. The pans have wide flares to accommodate the contours of the hearth loaves. By eliminating the outer pan assembly, there are no perforations to catch in the conveyor or in the indexing finger of the molder-panner. Also, the pans are lighter in weight, nest better, and are more sanitary and cheaper than the double-walled hearth bread pan.

Rolls—For many years, pan designers were extremely limited by the pan forming methods they had available. Pan manufacturers relied mostly on two basic methods that have been used by tinsmiths for hundreds of years; these are cutting metal sheet and bending the pieces. No one seemed to have much interest in developing new pan manufacturing processes or new designs. Bakers were reconciled to using traditional type pans, because they worked reasonably well with their old fashioned production techniques. Furthermore, they were making only traditional products for which they already had pans, and were not inclined to make the major investments which would have been required for replacing them with a new type.

As higher speeds and greater mechanization were forced on bun bakers by the pressure of competition, however, it became obvious that pans were, in many cases, the limiting factor in capacity and rate. Additionally, new bakery products were being developed which could not be made in the old pans. A few manufacturers of pans met the challenge and made continual improvements, using new processing methods, new materials, and new designs. Some of their competitors could not adapt to the changing conditions and disappeared from the business world.

One of the first goals was to maximize the use of oven space by eliminating wasted space in the pan. Both in individual loaf pans and straps, as well as in bun pans, cupcake pans, etc., there were non-product-containing areas which contributed nothing to baking efficiency or product quality, but were included in the design because of tradition or because of the requirements of outmoded methods of pan manufacture. This wasted space in the pan meant fewer pieces of dough could be included in a given area of hearth and, consequently, the production rate was reduced by an equivalent amount. The amount of wasted space is greater in roll pans than in loaf straps.

More rolls can be contained in a given area if the cups are square in outline. Some pans for brown-and-serve rolls are made this way, but consumers appear to have reservations about the appearance of the product. It was discovered that, as an example, the circular cavities used for certain kinds of soft rolls could be placed closer together while maintaining the rounded configuration. Sometimes the margin, or frame area, around the pan could be reduced. The cumulative effect can amount to an additional 20 to 25% of product in the same overall area. When cups or molds in cupcake frames and roll pans are over 0.75 inch deep, they are seamed into the panel. By eliminating the seaming of the molds, pan manufacturers were able to save as much as 0.5 inch between every row of cups in both directions. Normal seamed-in construction has a minimum spacing between cups of 0.625 inch. According to Inzerillo (1979), a completely new concept and technique was developed to obtain the minimum spacing which was made available to industry. Instead of bending the metal for the individual cups and then seaming the cups in a sheet which had holes punched in it, the cups were individually drawn with a rim of 0.0625 inch and welded together. The result is cups that have 0.125 inch spacing between them. Depending on the number of cups in a pan, the savings in space allow 20 to over 30% more cups in the same size frame, or an increase of the same amount in oven area.

The reader might ask, why not use close-packed geometry for the round cavities. This should allow some additional space saving, although the loss at the edges would take some of this back, depending on the size of the tray and size of the top of the cavity. The chief difficulty seems to be that such offset rows cannot be handled in current styles of panning machines or vacuum depanners.

Most pans for hot dog buns and the like do not have seamed in cups. They are made with drawn sheet and require about 0.75 inch, or a little less, of separation between the depressions. The minimum separation needed depends on the configuration—whether the molds are round, square, oblong, or cluster shaped. Simply reducing the separation is not a feasible alternative in these cases. The solution was to bring the cavities closer together by squeezing the metal upward between the rows to form a ridge. The ridge can reduce the spacing from the 0.75 inch originally thought to be the minimum to as little as 0.125 inch. If it is decided to reduce the width of the pan, the ridge would run lengthwise of the pan, and vice versa. It does not seem to be practical at the present state of the technology of metal shaping to run ridges both ways.

Pan manufacturers have learned how to make deeper depressions in drawn metal pans. For many years, the maximum depth was considered to be about 0.375 inch. Then, advances in the art made mold depths of about 0.55 commercially feasible. Further improvements in technique and equipment are leading to even deeper draws. The extra depth gives greater support of the dough piece during proofing and baking, so that buns are more uniform in size and shape. Since the drawn pans are made in seamless panels, there are fewer crevices to impede release and trap dirt.

In designing pans where some knitting (flowing together at the edges) of the dough is expected and desired during proofing and/or baking, either because clusters are needed for the packing operation, or because the consumer expects joined rolls, this factor will have to be taken into consideration when calculating the separation allowable between dough pieces.

Interaction with Product

Interaction of the pans with the products occurs as a result of the guiding effect of the container on the expanding dough, the support given the baked piece as it sets up, the rate of heat transmission to the product, and the shading effect on product surfaces, among other mechanisms. Some of these effects are not easily modified, but dimensions (i.e., relation of volume to height and width) and slope of sides can frequently be specified within fairly wide limits. Once the pans have been purchased, however, the large investment tends to strongly restrict the options of the baker in future planning.

When baking bread, deeper pans result in better appearing break and shred. This part of the loaf is higher, there is less collapse of the break and shred, and the sidewalls are stronger. With greater flare, release problems are reduced. Greater flare does not have much effect on loaf size. According to Torrens (1964), corrugations give stronger loaf sidewalls and have no effect on oven spring or pan release. The chief purpose of corrugations is to increase strength of the pan.

Interaction with Equipment

In the early 1970's, a committee of the American Society of Bakery Engineers completed a two-year test of open top bread pans having various structural features. The purpose of the study was to determine how well the pans withstood the wear and tear of actual bakery operations. Only tinplate pans were included in this survey, but it is believed the conclusions can also be applied, in a general way, to pans made of aluminized steel. The variables under test were: (1) 4X (0.022 inch thick) and 22 gauge (0.0315 inch thick) tinplate; (2) No. 5 (0.207 inch diameter) and No. 9 (0.148 inch diameter) wire in pan rim; (3) Plain pan sides and bottoms, and 45°, 90°, and quilted corrugated pan sides and bottoms; (4) Plain bottom edges and 45° and 90° corrugated bottom edges; (5) Pan tinplate Plain (unwelded), corner welded, and continuous welded to pan rim wire, also anchored 4X tinplate under pan strapping; and (6) Tempered pan strapping vs. untempered pan strapping, intended to reduce depanner bar damage. The construction variables of the test pans are shown in Figure 10.1.

The 22 gauge tinplate pans had flatter bottoms than the thinner tinplate pans without corrugations after two years usage, but they were no flatter than the thinner pans which included either the 45° or 90° corrugated feature on the bottoms of the pans. The thicker pan developed a slight bow due to its greater weight and there was also some crushing of the strapping at the points where the pan fell onto the depanner bars of mechanical depanners. Some of the report writers described the heavy pan as "a self-destructing pan."

After use for one year, the pans with No. 5 wire showed less denting of the top rim than pans with the thinner wire. This type of damage is usually due to pressure on pan conveyors, especially at 90° turns. Pans having two years service did not show as much difference between variables. A disadvantage of the heavier wire was a greater tendency of the top corners to open up.

When 90° corrugated bottom edges were compared with larger radius bottom edges, the latter pans seemed to have greater loss of original bottom dimensions. Balled corners, a feature generally used to make the corners of baked bread somewhat rounder, reduced the tendency of pans to develop sharp corners on the bottom—this should

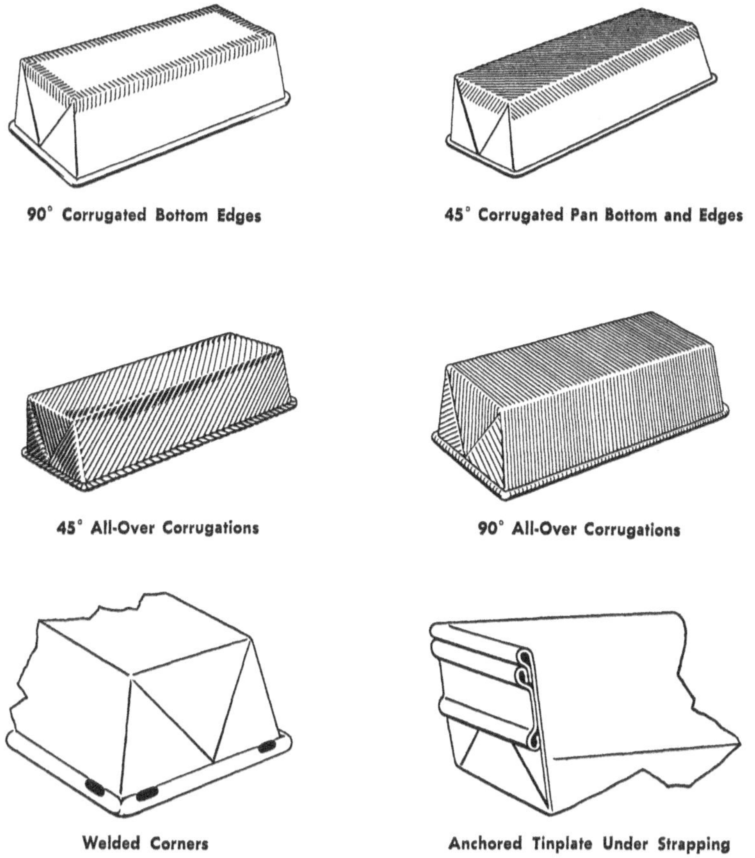

FIG. 10.1. CONSTRUCTION DETAILS OF THE PANS USED IN THE A. S. B. E. TEST DESCRIBED IN THE TEXT.

SOURCE: ANON. 1972.

increase pan life since sharp corners eventually lead to the development of holes at these points. It was noted that corrugation of some type, be it 45° or 90°, resulted in flatter bottoms on the test pans after extended use. In all cases, the 90° corrugated feature produced the flattest bottoms. Bottom corrugations of the quilted pattern type were about as effective as 45° corrugations. It was clear to the testing panel that any type of corrugation improved the sidewall strength as compared to plain sides. The 90° corrugation produced the strongest sidewall on the test pans; they showed less curving of inside pan walls such as results from manual stacking or mechanical depanning. The corrugating of bottom edges had effects similar to the sidewall corrugations. Welded corners were as effective as continuous welding in giving resistance to opening up of top corners, and anchoring of the 4X tinplate under the pan strapping was as effective as welded corners. Tempering of the strapping at the points where pan sets contacted depanner bars as they fell, was effective in reducing pan wear at these points.

The committee concluded that a combination of the following features would lead to optimum life of bread pans: (1) No. 5 wire in the rim; (2) 4X tinplate for the body of the pan; (3) 90° corrugated bottom edges; (4) Balled bottom corners; (5) 90° overall corrugations; (6) Welded corners or anchored 4X tinplate under the pan strapping; and (7) Tempered strapping if pans are to be used with automatic mechanical depanners.

When highly automated roll plants started to come into use, new demands were placed on pans. At the time this changeover began, single wire pans were common. These pans had a rounded top edge, usually formed by folding the rim of the metal around a thick wire. If these pans are handled in racks, or pass along a conveyor without contacting each other, they function satisfactorily. But, when these pans push against each other on conveyors, as they frequently do in automated systems, they tend to shingle, i.e, the leading edge of one pan slides over the rear edge of the pan in front of it. For example, pans may stop while conveyors continue to run as they assemble groups for entry into a proofer or oven. When this takes place, the single wire pans are subjected to forces which cause them to override one another, resulting in product damage or equipment damage.

An "anti-climb" feature was developed to eliminate this shingling. It consists of a second wire edge which was added below the first one, making a double tubular edge, the two wires being separated by a very short distance. The lower wire edge is commonly inset slightly from the upper one. Although this design did solve the shingling problem, it was expensive to make. Also, when a line of pans was under horizontal pressure in a jammed conveyor line, it was difficult to remove individual pans. A more recent development is the replacement of the two wires with a solid bar of approximately rectangular cross section around which the top rim of the tray is wrapped, giving a 0.5 inch flat vertical wall all around the pan. This type of construction allows a more uniform flow of metal around the rim core when the pan is made, and gives a smoother and tighter surface at the corners (Inzerillo 1979), so that the pan accumulates less soil and debris. The disadvantages are that the manufacturer has to have a separate tool for making each size of pan and the number and sizes of the molds are limited. In other words, the manufacturer has less flexibility in time and cost for making pans for specific requirements.

Another problem with the solid bar rim is that the rim absorbs heat during baking and proofing, and often cannot dissipate it before the pan returns to start another cycle. The rim doesn't cool down fast enough, so that pans reaching the panner are too hot. A solution was to make the rim hollow; it has a tubular cross-section rather than a solid wire center. Not as much heat is absorbed, and the rims have an opportunity to cool off before they start a new cycle. The pans retain the same outer configuration and practically the same strength, but they are lighter (Anon. 1974).

It is to be expected that pans will eventually wear out from the friction and damage that is encountered under normal conditions of use. If in some cases, it appears that excessive wear or damage is occurring, it may be possible to correct the problem by adjusting the equipment or relocating equipment. When this seems to be impractical, redesign of the pans can be considered. For example, it may be found that the guide rails continually abrade pan corners, eventually breaking through the metal. This leaves rough edges which can get caught in the machinery and torn away. Pieces of metal can fall into the product area or cause equipment damage. If heavy gauge strapping is extended over the wire edge, wear on the pan corners is essentially eliminated.

Straps and Other Multiples

Usually, two or more pans are held together in a set by metal strips for efficiency in handling and conveying. These pan sets, sometimes called "straps", serve to separate the pans an optimum distance. Spacer lugs are attached to the side and end bands to keep the sets at a proper distance from each other as they go into the oven.

A method for strengthening sets is the welding of the pan material to the wire rim, or bringing strapping over the wire. In most cases, the strapping is made of 18 gauge tin or aluminized steel. Torrens (1964) says one of the strongest sets available has a combination of 1.625 inch strapping over the wire and 1.375 inch strapping all around the bottoms and ends.

LIDDERS AND DELIDDERS

Lidders are used for pullman loaves (sandwich loaves) and a few other products. These machines fit into the production line between proofing and baking, automatically placing pullman pan covers on pans containing proofed dough. Pan covers, with open edge facing down, move into the machine on an accumulation conveyor at an upper level while the pans are moving along a conveyor at a lower level. The pan cover mechanism is synchronized to give the required rate of flow from the pan cover reservoir section. Covers move to a stop position, from which they are placed over the pans on a demand basis.

In older systems, the lids to bread pans could be removed manually or by magnetic appliances at the depanning station. The lids were then carried back to the lidding machine or to storage. Pullman covers should be either a stacking or nesting type for convenience in storing.

A recent development makes the lid an integral part of each tray in a tray oven. The operator has the option of running the oven "lids down" for pullman or sandwich bread or "lids up" for open top breads. This arrangement neatly eliminates the need for lidding, delidding, and storing and transferring lids, at some cost in complexity of oven design.

English muffins are also covered for part of their baking time with a lid which is a part of the oven or griddle mechanism.

PAN STACKERS, GROUPERS, AND CONVEYORS

In older systems, pan sets were taken off the depanning station by hand, placed on trucks which were pushed to the storage area by hand, and there removed and stacked by hand. When needed, they were removed from the stack by hand, placed on the truck, pushed or pulled to the panning machine area, and there transferred from the truck to the conveyor by hand. If forklifts were available, they could be used to transfer the trucks. This setup is obviously so slow and labor intensive that it had to be modified before any kind of automated system could be implemented.

Automatic stackers and unstackers were developed to ease the problem of pan handling. One system based on magnetic grasping of the pans functions at rates up to 40 per minute. The pan stacker receives pans from the conveyor system, stacks them to a predetermined height, and discharges the pan stack onto a storage conveyor or truck. The pan unstacker reverses this procedure by receiving pan stacks from a truck or storage system and feeding individual pans on demand into the conveyor system leading to the panning operation. When pan supply and demand are in balance, pans in the conveyor system flow through the stacker and unstacker without interruption.

The following description applies to one current type of pan handling system with automatic stackers. It is adaptable to handling both bread and roll pans. The heart of the system is comprised of a pan stacker, a pan unstacker, and a pan diverter. With its associated conveyors, it is designed to function with a minimum of operator involvement. A feature of the pan stacker is its continuous operation capability—it does not stop for discharge of a completed stack. Instead, the stacking operation continues at the top of the machine as the full stack at the bottom is automatically moved out. The new stack is then lowered into place to complete an uninterrupted stacking cycle.

This feature reduces the possibility of pan jams on conveyors, which can result from frequent starts and stops. The pan stacker is available with optional discharge conveyors to meet the requirements of various types of pan trucks or storage conveyors. Both the stacker and the unstacker are hydraulically powered for smooth, quiet operation. The stacker is available with an optional stack counter for variable stack loading of pan trucks. The stacker operates at a maximum rate of 50 pans per minute while the pan unstacker operates at an average rate of 40 bread or roll pans per minute.

The pan diverter sections of the system are air operated. When activated, the pan diverter directs pans to the appropriate conveyor without having to stop pans or conveyors while the diverting action is taking place. This 100% continuous operation can occur at rates of up to 50 pans per minute.

The pan handling system which has been described is equipped with a number of safety features that shut the system down quickly if pans or stacks are not being handled properly. This prevents the occurrence of massive pan jams which can damage pans and equipment.

Another company offers a bun pan stacker having the following features: (1) Stack height adjustable up to 44 inches; (2) Pans having dimensions of 8 to 32 inches wide and 14 to 32 inches long can be handled: (3) Stacks up to 35 pans per minute; (4) Left hand discharge, right hand discharge, or straight ahead discharge; (5) Hand wheel controls are front mounted and calibrated for accurate guide adjustment; (6) Operates continuously while changing trucks; (7) Has front mounted electrical controls for lowering and raising forks, resetting cycle, pre-setting amount of stacks per truck, high or low stack, on-off positions; (8) Pan truck safety latch holds truck in position until full; and (9) Heavy duty construction (Anon. 1976).

A "grouper" is essentially a conveyor which accumulates pans in a group of the size required for a certain operation, such as filling a proofer shelf or an oven tray. A commercial unit delivers a predetermined number of pans to the proofer loading conveyor in synchronization with the availability of a proofer shelf. It has dual conveyors consisting of low friction table-top chains powered by separate DC motors. A gate at the discharge end stops the moving pans until a group has been accumulated, at which time the gate lowers and the set is delivered to the proofer. The unit's gate section conveyor chain runs at half speed during grouping to reduce pan impact and wear on the pan bottoms. A clamp holds the first pan of the next group until the previous set is loaded and the gate can return to its up or stop position. Proximity sensors control the operation of the gate and clamp. The grouper cycles in response to a signal from the proofer. For very high speed operations, there can be added an optional traction assembly which employs variable electromagnets beneath the conveyor chain to increase the pan-to-conveyor friction. The magnetic pull of the magnets can be varied to adjust for pans of different weights (Anon. 1975).

Figure 10.2 shows a complete handling system for bun pans, including accumulators, groupers, indexers, and turners.

LOADERS AND UNLOADERS

In the least advanced and most advanced types of bakery operations, the question of specialized loaders and unloaders for the oven does not arise. Small retail bakeries and non-mechanized bakeries which are larger, simply use manual labor to put pans or straps into the oven and take them out. The most mechanized bakeries carry the pans through the oven on continuous conveyors, and special loaders and unloaders are not required. Most band oven operations do not need special mechanical devices for getting the product into and out of the oven.

For large tray ovens and, especially for traveling hearth (tunnel) ovens, specialized equipment for loading product into, and taking it out of, the baking chamber, is essential for efficient operation. There are different types of loaders and unloaders which are particularly suitable for products with special requirements. For example, pies require very gentle treatment, with no change in level, and, preferably no tilting as they move from the hearth to the removal conveyor. They are likely to be damaged by pusher bars or the like. These products can be loaded by "walking finger" unloaders, which use two sets of long thin bars, oriented across the direction of travel. One set of fingers, on which the pies rest, raises slightly and moves slightly forward, then moves downward to its base position during which movement it deposits the tins in a slightly advanced position on the second set of fingers. The second set of fingers picks the pies up and performs a similar operation. Then the first set repeats. The basic idea is that the fingers stay in about the same place from cycle to cycle, but the pans are moved along them in short steps. Good support can be given to the bottoms of the pies by these conveyors and movement is very gentle even though not continuous.

A second type of unloader, the roller bed unloader, is used if small pans are not to be handled and high production rates for larger pans are required. Its action is essentially as indicated by the name; pans roll across a series of rotating cylinders.

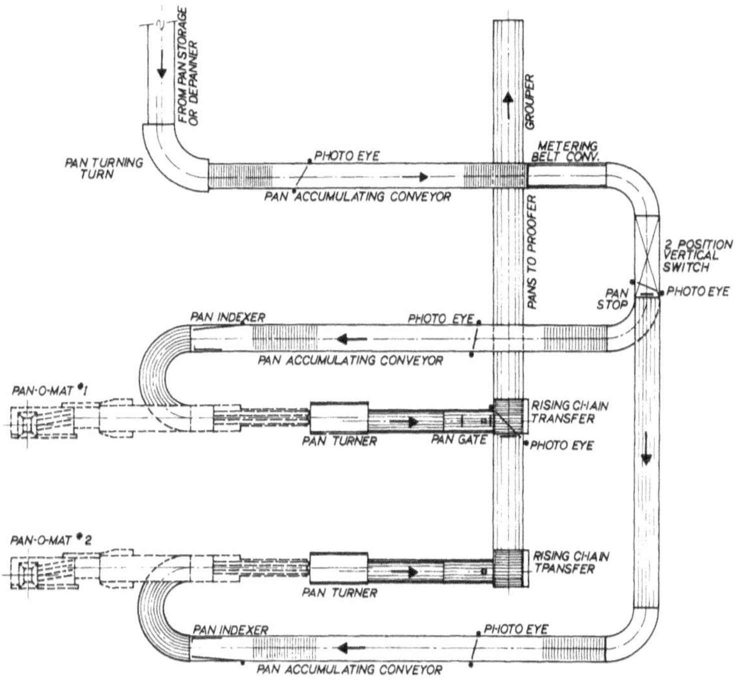

FIG. 10.2. COMPLETE HANDLING SYSTEM FOR BUN PANS.
SOURCE: VELTEN AND PULVER

The pie and foil pan unloader has a special fork lift mechanism that picks up a row of pans and gently transfers it to a take away conveyor.

A fourth type, the multiple flat-top chain unloader has an elevating conveyor surface enabling it to unload both panned product and hearth breads. Panned bread is transferred from the oven hearth in a conventional manner. The flat-top conveyor can then be hydraulically raised allowing hearth products to discharge directly down a slide plate to a belt conveyor underneath the flat top conveyor.

Manual slide plate unloaders are useful for hearth products and some pan products. The pieces simply slide off the oven hearth down the inclined plate to a take away conveyor.

A common type of loader consists of a pusher bar, the movement of which is synchronized with the travel of trays into the loading area so that a complete line of product goes onto the hearth at the same time. The pans are transferred across the front of the oven and along the pusher bar by conveyor action; when a complete line is on hand, the inflow is temporarily halted, and the pusher bar forces the entire line of pans into a tray as soon as a tray (or hearth space) becomes available. The loading platform may tilt upward to follow the travel of a tray as it passes the oven opening so that the oven conveyor does not have to stop. Unloading mechanisms either tilt the trays carrying the pans immediately before they reach unloading area or they use horizontal pusher bars.

An automatic unloader for a tunnel oven is shown in Figure 10.3.

Traveling hearth ovens can be provided with a number of types of loaders and unloaders. The following description applies to a system which can be converted to loading peels or pans. If peels are used, the boards enter the loader via the infeed conveyor. When a full group is in position, stops on the infeed conveyor rise to prevent further peels from entering and, at the same time, the infeed conveyor stops. With the peel boards in position, the "wing" wire mesh conveyor starts turning while at the same time this entire conveyor is moved back toward the peel. The wing conveyor is a mesh band conveyor extending the width of the oven which has an arc-shaped upper surface and a flat lower surface. Aided by a dough piece pick-up assist belt, the dough pieces are carried onto the wing conveyor and forward to its leading edge, at which time the mesh conveyor stops and the wing is moved towards the oven hearth. The speed of the carriage can be set by the operator to provide a wide range of spacings. Upon reaching its end position over the hearth, the carriage stops its forward motion and begins to retract, while the wire mesh conveyor re-starts, moving the dough pieces onto the oven baking surface at a speed about equal to the motion of the hearth. During this last operation, stop bars at the discharge end of the in-feed conveyor drop, allowing the empty peel boards to feed into the turn over chute and exit via the peel board return conveyor. At the same time, another line of fully loaded peels is entering the infeed conveyor for the next cycle.

When modified for loading pans, the wing conveyor is raised to a position that will allow it to pass over the pans being used. The pusher bar is, at the start, in the "up" position. Pans enter the loader in the same manner as the peel boards. When a group of pans is in position, the infeed conveyor stops and the carriage rides back over the pans. When the carriage frame and wing are over the pans, the air cylinder attached to the pusher bar moves the bar to the down position. The carriage frame with the wing and pusher bar now moves forward pushing the pans onto the oven hearth. With the pans loaded on the hearth, the carriage begins to withdraw, while the pusher bar retracts to the up position ready for the next group of pans which is now waiting on the infeed conveyor (Anon. 1980).

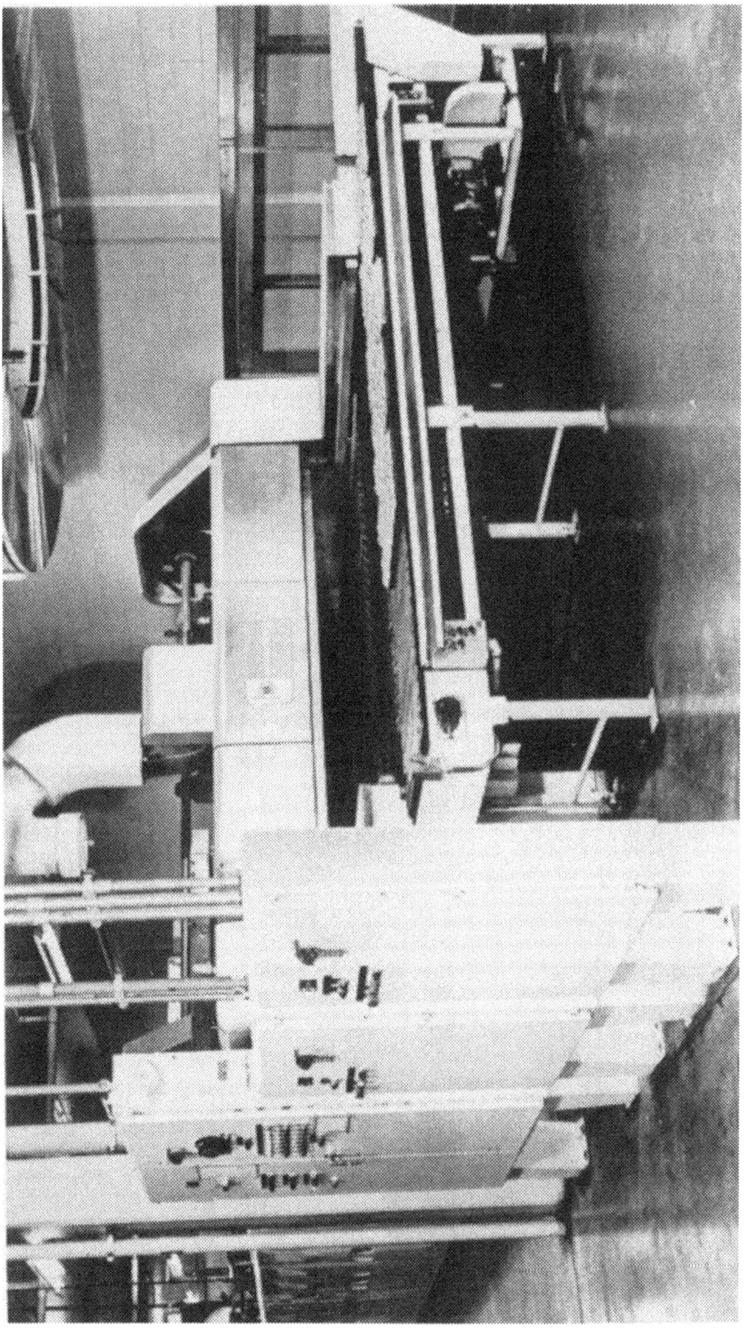

FIG. 10.3. DELIVERY END OF TUNNEL OVEN WITH AUTOMATIC UNLOADER FOR BREAD AND ROLL PANS. SOURCE: APV (BAKER PERKINS)

An automatic loader for sweet doughs (coffee cakes) is shown in Figure 10.4.

DEPANNERS

Manual depanning, as of bread loaves, is done by inverting the pan and dropping it on a frame which is open in the area where the loaves fall (see Figure 10.5). Automatic depanners are of two major types, the gravity models in which the loaf is dislodged by machines which first invert the pans and then drop them a short distance to allow the loaves to fall onto a conveyor, and vacuum depanners which rely on a multitude of suction cups to lift the loaves or rolls out of the pans. Vacuum depanners will function properly only with certain types of pans. One of the earliest patents described a perforated drum, with a vacuum therein, which did the lifting (Petersen and Heide 1963), but designs based on this principle were soon replaced by models using flexible vacuum cups.

In a typical vacuum depanning installation, the roll-filled pans are placed on the infeed conveyor, which has a positive indexing finger to properly locate them in the pick up area. In older models, a vacuum pick-up head driven by a continuous crank mechanism swings in an arc above the pan. The machine picks up 12 rolls simultaneously by fixing a flexible vacuum cup on their tops and gently drawing them out of the pan. Pressure is reduced in each individual cup through a hose connected to the pump. The vacuum is slight and the cup is very flexible so as to avoid distortion of the roll. Empty pans are then returned to a conveyor leading to the panning operation or to storage. Vacuum is supplied by a 3 hp turbine pump. Compressed air at 50 lb pressure is also required. These units are combined with packaging operations. A vacuum roll depanner is shown in Figure 10.6.

Other versions of vacuum depanners use a silicone rubber belt on which are fitted numerous vacuum cups to pick up the rolls. The belt surface travels at about the same rate as the pan beneath it. Since the belt surface is inclined, it gradually contacts the top of the bread, and as it continues on an upward path, lifts the loaf out of the pans. The pan conveyor has a magnetic rail to hold down the pans and, if required, can be fitted with a magnetic delidder. The vacuum cups then move up and around to deposit the loaf on a conveyor leading to the cooler while the pan proceeds at a lower level on to a conveyor leading to the panner or storage. When similar units are used on rolls, they may be provided with dampers to adjust the vacuum so that excessive force is not applied to rolls when most of the cups are closed off. Other models (for loaves) have a simple valve installed in each vacuum cup to eliminate

FIG. 10.4. TUNNEL OVEN WITH AUTOMATIC LOADER AND LOADING CONVEYORS.
SOURCE: APV (BAKER PERKINS)

air flow through inactive cups—this not only reduces the power requirements but also reduces the noise level. An average vacuum of 8 inches of water is recommended for some machines.

Conditions which are required to insure the best operation of the depanner-packer are (1) uniform make-up to insure rolls of the same size and shape, (2) good pan greasing to give easy release of the rolls, (3) uniform proofing to maintain even size, (4) uniform production flow from oven to cooling, and (5) uniform cooling. Typical maximum speed of a bun depanner is 35 pans per minute and of a bread loaf depanner is 400 loaves per minute. Roll machines deposit the desired number of rolls in the proper configuration in packages at the end of the traverse of the vacuum cup assembly.

Air jets directed into the corners of loaf pans are often used to loosen bread before it is depanned. If properly applied, compressed air can be very helpful in this regard, but if the jets are misdirected or the jet is too strong, the loaves can be damaged. It is also very important to thoroughly filter oil and other contaminants out of the compressed air.

PAN GREASERS

The grease applied to the pan to improve release of the baked product is, in effect, a part of the equipment even though some of it ends up in

FIG. 10.5. Tray Oven Showing Unloading Arrangement and Conveyor Leading to Hand Depanning Station at Far Right.
Source: APV (Baker Perkins)

the product as an ingredient—or, at least, as a component. There is no doubt that ideal practice is to use the irreducible minimum of pan grease since, though it may do no harm, it certainly does not improve the finished product. In order to reduce cripples, there is often a tendency to apply an excess of oil in order to make sure the product falls out of the pan, or can be easily lifted from the pan, at the depanning station.

A properly designed and maintained pan greaser can assist in getting good release with a minimum application of oil. There are several types of pan greasing machines available for bread. They fall into two main categories, the multi-nozzle units and the single nozzle units. The former type is designed to produce a series of overlapping round spray patterns to cover the entire interior surface of the pan. Because of the overlapping, some areas can receive excessive oil. Single nozzle units are said to be more efficient (Scott 1984). These units feature a single nozzle which yields a rectangular spray pattern conforming to the shape of the bread pan, eliminating the necessity for overlap. The single nozzle applies oil on the four sides and bottom of the pan.

Spray nozzles can be designed to meet specific needs. For example, equipment is available for applying pan oil to the ends and to a four inch square on the bottom of a perforated rye basket. The sides are not sprayed because the dough does not normally touch the sides. A properly designed spray pattern not only reduces the amount of pan oil required, but improves sanitation.

Some of the most important factors affecting the performance of the bread pan greaser are: (1) Nozzle placement—height and centering; (2) Placement of the guide rails; (3) Pan flow control; (4) Spray pattern; (5) Air pressure; and (6) Heat control (Scott 1984). For spraying regular loaf bread, the sprayer is normally set to give a 0.5 inch clearance between the top of the pan and the bottom of the nozzle protection bars. The clearance can be increased to 0.75 for sandwich and split top bread pans. It is important to make guide rail adjustments so that the pan will be centered under the spray, and thereby reduce the amount of oil which goes astray due to off center pans. The flow of pans through the greaser should be adjusted so there is a constant and even movement through the spray area. This requires frequent checking of the star wheel indexer or top trigger mechanism. The spray pattern is monitored daily by inserting into a test pan a folded paper conforming to the pan's inside contours. The oil droplets should give a uniform pattern over the exposed surface of the paper, not leaving some of the paper dry and soaking other areas. For testing purposes, the air pressure should be set at 70 psi for the multinozzle machine and 50 psi for the single nozzle units. During the test, the oil temperature should be 100°F for both sprayers.

If dispersion is not adequate, air pressure can be increased and another test made, but it should be kept in mind that excessive air pressure will cause misting and fogging, resulting in the sanitation problems which are always the result of airborne oil droplets. Poor dispersion is an indication of nozzle wear.

The pan greaser generally works better with oil at 100°F than with oil that is substantially cooler. Scott recommends the following levels of pan oil per 1,000 pans: For open top pans, 15 to 20 oz. for 16 oz pans, 20 to 25 oz for 20 oz pans, and 25 to 30 oz for 24 oz pans; For sandwich or pullman pans, 20 to 25 oz for 16 oz pans, 25 to 30 oz for 20 oz pans, 30 to 35 oz for 24 oz pans.

Some sample specifications for common blends of pan oil are given below.

Oil Composition	Minimum Smoke Point, °F	Iodine value	Gumming test, hrs
100% Soybean oil	345	130	48
20% mineral, 80% soybean	330	106	72
40% mineral, 60% soybean	325	78	96
Partly hydrogenated soybean oil	345	110	120
Partly hydrogenated SBO with coconut oil and mineral oil	320	52	312

Greasing of cake pans presents some of the same problems as bread pan greasing, as well as some different challenges. In addition to the usual nozzle type of sprayer, machines using spinning discs to apply the oil have been developed. Nozzle sprayers can be designed specifically for each pan cavity size and shape. Some greasing machines are mounted on the same chassis as the batter depositer, others are separate units. Smaller operations can use hand spray guns drawing the grease from a double acting pump mounted on the oil drum.

Cake pan grease is often formulated somewhat differently from bread pan oil. One popular cake grease consists of a fat phase of shortening and oil, with suspended cake flour or other cereal solids, and lecithin. Some of these mixtures are votated to promote an optimum

FIG. 10.6. A VACUUM DEPANNER SHOWING PICKUP HEAD IN FOREGROUND.
SOURCE: AMF

crystal structure. It is not clear how such a mixture can retain fat crystal structure during spray application.

About 2 to 3% of grease based on batter weight has been recommended for cake pans. Factors to be considered are grease pattern, force of the dough being deposited, temperature of the pans, water left in pans, excess crumb, and dumping cold cake.

COOLING EQUIPMENT

It is important, in fact essential, to cool many baked products before they are packaged. Loaf bread is removed from the pans shortly after it leaves the oven and cooled before it is sliced and wrapped. Rolls can be cooled before or after depanning. The situation varies with regard to sweet rolls and the like, which often go through some sort of finishing operation.

Putting a hot product in a closed container, whether of plastic film or fiber, will often lead to a greatly reduced shelf life because of moisture condensation and other factors. Cooling in ambient temperature systems, mainly by conveying product in dispersed array through air currents originating from fans, is very common and has been found to be satisfactory in many cases. It does involve lengthy delays in the processing line and long stretches of conveyors, and it exposes the products to contamination.

Most of the baked product cooling systems relying on ambient air movement use conveyors elevated above the processing area at a height sufficient to eliminate interference with operations taking place at the floor level. These conveyors often have multiple spiral passes vertically disposed. Other systems have multilevel straight-ahead conveyor systems at floor level.

To overcome the disadvantages of ambient air heat removal, accelerated cooling systems have been developed. Spiral, tunnel, and rack systems with refrigerated air are available. At least one manufacturer offers a compact, fully automatic, floor mounted rack model with washing and biological filtering of cooling air. Also, there is a tunnel rack cooler which is semi-automatic in operation and features a controlled atmosphere.

Buns can be cooled before or after depanning. Manufacturers of equipment which cools the buns in their pans say their system is better because: (1) When the rolls leave the oven, they have a thin brittle shell which is easily broken by depanning; (2) In-pan bun cooling facilitates placement of the depanner in line with, and immediately in front of the slicers, allowing positive alignment of the product with the slicer infeed; and (3) If the pans are recycled after the depanner, the pans will have more time to cool before they return to the molders. Manufacturers of bun coolers operating on products which have been removed from the pan say condensation can form on the bottom of the bun, leading to corrosion of the pan. It is doubtless true that both types of bun coolers are operating satisfactorily. Figure 10.7 illustrates the way an overhead in-pan cooler can be integrated into a pan handling system.

Transfer of heat out of low density porous products such as bread loaves and rolls is slow because of the porous nature of the product. Hot air has a tendency to become entrapped in the product. It has been found that cooling can be greatly accelerated by using vacuum. According to Fish (1980), modulated vacuum cooling systems are being used in many parts of the world to reduce the temperature of bread, buns, rolls, cake, melba toast, stuffing bread, and crouton bread. These systems use a combination of liquid ring pumps and water-cooled condensers with steam augmenters to create the vacuum. Cooling chambers can be in-line high speed coolers with single or double tier conveyors con-

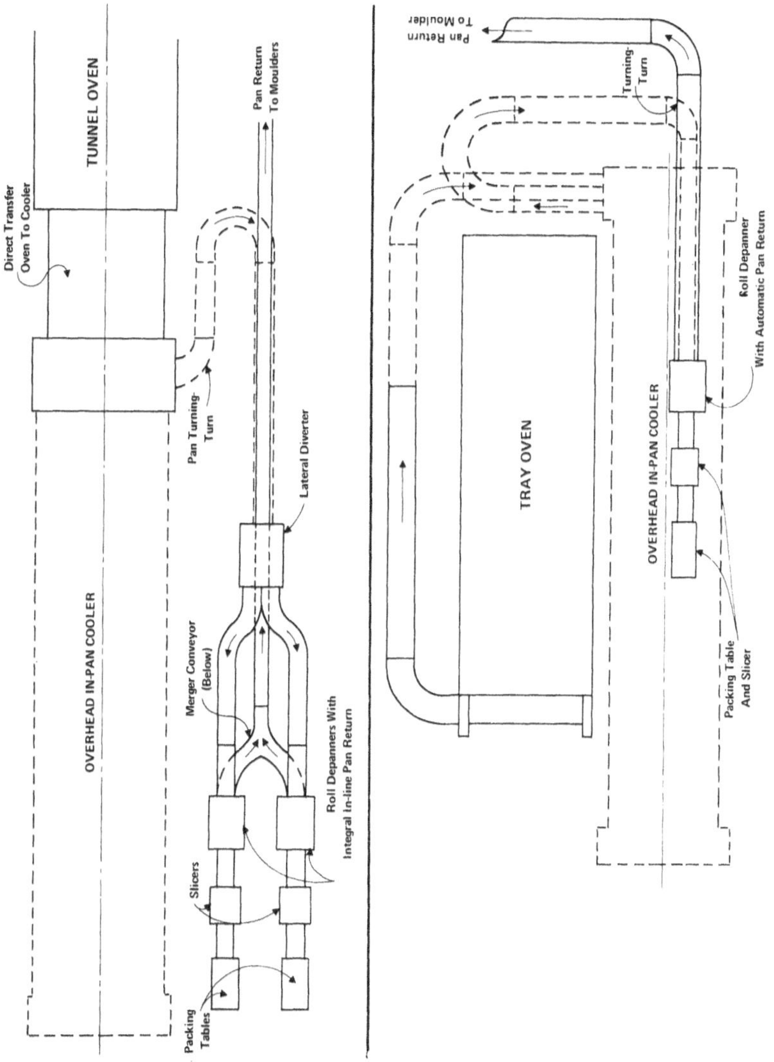

FIG. 10.7. SYSTEM FOR IN-PAN COOLING OF BUNS.

SOURCE: LATENDORF

nected to pivoting bridge conveyors at each end. The tunnel is clad with molded fiberglass panels supported by castellated beams. It is equipped with a washing system controlled by programmed circuit.

Operation of the vacuum cooler can be summarized as follows. Products are grouped in batches and carried to the infeed section; doors at both ends of the tunnel open, and products are carried onto the interior conveyor. After the batch has entered and the doors have been shut, a selected vacuum cooling cycle is applied. After the correct levels and duration of vacuum have been completed, air is readmitted through a valve fitted with a filter unit, the drain valves open, and the doors open. Then, the internal conveyor re-starts, transferring the cooled batch of products to the outfeed bridge conveyor and the cycle starts over.

Most of these vacuum coolers are employed to reduce the temperature of the product to 95° to 110°F at the wrapping machine, but lower temperatures can be attained. Fish (1980) claimed that the advantages of vacuum cooling include: (1) A dramatic reduction in cooling time, taking it to the order of minutes; (2) A reduction in baking times of up to 25% can be achieved because moisture is reduced through the application of vacuum and this can replace the moisture reduction which occurs in the last phase of baking; (3) Production can be increased; (4) A more stable product results; (5) An extension of shelf life is generally experienced; (6) Increased volume and extension of crust shelf life of French bread is observed; (7) Working conditions inside the bakery are improved; (8) Considerable reduction is possible in labor requirements; and (9) A space saving can be realized.

SLICERS AND ASSOCIATED EQUIPMENT

Introduction

Bread has been sliced on a commercial basis since the late 1920's. The first production scale slicers were based on the reciprocating (up-and-down) blade principle and would slice about 20 or 25 loaves of bread per minute. After the bread was sliced, it had to be placed in a cardboard tray or bound with a paper circlet to keep the slices together through the wrapping cycle. It took about two years for the manufacturers of wrapping machines to make improvements in their devices so that they could wrap sliced loaves not held together by trays or bands.

The first bread slicers were rather crude machines. They made a lot of noise and readily developed mechanical troubles. Because they were slow, there was a tendency to crush the loaves as they were pushed through the blades. A large amount of crumbs was generated, and it was necessary to cool the bread to about 90°F before it could be sliced.

Since wrapping machines could handle the output of two of the early slicers, initial improvement efforts were directed toward increasing the speed of the slicers. These efforts resulted in the development of slicers which could process 40 to 50 loaves per minute, or about the capacity of one wrapping machine. The principle of slicing the loaf with a set of parallel steel blades spaced about 0.5 inch apart and moving at high speed has remained the same since the earliest models. Some designs with reciprocating blades are still being made, particularly for slicing specialty breads and for retail bake shops, but all high-capacity machines for white pan bread now use endless steel blades for cutting.

Automatic slicers for bakery products can be classified into three categories based on the type of blade they employ: (1) Slicers using straight blades; (2) Slicers using continuous bands; and (3) Slicers using disc shaped blades. The latter type is substantially restricted to splitting buns and English muffins. Some "forking" devices have come into use for splitting English muffins; these do not use saw-type blades.

It is known that experiments have been conducted in slicing bread by lasers and high pressure water jets. Bread can be cut by these two forces, but lack of a practical method of focusing allows the jet or laser to spread as it goes through the object. Until this problem is solved, it does not appear that these two methods will find commercial application.

Bread Slicers

Reciprocating blade slicers—The back-and-forth movement of blades in this type of equipment is transmitted from the motor to the blade frame by a crankshaft and levers mechanism. The vibration and noise created by this mechanism was considered one of the most objectionable features of reciprocating slicers. Machines of modern design are constructed so as to minimize the noise and vibration, however, and these factors should no longer be regarded as major objections to their use.

The blades are held in a rectangular metal frame, which may be made of cast magnesium. Each of the blades is held taut by a heavy spring. When a blade is to be inserted into the frame, its spring is compressed by a special tool, allowing the blade pin to be slipped into the appropriate retaining notch. Each slice thickness requires a separate frame. Standard frames are available for several slice thicknesses, such as 0.375, 0.4375, 0.5, and 0.5625 inches. The time required to change a frame is about 5 to 8 min. Figure 10.8 shows a partially diassembled reciprocating slicer with frame being inserted.

FIG. 10.8. DISASSEMBLING A RECIPROCATING SLICER. MECHANIC HOLDS FRAME WITH SLICER BLADES.
SOURCE: BATTLE CREEK BREAD WRAPPING MACHINE CO.

Different models vary in the orientation of the slicer blades and the direction the loaf moves during the slicing process. In the slicers used in many retail shops, the loaf either moves straight through slanted saws or on a slanted path downward through vertical saws. Other versions have an elevator mechanism which moves the bread straight upward through horizontal blades. Each style has advantages for specific purposes, primarily for its adaptation to a particular packaging method. In one popular model, the knife frames are tilted 20° toward the top crust of the bread so as to provide angle cutting and to accommodate the natural flow of loaves through the saws.

Knives in reciprocating slicers attain a maximum velocity of 35 ft per min. They will slice from 10,000 to 20,000 loaves of bread without resharpening. Much longer service can be obtained from continuous bands. There is also no way of providing automatic honing on reciprocal slicers, and the parts of the blades which cut the crust undergo more abrasion than the other parts, so that wear is not uniformly distributed over the blade length.

A disadvantage encountered in using reciprocal slicers for soft white loaves is that the surface of the bread slice may exhibit slight waves or corrugations. This is usually objectionable to the salesman and may even be so to the consumer. The effect is much less noticeable on denser varieties such as rye and whole wheat breads.

When band slicers are used for cutting raisin bread, the sticky substances from the fruit are collected by the blades and deposited on the drums. Eventually, the deposit may become thick enough to interfere with the operation of the machine and the bands may break. Any bread with a soft and sticky crumb can cause a similar accumulation of material on the band. Although there are also difficulties in slicing these breads with reciprocating machines, the problems are not as severe, and their blades will not break from product buildup.

Reciprocating slicers are also used for slicing hard crust breads such as French or Italian, heavy rye and health breads, and iced loaves. They are also useful in low volume applications such as retail bakeshops where loaves are sliced at the customer's request, etc. On the whole, machines of the reciprocating type can be said to be more versatile but less efficient than band slicers.

Some special devices include slicers which accept the simultaneous feeding of two loaves (placed end to end) into the slicing head as a means for increasing output of hard-to-slice breads, conveyors which allow removal of the heels, slicers which allow feeding of alternate loaves of two different kinds and then re-assemble halves so that each loaf contains two different types of bread (e.g., whole wheat and white), and conveying half loaves to the wrapper.

Slicers using continuous blades—Band slicers are the preferred type for slicing uniform loaves of white bread at high rates of speed. They are also usable for many other types of bakery foods, but, as the products assume irregular shapes or have dense moist crumbs or contain sticky ingredients, their efficiency decreases. Some of the recognized advantages of band-type slicers are: (1) Minimum vibration; (2) Greater cutting speed; (3) Less frequent need for blade sharpening; and (4) Continuously variable slice thickness—within limits.

One type of endless blade slicer requires that the bands be placed around steel cylinders and crossed in the center so that they form a figure eight as viewed from the side of the machine. The bread contact area is at the point where the blades cross. Another type, which will be discussed later, uses four drums to give an approximately parallel orientation of blades in the slicing area. Since the blades must pass over the drums in a flat position and then turn to present their serrated edges to the bread, it is necessary to move them through a 90° angle between the time they leave the drum surface and the time they contact the bread. This turning function is performed by hardened steel guides located above and below the line of travel of the loaves.

Three types of devices are found in all band slicers, (1) rotating drums which power the bands, (2) slicer blades, and (3) blade guides (see Figure 10.9). The first band slicer had two drums with relatively close centers and an overall height of about 6 ft. After these machines had been in operation for some time, it was found that the drums were too close together, causing the blades to twist too abruptly as they turned through a total of 180° between the upper and lower drums. The blades exerted too much pressure on the band guides and, as a result, the guides and bands overheated and suffered excessive wear and breakage. Experimentation with various band lengths resulted in the conclusion that drum centers about 7 ft apart would make the band twist sufficiently gradual that excessive heat generation in the knife guides would be eliminated.

Slicers with drums placed relatively far apart also have the advantage that the longer blades require replacement or resharpening less often, saving labor costs and material. In addition, the blades are more nearly parallel at the position where they cross, creating better cutting conditions. The disadvantage of the arrangement is that the machine becomes quite tall. Unless an unusually restrictive head space situation exists, however, a height of 8 ft or so should not create any major placement problem. The floor space is not increased appreciably and this is usually the critical factor.

Some manufacturers have decreased slicer height while maintaining the increased band length by placing the blades at an angle to the floor. The chief operational difference between the vertical and the slanted arrangement is that vertical alignment permits an almost perpendicular approach of the blades to the loaf, resulting in the exerting of a constant force on the bread at all time. With offset drums, the blades

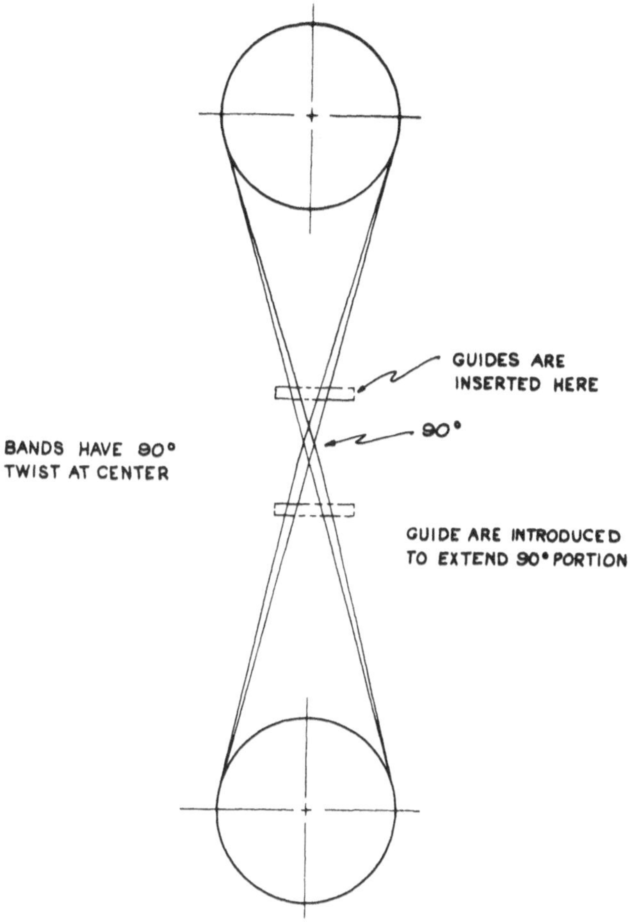

FIG. 10.9. TYPICAL ARRANGEMENT OF DRUMS, BANDS, AND GUIDES IN CONTINUOUS BLADE SLICERS.

contact a corner of the bread first and the force on the loaf continually increases until the blade crossover point reaches the center of the loaf. In most practical situations, it is debatable whether or not a difference between the two methods would be detectable. Machines designed with offset drums may have the product approach conveyor (or slide) changed from the horizontal to a descending slope as the slicing zone is reached so that the blades are almost at a right angle to the line of travel of the loaves (see Figure 10.10).

FIG. 10.10. A CONTINUOUS BLADE SLICER USING OFFSET DRUMS.
SOURCE: AMF

In another machine designed for use on very soft bread, four drums are positioned to give the blades an obtuse angle and keep them in nearly parallel alignment in the slicing zone. Balanced forces, low blade speed, and reduced rate of loaf travel tend to minimize tearing, crushing of side walls, and keyholing. As in most machines, thickness of the slices is adjusted by a simple band wheel control which moves the blade guides farther apart or closer together. The band guides are attached to a so-called lattice which permits adjustment of the separation in a non-stepwise manner.

The knife speed in modern slicers is usually from 900 to 1,200 ft per min. There has been a trend to lower blade speeds since this reduces loaf distortion and temperature rise in the bands. Hot bands accelerate dough buildup on bands and drums. Blades are capable of slicing 500,000 to 2,000,000 loaves of bread between resharpening periods if they are properly rehoned at intervals.

Most band slicers include automatic honing devices and these are often activated while slicing is in progress. Two types of hones are the rotary and the fixed. Different grit sizes and different hardnesses of the binder are obtainable. Grits up to 600 are available in rotating hones, while the nonrotating variety is offered in fine, medium, or coarse grits. Automatic hones are often activated by a timing device which applies the hones after every hour or half hour of slicer running time. Overhoning causes excessive wear and reduces band life.

In most slicers, the hone is located about 5 or 6 inches below the point where the band meets the upper drum. At this point the band is vibrating with sufficient amplitude that the hone contacts all parts of the cutting surface. In the absence of this vibration, a stationary hone would not be able to process the complex beveling in a modern blade.

Blades—The useful service life of a blade is determined mostly by the quality of the steel from which it was fabricated. Most, perhaps all, slicer bands are made of chrome-vanadium-steel conforming to close specifications for uniformity, hardness, dimensions, smoothness, accuracy of the rounded edges, and the camber or straightness of the steel over the 20 ft or so of length required for a band. This relatively high priced alloy contains chromium to improve its hardness, so the the edges will maintain their sharpness for a relatively long period of time, and vanadium to increase its toughness so the band can undergo repeated twisting and bending without breaking. A recent improvement has been the introduction of blades with hardened tips.

The effectiveness of a blade in performing its intended function is determined by the design of the cutting edge and the accuracy of the manufacturing method in implementing that design. The standard shape is a scalloped tooth of specific dimensions and angles. The depth of the curve affects the smoothness of the sliced surface, deeper scallops giving better penetration and less crushing of the loaf but also causing more tearing so that the sliced surface is rougher. The type and direction of beveling also affect the smoothness of the slice.

According to Fitzmaurice (1970), blade types are generally differentiated into three categories on the basis of the relative softness of the bread which they are best adapted to slice. Firm breads (e.g., French, rye, and whole wheat breads) require a 0.5 inch scallop having a cross-ground primary bevel. This bevel tends to produce relatively rough surfaces. A variation in which the cross grind marks have been removed by parallel grinding provides a smoother bevel with a reduced tendency to tearing the crumb. Conventional pan loaves with fairly soft crusts and a medium firm texture are best sliced with blades having a secondary bevel applied to each side so as to remove some metal at the cutting point. Even smoother slicing results when the second bevel is ground parallel to the cutting action. Very soft breads, such as continuously mixed bread, require a four-beveled cutting edge in which all the bevels are ground parallel to the cutting action. This provides the thinnest possible cutting edge.

Some adaptations are used to modify continuous slicers for better processing of raisin bread and other loaves with soft sticky characteristics. The usual procedure is to continuously dampen the blades with a wet sponge or the like. Frequent cleaning of the bands, drum, guides, and sponges is absolutely essential. Teflon coating of blades has been used for the same purpose, but Teflon is very subject to mechanical erosion and the extremely abrasive conditions encountered by slicer blades causes the material to wear off very quickly.

Blades work best when they are kept under a constant tension which keeps them fully extended, but excessive tension can cause undue wear on the drums and adds to band breakage. Devices have been developed which automatically adjust and maintain the tension by means of hydraulic cylinders acting on the axle of one of the drums (Jongerius 1987).

Blade guides—Blade guides used to be made of chrome plated steel with two opposing fingers contacting the blades. Tungsten carbide, or sometimes ceramic inserts, are now used in some slicers, to improve wear resistance and decrease heat generation. The fingers have been offset to allow crumbs to flow through better (Wright 1988).

A typical slicer—The following description is of the AMF Slice-Master, but many of its specifications are applicable to other slicers. In the AMF slicer, bread is received from racks or a cooling conveyor and deposited on a horizontal woven metal belt which transports it up against a rigid stop. An elevator mechanism releases loaves one at a time and delivers them into the flights of an infeed conveyor. The flights move down the infeed conveyor and then retract so that the gentle pressure of the accumulating bread moves the leading loaf toward the slicing area. A cushion of two loaves is set up between the blades and the positive feed.

Blades rotating around drums in a figure-eight pattern tend to pull the loaves forward, thus minimizing the need for force feeding. The sliced loaves are brought by a flighted discharge conveyor to the packaging machine. The blade drums are all steel and chrome-surfaced. They are precision ground and balanced. Supporting shafts are fitted with removable screw plugs on the operator's side to facilitate blade changes.

A heel control adjustment permits changes to be made to equalize heel thickness after the bread rails have been set to the proper loaf length. A slice thickness control handle allows changing slice thickness from 0.375 to 0.625, the current setting being indicated by a pointer and scale having graduations of about one-sixty fourth of an inch.

Disc Slicers

Slicers with horizontal disc blades are used primarily for slicing hamburger buns and hot dog rolls so that the bun can be readily spread apart in more or less equal parts for inserting the meat. These slicers consist of two motor-driven discs with serrated edges. The saws are disposed in a plane horizontal to the bun conveyor and are separated by a short distance (e.g., 0.12 inch) so they leave hinges in adjacent edges of the pairs of rolls which they are slicing. That is, each roll opens on the "outside" edge. The distance between the blades and the distance between the blades and the table or conveyor (height) are usually adjustable.

In the different designs of this general type, the characteristic operating element (toothed disc) remains the same while the methods of orienting the rolls and moving them through the slicing area differ. For example, a series of belts, moving in the same direction as the incoming conveyor but raised an inch or so above it, grasp a cluster of buns as they travel down the conveyor belt and carry them past the blades. There are two vertical knives immediately in front of the saws for splitting the clusters so they can pass by the spindles of the saws. Such equipment will process up to 35 clusters of 8 buns per minute.

Another device carries an entire pan of individual buns or clusters, on or off pans, through a slicing head where the circular blades slice each bun. A special rimless pan similar to a cookie sheet must be used. An advantage is obtained by having the buns kept in a fixed relationship when the entire pan is conveyed through the slicer. Rates of 1,200, 2,400, and 3,200 doz buns per hr can be obtained when individual buns and clusters are sliced in the pan, while a rate of approximately 1,200 doz buns per hr is possible when slicing clusters out of the pan.

Still another arrangement uses a circular table rotating at high speed to bring buns to its outer periphery where they pass under a revolving paddle wheel called an unscrambler. The wheel rotates in a direction opposite to that of the table and separates the buns so that they can be individually fed between two horizontal belts. The belts carry the buns past a disc saw where they are sliced before being ejected onto a packing table. This equipment will slice up to 1500 doz sandwich buns per hour. It is less satisfactory for hot dog buns because some of the pieces go through the slicer on their sides and receive cuts on their top or bottom. Other devices include air jets to partially open the bun as it is cut, and to cool and clean the knife blade.

Bread Cubers

The slicing of bread into cubes for stuffing mixes and croutons can be done with special arrangements of bands. One model uses eight drums to cut up to 70 loaves per minute into cubes 0.5 inch on each side, with automatic conveying of the loaves, strips, etc. from one operation to the next.

Splitters for English Muffins and the Like

It is possible to slice English muffins with the same kind of disc knives used to split hamburger buns, and some are sold with this kind of split. Nearly all large manufacturers use scoring machines which leave the two halves clinging together except at the edge. This simulates to some extent the fork separation that connoisseurs of English muffins are said to prefer. In a typical machine, a shallow razor cut is made around the mid-perimeter of the muffin by a continuously moving chain knife operating on a clockwise rotating muffin (Clock 1988). After being scored, the product passes through a set of four wheels having many knife blades. The wheels turn at different ratios and serve to perforate the muffin through its entire diameter while leaving enough cell structure between the tines to hold the halves together. Both scoring and splitting speed, as well as depth of cut, is variable to meet the baker's requirements. A variation of this machine includes a web or hinge slicer with automatic or manual provision for backing the scorer and splitter elements away from the product flow so as to permit use as a single file conveyor or hinge slicer. Other types of forkers use a straight line of tines which move back and forth on each side of the muffins as they move down a conveyor. Tines can be coated with Teflon or other nonstick coatings.

It is said that the consumer wants the periphery of the muffin to be sliced inwardly about a quarter of an inch all the way around, so as to prevent unsightly raggedness along the edges of the muffin when it is pulled apart. A device to perform this operation, while at the same time scoring it all the way around its periphery, is described in a patent by Hanson (1986).

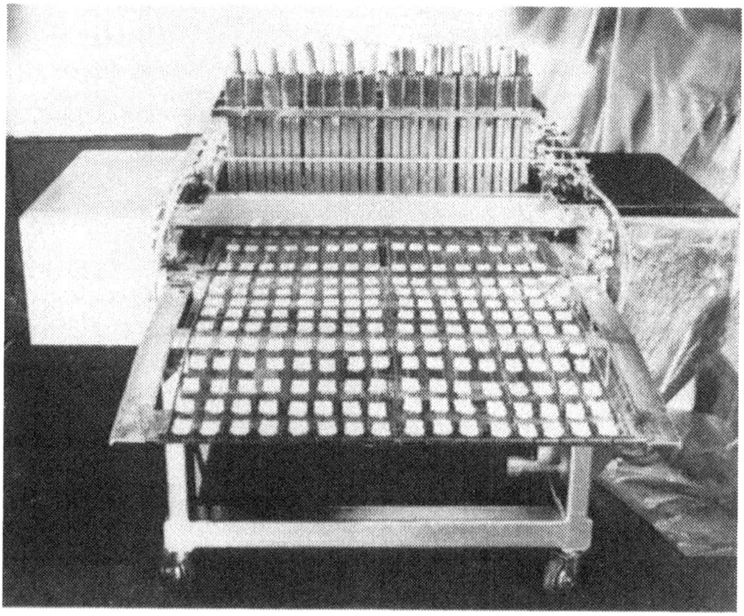

FIG. 10.11. SLICER FOR MELBA TOAST. MAGAZINES FOR VERTICALLY ORIENTED SMALL LOAVES AT REAR.

SOURCE: CLOCK ASSOCIATES

There are available melba toast and bagel slicers which make many thin slices from loaves or bagels held in vertical magazines (see Figure 10.11). They incorporate a high speed band blade and an O-thane or wire flex belt carry away conveyor. Depending on the number of input product magazines, the unit can produce up to 320 pieces per min (Clock 1988).

BIBLIOGRAPHY

ANON. 1972. Report on project of research and study committee on bread pan structural feature study. Am. Soc. Bakery Engineers Bull. *194*.

ANON. 1973. The whys and wherefores of tinplate's retirement. Ekco Bakery Engineering Review *3*, No. 1, 1-4.

ANON. 1974. The case of the hot pans. Ekco Bakery Engineering Review *4*, No. 1, 1-4.

ANON. 1975. The Teledyne Readco Grouper. Teledyne Readco, York, PA

ANON. 1976. Model M-44 Bun Pan Stacker. Alto Corp., York, PA

ANON. 1980. Operation of the Pan and Peel Oven Loader. Latendorf Conveying Corp., Kenilworth, NJ

CLOCK, T. Q. 1988. Personal communication. Clock Associates, Portland, OR

CRALL, R. D. 1980. Handling pans on air. Proc. Am. Soc. Bakery Engineers *1980*, 53-61.

FISH, A. R. 1980. Vacuum cooling. Proc. Am. Soc. Bakery Engineers *1980*, 120-126.

FITZMAURICE, D. T. 1970. Selection and care of slicer blades. Bakers Digest *44*, No. 3, 52-53.

HANSON, D. R. 1986. Machine for prescoring English muffins or the like. U. S. Patent 4,581,970.

HARTMAN, W. W.. 1948. The Story of Bread Slicing. Maine Machine Works, Los Angeles, CA

INZERILLO, A. N. 1979. Pan design. Proc. Am. Soc. Bakery Engineers *1979*, 64-71.

JONGERIUS, S. C. E. 1987. Slicing device for bread or the like. U. S. Patent 4,694,715.

LECRONE, D. S. 1980. Automated roll slicing and handling into packaging. Proc. Am. Soc. Bakery Engineers *1980*, 129-135.

MARCKX, E. I. 1971. Bread and pastry processing apparatus. U. S. Patent 3,614,933.

PETERSEN, C. W., and HEIDE, H. A. 1963. Depanning apparatus. U. S. Patent 3,099,360.

SCOTT, E. A. 1984. Product release agents. Proc. Am. Soc. Bakery Engineers *1984*, 157-165.

TORRENS, E. J. 1964. Pan care and maintenance. Proc. Am. Soc. Bakery Engineers *1964*, 147-153.

WRIGHT, S. 1988. Personal communication. Hansaloy Corp., Davenport, IA

CHAPTER ELEVEN

FRYERS AND ASSOCIATED EQUIPMENT

INTRODUCTION

Not all "bakery" products are cooked in ovens. All doughnuts, a large percentage of the crusts made into frozen pizzas, numerous snack products, and some other foods based on cereal ingredients are fried. The unifying principle in ovens and fryers is that they are intended to transfer heat into the product. In ovens, this is done by conduction from the pan, convection from hot oven gases, and radiation from hot oven parts, while in fryers the heat is transferred to the dough piece by conduction from a liquid transfer medium, hot edible oil. In ovens, the product is supported in a pan or on a hearth during the whole time it is being baked, while frying products are generally unsupported (though they may be submerged) except by their buoyancy in the oil.

Radiation probably contributes very little to to heat transfer during the frying process although convection currents in the oil contribute a great deal. Another important difference between baking and frying is that, in the latter process, some of the heat transfer medium (edible fat or oil) becomes part of the finished product.

GENERAL CONSIDERATIONS

Block (1964) states that frying differs from other heat-processing methods in the following principal respects.

1. Cooking is accomplished in a relatively short period of time, generally within five minutes, due to (a) a great temperature difference between the heat source and the food, and (b) the individual food unit which is being cooked is usually small, often less than one ounce in weight.

2. The frying fat becomes a significant component in the end product—varying from as little as 10% by weight of the end product in breaded fish sticks to 40% or more by weight in potato chips.

3. Fried products are usually crisper on the outer surface than other heat-processed foods.

4. The heat-transfer medium (i.e., the hot fat) is subject to changes in composition and often in performance characteristics during its process life.

5. There are unique mechanical problems involved in commercial frying operations.

As a consequence of their cooking process, doughnuts differ from most other bakery products. This is because: (1) No baking pan or mold is used to determine the finished shape; (2) The frying or "kettle baking" takes place in a very short period of time; and (3) During the frying, 15 to 20% of the composition of the finished product is added in the form of shortening from the fryer (Belshaw 1976).

Many models of batch fryers and continuous fryers are available. Most batch fryers are used for purposes other than frying dough products (frying chicken, for instance) but a significant fraction is being used for doughnut frying, and a few for specialty dough products of different types. Probably more continuous fryers are being used for potato chips, manufacturing breaded meat products, and frying snack products than are being used for doughnuts and other dough products, but it is the latter types which will be discussed in the subsequent sections.

CONTINUOUS FRYERS

One expert states that the four requisites for efficient automated frying systems are accurate temperature control, efficient heat transfer, minimum shortening contamination, and rapid shortening turnover. In making this assessment, Belshaw (1976) was primarily concerned with doughnut fryers.

Most continuous frying systems consist of at least five more-or-less independent sets of equipment: (1) The tank or trough containing the frying medium and providing a means for draining fat from the exiting product back into the kettle; (2) A conveying means for moving the food into, through, and out of the fat; (3) The fat system, which pumps and filters the frying medium and replenishes it from a bulk supply as needed; (4) A heating unit with a control system for generating thermal energy as required and transferring it to the fat; and (5) An exhaust system for removing the hot vapors emerging from the fat. The fat itself is an essential and interacting part of the whole system, and its characteristics strongly affect the functioning of the equipment and the results of the operation as reflected in the quality of the finished product.

Figure 11.1 is the layout of a continuous doughnut frying plant showing the relative locations of the various units to be discussed in subsequent sections.

Conveying mechanisms for transferring the product from the entrance to the exit end of the oil trough include the following types: (1) Spacer bar conveying, in which the frying piece floats between transverse bars that push it over the surface at a rate adjusted to give the

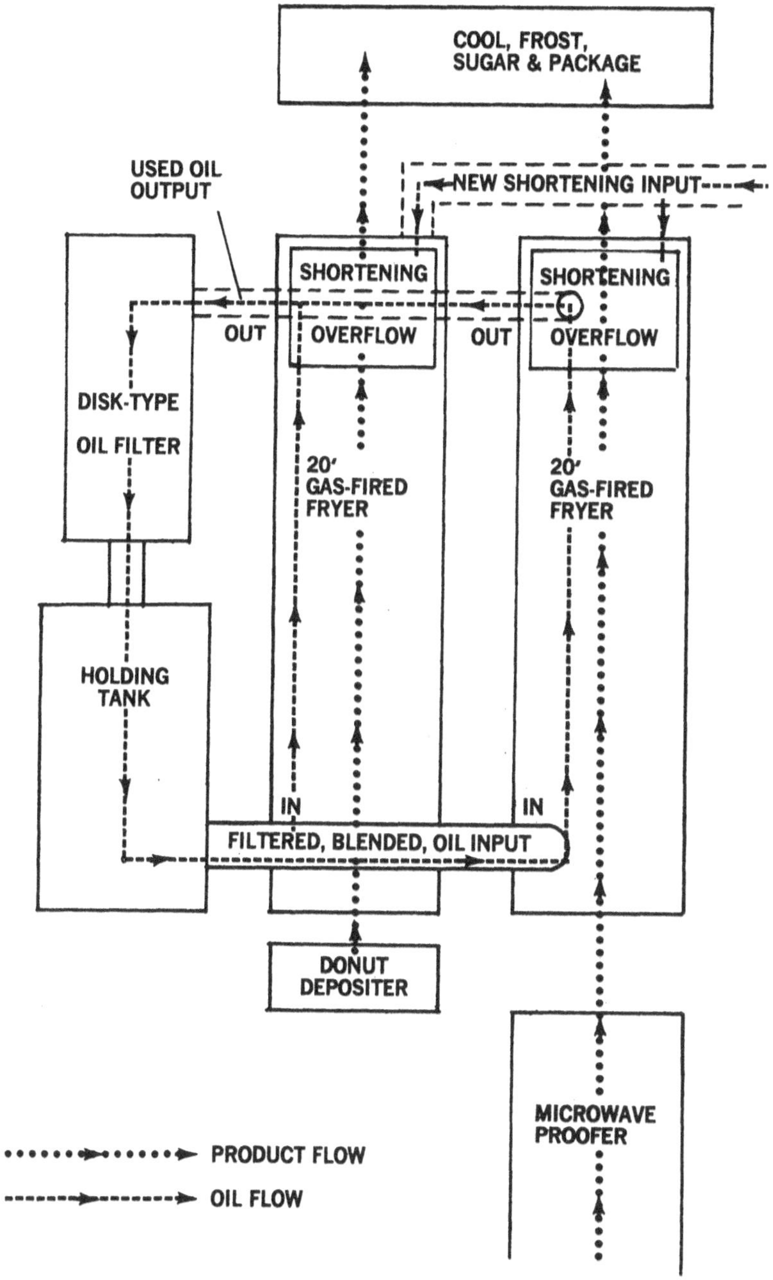

FIG. 11.1. FRYERS AND RELATED EQUIPMENT IN A DOUGHNUT PLANT.
SOURCE: BAKING INDUSTRY

desired cook by the time it emerges from the end of the vat; (2) Fat-current conveying, in which the product is carried along with the current of fat being pumped through the length of the fryer; (3) Drop plate conveying, in which the product rests on shelves moving below the fat surface until the pieces develop enough buoyancy due to internal expansion to rise to the surface and contact the upper conveying means; (4) Submerger conveying, in which a mesh band moving just underneath the surface of the fat contacts the upper surface of the buoyant product and carries it along; (5) Conveyors for nonfloating products which utilize baskets, belts, or other holding devices with no covers traveling well below the surface of the fat; (6) Restrained conveying, in which products are carried between two horizontal mesh belts: and (7) Compartmentalized conveying, in which the pieces are contained in small molds that help shape the product as it cooks.

The usual conveying means for doughnuts is the spacer bar type, although some varieties of doughnuts require submerger conveyors. For many other products, it is some sort of continuous mesh belt moving in both directions beneath the surface of the fat, usually with the discharge end emerging from the fat to dump the product on another wire mesh belt for draining and cooling. The methods for transferring the cooked product out of the fat often involve mechanisms of considerable complexity. The raw product is often dumped directly into the hot oil, but certain items require more elaborate handling, such as that in Figure 11.2, which shows a reciprocating conveyor depositing lines of pies on the infeed of a fryer.

Fuels and Heaters

Gas, oil, and electricity are being used as energy sources for food fryers. Dual-fuel systems convertible from oil to gas are available. An example of a doughnut fryer equipped for using alternate energy sources has oil and gas supply lines leading into a pair of oil/gas "combustors" feeding fire into flame distribution manifolds from which the heat is diverted into a number of tubes crossing the vat.

Heat sources may be located within the kettle or vat, or they may heat fat in an outside reservoir for transfer to the frying area. Tubes containing gas burners may be immersed in the fat, and electric immersion heating coils can also be used. If an outside heat exchanger is being used, the oil is pumped from the kettle through tubes in a chamber where gas is burned; modifications of this system are also available.

According to Belshaw (1976), heat is applied to automated doughnut fryers by (1) bottom-fired gas strip burners, (2) infrared gas heat on V-bottom, (3) tube-fired, atmospheric burners, (4) tube-fire premix

FIG. 11.2. PIES BEING ALIGNED ON INFEED OF FRYER BY RECIPROCATING CONVEYOR.
SOURCE: DAWN EQUIPMENT

burners, and (5) electric tubular heaters with lift-out heater assemblies. The simplest heating system consists of gas flames directed against the bottom of the frying kettle. If the bottom of the vat is flat, solid matter such as dusting flour and pieces of batter accumulate in the bottom and form an insulating layer between the heat source and the fat, resulting in a decreased rate of heat transfer, a loss of control, and an accelerated rate of shortening breakdown due to localized overheating and contamination. By constructing the tanks with a V-shaped bottom, this situation is alleviated to a certain extent since the solid material settles into the lowest part of the vat, below the level at which the heat is being applied. An improvement on the V-bottom design uses a shortening circulating pump to flush the solids to channels at the side of the tank.

Tubular heating systems transfer heat by passing burning gases through tubes that run through the vat from side to side. The advantage of this design, as compared to direct bottom heating, is that a cooler zone exists below the heating tubes; solids settle into this area away from the high heat and turbulence so that they do not significantly interfere with heat transfer and do not deteriorate as fast. Tubular systems also provide a relatively large heating surface, giving a lower heat ratio per square inch, which can be further improved by installing baffles inside the tubes to increase their heat transmitting efficiency.

Two types of heaters are used in tubular systems. In the atmospheric burner system, gas is fed in a thin stream to an orifice where it draws air into the burner. The gas and air mixture passes through the burner ports and the flame is drawn into the fire tubes within the frying tank. Belshaw states these atmospheric burner systems are simple to install, easy to maintain, and provide good control as well as an acceptable level of efficiency. The second type of tube heater is based on a premix system where the gas and air are blended in the proper ratio and the mixture fed under slight pressure to burner orifices where it is ignited. The mixture burns with a high degree of efficiency and requires no secondary air to complete the combustion. The premix system and power burners are rated superior in efficiency to any other gas-fired system.

Tube-fired systems have some inherent advantages. They create a cold zone, they have a large heat transfer surface in relation to shortening volume, and they have a high heat transfer efficiency. Two disadvantages are the difficulty in cleaning and the need for a larger quantity of shortening for the same output of product.

Electrical heating systems are convenient, though seldom economical so far as energy cost is concerned. They are used mostly in batch fryers and small continuous fryers. Electrical heating elements can be installed in such a manner that the whole assembly can be raised up out of the vat to make cleaning easier. In other designs, the entire heating complex can be removed quickly and completely for replacement or cleaning. In these systems, electricity is used to heat resistance elements contained inside small stainless steel tubes assembled in grids. These grids are fixed in the fryer just above the bottom of the vat. Location of the grids can be varied so that heat can be concentrated in some areas and minimized in others. Electrically heated fryers require relatively small amounts of fat. One manufacturer recommends using three-phase 380 to 440 volt current for these elements.

The oil can be heated in a separate heat exchanger located at some distance from the fryer and transferred to the vat by pumps and piping. An example of a heat exchanger for fryer oil is shown in Figure 11.3. Heat exchangers have been designed for direct and indirect heating of

FIG. 11.3. HIGH EFFICIENCY HEAT EXCHANGER.
SOURCE: HEAT AND CONTROL

the fat. Capacities up to at least 14 million Btu are offered in horizontal direct-fired gas, oil, or propane models; burner systems are completely automatic. Indirect systems have been designed to use chlorinated hydrocarbon heat transfer fluids for carrying the energy from an outside heat exchanger to tubes traversing the oil vat.

In one of the simpler designs of heat exchanger, the fat is circulated through tubes traversing a chamber in which gas is burning. The constant circulation of the fat tends to even out temperature differentials. The vat can be shallower, since it is not necessary to provide a space for heating elements or for collection of debris. This reduces the amount of fat which is required in the vat. On the other hand, additional fat is required to fill up the heat exchanger and the piping to and from it. Debris can be filtered out continuously as the oil is brought back to the heat exchanger.

A commercial low pressure gas-fired cooking oil heat exchanger is available in sizes up to 1,500,000 Btu per hr. It features completely inspectable all stainless steel tube construction designed to meet USDA requirements. Heat is supplied by multi-port inspirator type low pressure gas burners of high thermal efficiency and controlled by a proportioning temperature control.

A patent granted to Bullock (1984) describes one such system. Since the oil is carried through the heat exchanger in very turbulent flow, heat transfer is very rapid and the temperature of the oil flowing through the pipe and the oil at the wall of the pipe will differ by only about 18° to 22°F, compared to temperatures at the surface of the heating element in vats of about 600° to 1,200°F. The lower temperature in the heat exchanger should cause less degradation of the fat.

Although use of an external heat exchanger solves a number of the problems resulting from the placement of heating elements in the vat, it creates others. For example, there is the cost and extra maintenance involved in installing another piece of fairly complex equipment.

Conveyor Systems

For cake doughnut fryers, conveyors generally consist of slatted drop plates on which the dough pieces fall immediately after they are cut. This submerged plate moves in registration with the surface flight bars until the doughnut becames light enough to float, after which the dough piece is moved by the flight bars. The conveyor is coordinated with the cutter so that the doughnuts will be dropped in the proper position between the flight bars. In modern designs, this is accomplished by using a separate cutter drive, a mechanical adjustment to time the cutter signal from the conveyor, and a time-delay relay permitting an adjustment of the interval between the signal from the conveyor drive to the cutter head drive.

Speed of the flight conveyor should be adjustable to give the required frying time. The rate at which the leavening systems of cake doughnuts react depends upon the amounts and types of leavening chemicals used and on the temperature. The time at which a doughnut should be turned over is, in turn, dependent upon the rate of reaction of the leavening system, as well as on some other factors. If the doughnut is turned too soon, it will tend to be low in volume. When turned too late, it may become misshapen, low in moisture, and high in fat. If the conveyor timing is not set in registration with the cutting mechanism, the dough piece as it is falling may strike the flight bars causing misshapen product.

Some conveyor footage must be added to cool the product after it is fried and before it is coated or packaged. It is recommended that the doughnuts be cooled to between 88° and 92°F before packaging. One purpose of this cooling is to slow moisture migration from the crumb to the crust so that moisture droplets will not condense on the package surface. Another purpose is to allow the doughnut to become firmer so there is less chance it will be damaged in the packaging operation.

Fried pie fryers—at least some of them—carry the pies between two mesh belts. The pie is completely submerged throughout the frying process. Pizza crust fryers generally use single belt submerging conveyors since the buoyancy of the expanded dough creates enough friction with the belt to carry it through the fryer.

Various other types of conveying systems have been devised to carry kinds of foods requiring special treatment, but these have not generally been applied to dough and batter products. An example is the rotating drum conveyor described by Bullock (1984). The drum has internal shelves on which the product is placed. During part of the rotation, the food is immersed in hot oil.

Control Systems

There are three types of controllers used for ovens, driers, and fryers: pneumatic, electric, and electronic. Pneumatic controllers receive a mechanical incoming signal and use a relatively complicated air-pressure balancing system to produce a pneumatic force that operates the heater valves. The sensing element can be a liquid-filled bulb connected by a capillary tube to the control instrument. A system of levers multiplies the slight motion of the liquid to an extent sufficient to activate the balancing section. An electric controller also receives a mechanical incoming signal, and uses a simple electric resistance slide wire balance to generate the electrical impulse that operates the gas valves. The electronic controller receives an incoming electric-voltage signal from a thermocouple and processes this through an electronic circuit to modulate the electric current that operates the valves. Most, if not all, modern equipment is based on electronic control.

On-and-off and modulating control systems can be used for controlling the input of heat to the fat. On-and-off controls consist of a thermocouple which activates a thermostat switch; they can maintain the temperature within a range of about 10°F. The more expensive modulating-thermal-control systems can adjust the amount of gas flowing to the burner as required to maintain the temperature within plus or minus 2°F.

Temperature Variation in Zones

There has been a gradually increasing recognition of the value of having different heating zones in fryers, especially in long conveyor fryers for doughnuts (Belshaw 1976), similar to the technology development in tunnel and band ovens. In large doughnut fryers, the demand for heat is greatest in the region where the dough pieces enter the fat and, again, where the pieces are turned over. Also, the increased width and length of fryers being installed today make it impossible to get an even distribution of heat throughout the vat if the temperature is controlled and supplied by one set of heaters. When heaters and temperature sensors are located in areas of maximum demand, a rapid response can be obtained as the need for additional heat occurs.

Most modern fryers have at least two zone temperature controls, so that the infeed end of the vat can be set at a different temperature than the exit end. Leavening system reaction, color, shape, and volume are controlled mostly by the adjustments at the infeed end, while the exit end adjustments control bakeout and moisture loss.

Exhaust Systems

A large amount of vapor is generated by fryers. This vapor is mostly steam but it also includes droplets of fat, gases given off by the frying food, and decomposition products of the fat. Even small batch fryers should be vented, but it is unacceptable to allow the vapor from a continuous fryer to escape into the processing area. Exhaust systems must collect and dispose of the hot gases in a safe and sanitary manner. The fryer may be completely enclosed in the exhaust system or an overhead hood may be used (see Figure 11.4). The totally enclosed design contributes very little total air movement within the room, but hoods exhaust large amounts of air from the processing area along with the steam, and it may be necessary to provide replacement air sources in order to compensate for the loss. All exhaust systems must be designed so as to prevent condensed materials such as fat from dropping back into the fryer.

It is inadvisable to vent the hood exhaust directly to the environment. Cooker vent scrubbers are available for oil emission control. Some of these are based on the cyclone separator method. Water spray is mixed with the cooker gases as they enter the centrifugal blower, where violent turbulence disperses the water and then throws the oil-water droplets to the blower walls by centrifugal force. Separation is completed in a tangential entry centrifugal separator that follows the blowers. There are no filters involved. Such units are quite effective in removing the oil, but other devices are needed for odor control.

For removing odorous compounds and other organic substances, thermal oxidizing units can be fitted to the exhaust system. They mix contaminated fumes with an intense flame in a combustion chamber. Sufficient time and turbulence allow the organic materials to be completely converted to carbon dioxide and water vapor. Figure 11.5 shows an odor control unit combining an oil demister and a system for condensing the odor laden vapors and returning them to the fryer's burner for incineration.

Examples of Commercial Equipment

In the following paragraphs are given details of a fryer suitable for cooking both yeast raised and cake doughnuts. This fryer is available

FIG. 11.4. PRODUCTION FRYER WITH EXHAUST HOOD WHICH IS LIFTED HYDRAULICALLY ALONG WITH CONVEYOR.
SOURCE: DAWN EQUIPMENT

FIG. 11.5. ODOR CONTROL SYSTEM.
SOURCE: HEAT AND CONTROL

in lengths of 12, 16, and 20 ft and widths of 50 and 59 inches. The largest model has a hot fat capacity of 151 gal if the heating tubes are below the conveyor and 130 gal if the tubes are between the conveyor belts, while the largest size has capacities of 358 and 275 gal, respectively. The kettle and all stationary parts within it are constructed of Type 304 stainless steel, while the frame is sanitary heavy gauge tubular steel.

There are independent heat controls for infeed and discharge sections, a shortening high temperature cut-out at 410°F with built-in fail safe operation, and a shortening low level shut down of the combustion system. There is independent electronic flame supervision on the infeed and discharge sections.

This fryer has a traveling baffle conveyor with a choice of 3.75 or 4.62 inch flights, and, optionally, either a full submerging conveyor or partial front and rear submergers. The conveyor system is compensated for thermal expansion, and has self-aligning chemically inert bushings on all bearing surfaces. There is a conveyor synchronizing device. Ratchet assemblies enable the operator to make timing adjustments to both the conveyor and turner while the fryer is running. There is an automatic line-up device for use with automatic proofers.

The electrical system requires 230/460 volts, 60Hz, three phase and a 110 volt, 60 Hz, one phase control circuit. Power required is 1.5 hp for the control circuit, 0.1 hp for the fat pump, 1 hp for the conveyor drive, 1 hp for the air pump, and 0.2 hp for the flue exhaust blower.

Gas requirement for the smallest fryer is 500,000 Btu per hr of natural gas (6" W. C. minimum) and for the largest fryer is 1,000,000 Btu per hr. Air required is 9.5 cfm at 20 psi.

Another frying system made in sizes of 250 to 10,000 lb per hr is described as follows. A sealed tubular frame and insulated tank with hermetically sealed outer skin are featured. Frying oil temperature is held at a pre-set operating level of approximately 400°F by a temperature sensor and controller and never exceeds 500°F. Since the heat is electric, the cooking oil is never exposed to open flame or intense heat. The frying food products travel on an endless conveyor or belt just below the oil surface. A screw conveyor in a v-shaped trough at the bottom of the fryer continuously keeps the oil free of submerged food particles. Material discharged by the screw conveyor together with surface oil is skimmed from the exit end of the fryer and pumped to a fine mesh vibrating screen from which it is circulated through a five micron filter.

Fryers designed for corn or tortilla chip snacks generally consist of a product conveyor belt, a tank containing cooking oil maintained within an optimum temperature range, a gas train which includes a pilot burner and a flame sensor, and a control panel. The extent to which the product is cooked is preferably determined by changing the time it stays in the fryer (through adjustments of the conveyor's variable speed drive) rather than by varying the temperature of the oil. In one unit, angled stainless steel paddles meter chips to a submerger conveyor that holds them beneath the cooking oil until they are released at the discharge end. Heating of the oil is accomplished by the usual methods, i.e., direct heating of the tank, immersion tubes, or external heat exchangers. If the oil is circulated externally, fines can be continuously removed by an in-line catch box, and oil can be automatically added as needed by a pneumatic oil level control system. Processing rates of available sizes range from 350 to 2,000 lb per hr of tortilla chips.

A large fryer for pies is shown in Figure 11.6.

SOURCE: DAWN EQUIPMENT

FIG. 11.6. A FRIED PIE COOKER.

Frying Temperatures and Times

Frying time, frying temperature, and fat level in the vat are three variables that can be controlled by the equipment operator. Fat levels that are too low can produce misshapen or flat products, rough surfaces, and high fat absorption. If the fat levels are too high, the doughnut may crust over before it reaches the turner or it may float over the flight bars and fail to turn properly. The surface of the fat would ordinarily be kept within about 2 to 3 inches above the top of the drop plate.

For yeast leavened doughnuts, frying temperatures in the range of 375° to 395°F have been recommended (Braden 1976). Optimum frying times depend upon the weight and configuration of the dough piece, among other factors. For regular round doughnuts weighing 18 to 19 oz per doz, unfried weight, an average total frying time of 110 to 120 sec is considered normal. This allows about a minute on each side. Higher temperatures within the normal range may encourage greater expansion, but excessive temperatures firm up the skin too fast so that maximum expansion cannot occur. Frying at temperatures which are too low causes a slower coloring of the crust and allows too much fat to penetrate the crust before the interior is cooked. Settling, or collapsing of the white ring, is due to a frying time which is too short. Frying times which are too long lead to excessive fat absorption.

"Rise time" is the time it takes for the doughnut to come to the surface, measured from the time it drops from the cutter into the fat until it appears on top of the fat. It is usually about 3 to 7 secs depending on the size and type of doughnut. Variations in rise time can indicate problems with the leavening system, or other processing difficuties..

In the cooking of fried pies, some additional problems are observed. The crust must be thoroughly cooked to the desired medium brown color and the filling must be cooked as much as necessary before significant steam evolution occurs in the interior of the pie. Boilout or bursting of the pie crust leads to contamination of the frying fat with rapid breakdown. Of course, the pie itself must often be rejected when this occurs. One operator (Taylor 1971) recommends 3 min 22 sec at 385°F as suitable conditions for a 4.5 oz fruit pie. Other authorities (Harding and Starwich 1965), specified 3 min 45 secs for 3.5 oz pies frying at 355°F, and 4 min 15 sec for 4.75 oz pies frying at 380°F. These conditions are used with submerging conveyors.

Fryer Sanitation

The elements of a total sanitation program are (1) master cleaning schedules, (2) detailed cleaning procedures, (3) fundamentals of cleaning, (4) pest control program, and (5) inspection program (McMurray 1976). The master cleaning schedule is an itemized list of equipment and utensils with predetermined intervals for cleaning. Detailed cleaning procedures are written out for virtually all utensils and machinery. Since the primary emphasis of this discussion is on fryers, a sanitation check list from the cited article for a continuous fryer will be reproduced in modified form.

****

SANITATION/FRYERS—CHECK LIST
CONTINUOUS EQUIPMENT

Cleaning required weekly, more often if production volume dictates. Time to complete, approximately 4.5 hours

PROCEDURE; 1. Remove and clean exhaust hood filters. 2. Pump shortening from reservoir into fryer; close valves. 3. Unhook inlet and outlet pipes of reservoir. 4. Hook up drain hose from fryer to drain. 5. Wash reservoir and wipe down dry. Close outlet valve. 6. Turn heat off and filter hot shortening to reservoir. 7. Attach hand hoist to conveyor, disconnect clutch, and lift conveyor out of fryer. Scrape and clean where necessary. 8. Scrape out sides and bottom of fryer. Remove overflow pipe. Install plug. Open drain valves. 9. Lower conveyor back into frying position. 10. Rinse fryer and conveyor. Flush with clean water. 11. Close drain valve of fryer and start to fill with water. When water covers tubes by one inch, turn heater on. Pour in cleaner solution. Allow to soak 45 min at 200°F.

****

Microwave Assisted Frying

In an earlier chapter, the use of microwave heating to speed up the proofing of yeast-raised doughnuts was mentioned. Doughnuts which normally take 25 to 45 minutes to proof, can be proofed in about four minutes if heated by microwaves. It was also mentioned that microwaves have been used, or tested, for reducing the moisture in baked crackers so that their time in the oven can be reduced. Microwave heating has been used for much the same purpose (moisture reduction without excessive color development) in finishing potato chips. It is not a large intellectual leap to consider the use of microwaves for heating of other fried products, such as doughnuts.

Stein (1972) discussed the cooking of cake doughnuts in a fryer that had an input of microwaves as well as heat from the conventional heat sources. In this type of microwave frying, when doughnuts are heated in the frying oil on the first side, microwave heating is also applied. This heats the interior of the doughnut so that its temperature rises much more rapidly than it would in the absence of the microwaves. The doughnut expands to its fullest while the first side is being fried, and there is no dense center portion. The total volume of the product is larger as it comes out of the fryer. Fat absorption is reduced by 25%. However, Stein says these results are obtained only on blooming-type doughnuts which are characterized by a relatively thin crust, low density, an open crumb, lower fat absorption, and a greater expansion.

The microwave equipment used in the preceding experiments was a large conventional continuous doughnut fryer with a microwave cavity mounted only on the first half of the fryer. The power packs were placed at remote locations but were controlled from a single control panel.

Infrared Frying

Contact frying, or a simulation of contact frying, represented as being suitable for pizza, pancakes, and sandwiches, can be effected by infrared irradiation, according to the patent of Dagerskog et al. (1986). The invention consists essentially of a single solid endless thin steel conveyor belt which is heated from below by short wave infrared radiation. Automatic control is provided by photocells and thermocouples connected to computerized heat modulating equipment.

FRYING FAT

The frying fat is an integral part of the cooking system for fried dough products. It also becomes one of the ingredients and it acts as a binder for powdered sugar (or similar materials) applied to the surface, but this discussion will be concentrated on its role as the heat transfer medium. It has become well understood over the last three or four decades that the specifications for frying fats can be quite different from those for ingredient shortenings.

Fat Absorption during Frying

Fat absorption of 2.5 to 3.5 oz per doz for yeast leavened doughnuts weighing 18 to 19 oz per doz raw weight is considered normal. Richer formulas generally absorb more shortening. A large doughnut operator frying many varieties (cake and raised), found that an average of 20 to 21% shortening was absorbed based on the total weight of the batters or dough. There is considerably more absorption when doughnuts with ragged surfaces are being fried; "old fashioned" doughnuts take up about 24 to 26% fat, while crullers may go as high as 28 to 32% (Goodsell 1984).

For a 4.5 oz fried fruit pie, cooked at 385°F for 3.3 to 3.5 min, an acceptable range for fat absorption is said to be 3.5 to 5 oz per dozen.

Specifications

A high smoke point and excellent stability to hydrolysis and oxidation at elevated temperatures are essential qualities for fats used for frying doughnuts, honey buns, etc. In most markets, the frying fat is expected to contribute very little flavor of its own to the finished product, although consumers in some regions of the country are accustomed to the flavor of lard. Hydrogenated cottonseed oil is one of the most common frying fats, and it is bland.

The fresh fat, as it is received by the doughnut manufacturer, may not be completely satisfactory for frying becauses its heat transfer characteristics—partly a function of the viscosity—are not optimal. If the fryer is replenished from time to time with small additions of fresh fat, an equilibrium state can be maintained in which the finished product always has satisfactory properties. When the frying equipment contains only fresh fat, it must be "broken in" by heating the fat at frying temperatures in contact with product until a free fatty acid content of about 0.4% is reached, or until the product is being cooked as desired. The increase of viscosity is probably due mostly to the accumulation of polymeric substances which develop from oxidized fatty acids.

Fat is decomposed to free fatty acids, glycerol, etc., by steam from the product, by oxygen from the air, and by heat. Oxidation also forms hydroperoxides which can transform into volatile flavor compounds or polymerize to gummy substances which accumulate on the bottom or around the rim of the frying vat. These long chain, poorly characterized materials, are probably responsible for foaming, as well.

Excessive deterioration is evidenced by smoking, darkening, and foaming. Methyl silicone is often effective in reducing foaming. Filtration is effective in removing particles that can char and discolor the finished products, but it does not reverse oxidation which has already occurred and it does not affect the free fatty acid content or the viscosity (Moyer 1965). Some types of filtration, where the fat contacts adsorbent compounds, will reduce free fatty acids. Copper ions greatly accelerate the development of oxidative rancidity, so fats should be protected from contacting copper, brass, or bronze utensils or fittings at all times.

Some typical requirements for frying fats are (Fischer 1976):

Free fatty acids (% as oleic)	0.05% max.
Smoke point	400°F min.

Wiley melting point .... 112°F + or-2°
Iodine number for vegetable fat .... 70 + or-1
Iodine number for animal fat .... 55 + or-1
Peroxide value (meq/kilo) .... 0.5 max.
Stability (AOM hr) .... >90
Odor .... Bland
Color (Lovibond) .... 20 Y, 2 R
Solids content index, percent:

50°F.....................37 to 43
70°F.....................25 to 29
80°F.....................20 to 24
92°F.....................15 to 17
104°F.....................10 to 12

Crystallization rate (percentage solids when melted shortening is cooled for the given time at the given temperature):

92°F for 30 min............1 to 7
92°F for 120 min..........13 to 16
80°F for 30 min...........14 to 17
80°F for 120 min..........16 to 19

The free fatty acids specification gives an indication of the effectiveness of the refining operation applied to the shortening and of the deterioration it has undergone since then. Melting point and solids fraction index are related to the eating quality of the doughnut and stability of the coating; an SFI which is too high may cause the doughnut to be waxy in texture and could result in poor adhesion of the coating sugar. An SFI which is too low may result in a doughnut which is too greasy and which lets oil penetrate the coating. Smoke point is important because it may indicate poor performance if too low.

Fat will always break down to some extent during frying. Generally speaking, the higher the process temperature the faster the shortening deteriorates. This deterioration is accelerated if the fat is maintained at the frying temperature with no product going through the line. Any condition which brings air into the fat—bubbling, spraying, spattering—will cause increased breakdown. Contact with brass or copper at any point will greatly accelerate oxidation.

Filtering and Other Cleaning Operations Applied to Fat

Various types of filters and other reconditioning equipment can be employed to prolong the useful life of frying oils by removing small particles and other debris which accelerate the development of free fatty acids and add objectionable dark specks to the fried object. An example of a filter press is shown in Figure 11.7. When an operator

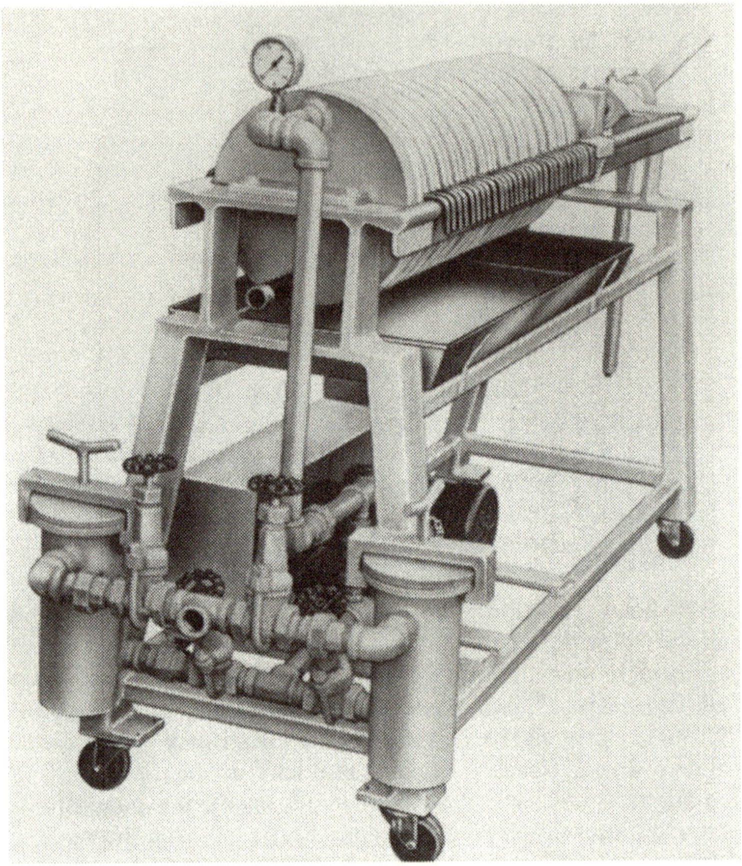

FIG. 11.7. A 30-DISC 12-INCH DIAMETER FAT FILTER WITH IRON PUMP AND DUAL BASKET STRAINER.

SOURCE: STAR FILTER

determines that the oil in a fryer needs to be purified, the shortening is first pumped into a holding tank which is heated to keep the oil liquid. Normally, the fryer will be thoroughly cleaned at this time as part of the overall procedure. The hot shortening is pumped through a series of circular filter plates held in the press frame.

A common type of filter pad is a disc of nonwoven cotton cloth, but other materials can also be used. Filter aids, such as diatomaceous earth, are often mixed with the oil in the holding tank to absorb fatty acids, odorous substances, and other undesirable chemicals as well as to improve the efficiency of particle retention. After the filtration step has been completed, the filter discs with their adhering coating of diatomite and entrapped residue, are discarded. The oil that has passed through the filter press may be withdrawn into drums, another holding tank, or a clean fryer. If the product that is being fried takes up a large amount of oil, requiring the frequent replacement of the frying media with fresh oil, it may not be necessary to discard any of the "old" fat if filtration is properly applied (Burns 1981).

Another type of filtering system uses a roll-stock filter cloth. In this equipment, oil flows by gravity through a concave filtering area formed by the conveyor belt that holds the filtering medium. An oil-level control, coupled to the conveyor drive, automatically advances the filter medium when the flow through it is reduced below the incoming flow rate. Oil that has passed through the filter cloth is collected on the sloping pan bottom and continuously discharged through a drain connection. A hood with a vent stack opening covers the entire filter area, and a clean-in-place spray ball is located under the hood for connection to the fryer's CIP system. USDA-approved filtering media for this equipment are available in 250 yd rolls in a range of porosities suitable for different requirements. The used filter pad can be collected in a portable disposal cart or other container. An example of a very high capacity continuous filter is shown in Figure 11.8.

One type and model of filter used in cleaning frying oil has the following characteristics. It is a 20-disc round plate and frame fabricated filter with 18 inch diameter plates and frames and 1.5 inch wide sludge frames built to USDA guidelines on an 87 inch long stand. It is equipped with a 3 hp motor and Viking pump. The filter is dressed with #218 filter media between each plate and frame and changed every day after eight hour shifts during whick 4,500 lb. of product is cooked. The single stage media is either 100% rag content pure cellulose or 100% pure nonwoven cotton. It removes particles down to 2 or 3 microns.

One manufacturer describes their line of fat filters as follows. These square plate and frame filter presses are completely constructed of plates, bars, and forgings. Standard sizes run from 13 to 60 inches square. Standard construction material is #304 stainless steel, but carbon steel and #316 can be used. Standard working pressure is 75 psi. Maximum working temperature is 425°F. The filter frames are of open construction with support bars between ports. Both sides of the plates are covered with heavy perforated sheet metal and have strong internal reinforcing and screen support bars to give the maximum clear filtration area. No cloth is required to support the filter media. Front and back heads are made of forged steel and are heavily clad

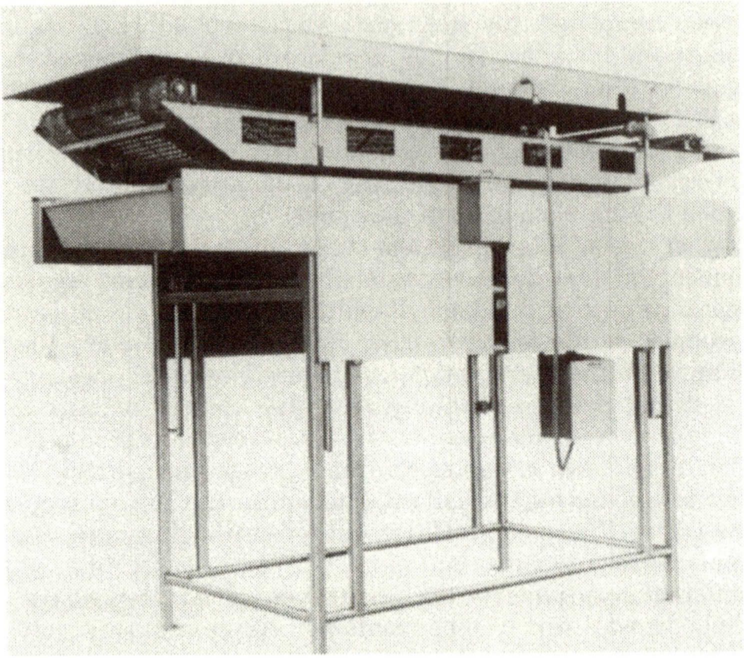

FIG. 11.8. CONTINUOUS OIL FILTER.
SOURCE: HEAT AND CONTROL

with stainless. Standard unit construction is closed delivery, 2 or 4 ports, top or bottom inlet. All units are suitable for filtration using paper, canvas duck, synthetic fiber, or submicron filter media together with filter aid and diatomaceous earth.

These devices can be augmented with a drum type cooking oil pre-filter, especially if the foods being fried shed a lot of fairly large particles. The pre-filter passes all the oil in the circulating system through a revolving mesh screen drum. This removes large particles prior to final filtering. A continuous pneumatic system blows fines from the screen to a conveyor, which carries them to a steel drum or other receptacle.

Filters are also available for batch frying operations. One manufacturer supplies models with 100, 200, or 250 lb capacity. A typical arrangement is to have the filter, with its motor, pump, and tank, on casters so it can be rolled into place near the fryer when required. This allows one filter to be used with several fryers. It is claimed that less than 10 min are required for draining the shortening and returning it to the kettle. Disposable filter pads are used.

Heat Economizers

Manufacturers of frying equipment have offered equipment which conveys incoming product through the steam and vapor emerging from the vat (see Figure 11.9). During its passage above the vat, the raw product absorbs some of the oil and becomes heated, reducing the amount of heat and oil it requires when it reaches the fryer. In addition, the effluent gas continuing on to the exhaust is reduced in volume and its oil content is reduced, making it easier and cheaper to cleanse. These economizers have been applied mostly to potato chip fryers, but they may have some value in dough product frying.

NEW PRODUCT POSSIBILITIES

In developing new bakery products, most technologists shy away from using frying as a cooking method. There are several reasons for this attitude. Baking is, in many ways, simpler than frying. Since ovens are easier to automate, generally cleaner to operate, and usually cheaper to purchase than fryers, they are often the first and last choice of the technologist when a new product is being designed. Some opportunities may be missed if frying is not considered as an alternative, however. The special flavor contributed by frying, so characteristic and appealing in doughnuts, should not be ignored when projecting the marketing potential of new products. Crust texture and appearance of fried bakery-type items may have unique advantages which the consumer could appreciate—the upsurge in fried crusts for frozen pizzas is certainly a case in point. It may be that the ability to cook unsupported dough pieces, such as doughnuts, gives the technologist opportunities to process foods which are not suitable for baking in pans.

The author has suggested from time to time the possibility of frying small snack crackers, based perhaps on the oyster cracker shape or on the small round or fish-shaped crackers now being offered as premium snack crackers. Unique flavor and better crust color could be expected. The texture should be flakier and the appearance more rounded. Frying should particularly benefit cheese crackers. Similarly, small bread rolls could be fried to give new and very acceptable products. Certain types of cookies and pretzels might be adaptable to frying. If a crisp crust and soft center is desired in cookies, is it possible to batter and bread a chewy cookie dough and fry it? Sort of a chicken-fried chocolate chip cookie.

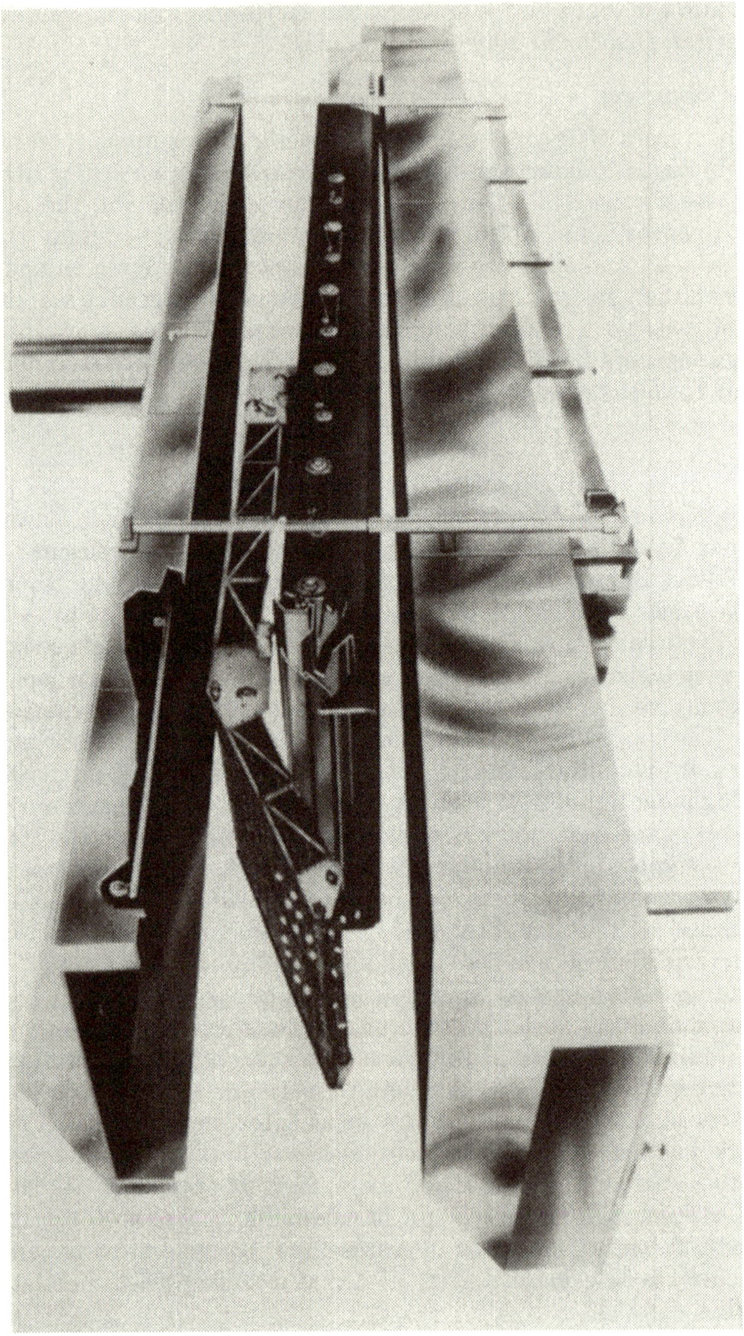

FIG. 11.9. OIL AND HEAT ECONOMIZER. CONVEYOR SYSTEM MOUNTED INSIDE COOKER HOOD TO PREHEAT INCOMING PRODUCT WITH FRYER VAPORS.

SOURCE: HEAT AND CONTROL.

Honeybuns, a type of cinnamon roll, is a fried product which is very popular. Is there an opportunity to fry more complex sweet rolls or even coffee cakes? Small fried pies are eaten by the millions every day—is it possible to fry large round fruit pies? What happens if you fry a cream cracker?

One point which emerges from the preceding discussion is that each bakery laboratory or pilot plant should have at least a small batch fryer for test cooking even though every operation in the existing plant is geared toward making baked products.

BIBLIOGRAPHY

BELSHAW, T. E. 1976. Cutting and frying equipment. Proc. Am. Soc. Bakery Engineers *1976,* 112-120.

BLOCK, Z. 1964. Frying. *In* Food Processing Operations, Vol.3. M. A. Joslyn and J. I. Heid (Editors). AVI Publishing Co., Westport, CT

BRADEN, B. W., JR. 1976. Yeast-raised doughnuts. Proc. Am. Soc. Bakery Engineers *1976*, 127-131.

BULLOCK, R. F. 1984. Fryer with oil circulation and conveyor. U. S. Patent 4,478,140.

BURNS, N. R. 1981. Personal communication. Star Systems Filtration Div., Timmonsville, SC

DAGERSKOG, M., GANROT, A. G., and JONSSON, K. O. G. 1986. Method and apparatus for frying. U. S. Patent 4,565,704.

FISCHER, L. G. 1976. Cake doughnuts. Proc. Am. Soc. Bakery Engineers *1976*, 121-126.

GOODSELL, G. R. 1984. Cake doughnut production. Proc. Am. Soc. Bakery Engineers *1984*, 119-131.

HARDING, M., and STARWICH, B. G. 1965. Formulation and production of fried pies. Proc. Am. Soc. Bakery Engineers *1965*, 290-298.

MC MURRY, T. K., II. 1976. Sanitation and safety. Proc. Am. Soc. Bakery Engineers *1976*, 132-144.

MENDOZA, F. C. 1987. Oven for preparing fried food products. U. S. Patent 4,711,164.

MOYER, J. 1965. Selection, maintenance, and protection of frying fats. Proc. Am. Soc. Bakery Engineers *1965*, 273-279.

SIMS, R. J., and STAHL, H. D. 1970. Thermal and oxidative deterioration of frying fats. Bakers Digest *44*, No. 5, 50-52, 70.

STEIN, E. W. 1972. Application of microwaves to bakery production. Proc. Am. Soc. Bakery Engineers *1972*, 46-51.

TAYLOR, J. C. 1971. Fried fruit pies, production methods. Proc. Am. Soc. Bakery Engineers *1971*, 182-191.

CHAPTER TWELVE

APPLICATORS FOR ADJUNCTS

INTRODUCTION

This chapter contains descriptions of finishing and decorating equipment. Depositers of jellies, marshmallow, and other adjuncts, applicators of sugar, salt, colored particles, chocolate vermicelli, and the like, injectors of fillings such as whipped cream and Bavarian cream, pizza toppers, spreaders of icings, frostings, and cremes, sprayers for washes and oils, and enrobers for chocolate and other fat based coatings are included. The general term "finishing" can be applied to all of these operations, and in this book the word "adjuncts" is used for the materials which are added because it seems to be the only common term broad enough to encompass all the many materials of very diverse qualities to be discussed here.

DEPOSITORS FOR PARTICLES

Powders and Granules

Topping applicators are used with all sorts of bakery goods, even bread and rolls, but they are especially prominent in cookie production where they can be found at many places along the production line. Salters are usually placed after the cutter and in front of the oven. Sprinklers for coconut, nonpareils, and the like are often placed after a marshmallow depositer and before the skinning conveyor. This ensures maximum stickiness of the marshmallow and therefore, good retention of the granules. Most of this equipment is portable, so that it can be put on or brought off the line as needed.

General purpose depositers—Manufacturers offer machines which can be adapted to deposit particles of different materials. It may be necessary to use change parts, such as screens with different size apertures, to adapt the equipment to a particular ingredient. These machines are often among the simpler designs, since the more complex machines tend to derive their marketing advantage from a more efficient way of depositing a specific ingredient, and the design features leading to this efficiency often limit their applicability to other ingredients. As a result, general purpose depositors are often based on the principle of rolling, pressing, or brushing particles through a screen placed at the bottom of a hopper. Others use interchangeable types of rotating shafts to dispense the material.

Some of the specifications for an industrial dispensing machine which can be adapted to deposit several different materials are as follows. The rotary dispensing shaft for dusting is a knurled rod in a choice of extra fine, fine, medium, coarse, or extra coarse to match the average particle size of the material to be dispensed. For depositing coarser materials, a grooved rod, with shallow, medium or deep corrugations, is used. When seeds are to be sprinkled, a drilled shaft, with shallow, medium, or deep holes, can be obtained. Two nylon brushes outside the hopper are counter opposed to the rotary dispensing shaft. Both brushes are secured in metal holders and are adjustable by thumb screws for regulating their contact with the dispensing shaft. Brushes are channel locked; a construction designed to prevent shedding of bristles. For larger particles, one of the brushes is replaced by a deflector plate or a pattern control plate. One agitator blade oscillates inside the hopper. It is linked to two movable walls inside the hopper. This combination improves flow of the material being dispensed and reduces or prevents bridging.

There are four adjustments which can be made on the above described machine to aid in obtaining the desired rate of deposit of the topping. The rate of rotation of the dispensing shaft can be varied by adjusting the motor controller unit. Four thumb nuts adjust the outside brush holder so the contact of the brushes with the rotary dispensing shaft can be modified. Thickness of the flow is controlled by the two density adjusting plates bolted to the outside of the hopper. These are positioned by elongated bolt slots and wing nuts. By loosening the wing nuts, these plates may be moved up to increase the flow, or downward to reduce the flow. Width and pattern of the deposit are regulated by block offs and slide adjusters operated manually from the end of the hopper away from the motor (Anon. 1972).

Flour dusters—Dusting flour is the material for which the greatest number of dispensers are made. Almost any dough handling line has a space for a flour duster, and most lines have several of these devices. Flour is a rather difficult powder to deposit uniformly because of its tendency to hang up in hoppers and to clog sieves. One important goal the designer must keep in mind is the prevention of accumulations of flour being dumped on the dough surface; these are very deleterious to dough properties. The usual type of mechanism placed in these machines is a rocker arm assembly moving above a screen. Often, there will be a rotating spindle below the hopper outlet to further break up the flour and spread it more or less uniformly over the dough sheet. Rates of deposit can be adjusted by changing the distance moved by the rocker arm. The width of the slit beneath the hopper can often be adjusted by moving metal slats in or out.

Starch dusters which surround dough pieces with a cloud of the powder in a slightly vacuumized chamber are also available.

Salt dispensers—Salt is sprinkled on many types of products: soda crackers, pretzels, and salt sticks being a few examples. Size of the granule determines to some extent the type of depositer which is needed. If it makes little difference to the consumer whether the granule is large or small, or if there is a wide distribution in particle sizes on the product, then the shaking depositers are adequate. If it is important to keep the particles as large as possible, an apparent advantage for salt sticks and pretzels, then a more defined pattern is desirable, suggesting that a more complex type of machine may be required. This is also thought to be true for topping salt applied to soda crackers.

Simple mechanical sprinklers may use a vibrated or reciprocated screen or a rotating grooved bar to deposit sugar, salt, or spices from a stainless steel hopper on to the baked product. Rotating brushes to push the powder into the dispensing element and to bring it out again are common additions to the device. The brushes help prevent the clogging which would certainly affect the amount of powder going through a screen. Pressure of the brush can often be regulated to modify the rate of deposit. Often, there is an adjustable damper regulating the infeed. The powder commonly falls first onto a distributor plate where it slides for a brief time before it falls off onto the product.

The topping salt used on crackers is often a coarse screening of natural flake salt. These flakes are preferred because of their white color and dissolving characteristics, but their irregular corrugated shape and fragile form complicate the dispensing operation. The particles tend to intermesh and dispense as loose aggregates of several particles, producing clumps on the cracker surface, rather than an even distribution.

Mechanical roller type salt dispensers usually incorporate one or more mechanical aids such as brushes, serrated rollers, or vibrators to break up aggregates. These devices almost always cause attrition of the fragile flakes, however, producing fines that not only disrupt flow through the dispenser but also dissolve quickly when they contact cracker dough. This is to be avoided because rapidly dissolving salt fines can cause objectionable brown spots and blisters on the finished cracker. Coarse flake salt requires a dispenser that will feed it uniformly without resorting to mechanical action that can crush or grind the salt. The electrostatic salt dispenser was developed to meet this requirement. Initially, it was used for adding salt to potato chips but it was later found to be suitable for baked products.

The electrostatic dispenser maintains uniform spacing of particles during depositing by inducing a negative static charge on their surfaces, causing the particles to mutually repel each other. In a side-by-side test on two cracker lines, this dispenser was compared by Strietelmeier (1969) with a mechanical depositer. Two comparisons were made: (1) Actual concentration of salt at each sampling point and across the roll to determine how much variation occurred within a given sampling period, and (2) A salt particle count within two predesignated areas on representative crackers taken from each sampling point to determine the degree of dispersion on individual crackers.

The range of salt levels for the electrostatic salter line was 2.2 to 3.6%, while the conventional salter line showed a range of 2.0 to 4.2%. There were significant variations for both salters at all three sampling points. Across the width of the sheet, eight of the fourteen sampling points showed significant variations in salt deposited by the electrostatic machine as compared to eleven points for the conventional salter. So far as uniform distribution on the cracker surface was concerned, it was apparent from the larger number of particles on the crackers obtained from the line with a conventional salter that a severe degree of particle breakdown had occurred. The standard deviation of difference and the coefficient of variability of salt distribution on the electrostatic line were considerably less than on the line with the mechanical salter.

Some salters for biscuit lines are combined with a panner conveyor so as to form a link between a rotary cutter for cookies or crackers and the oven band while, at the same time, applying salt to the dough pieces. An example of an electrostatic salter combined with a panner conveyor is shown in Figure 12.1). Excess salt falls through a mesh band to a salt recovery system; fully automatic recycling systems can be used.

Pneumatic salters rely on air currents to transport salt to the dispensing area and to disperse it over the dough. Similar equipment is used for spices. In one type of equipment, an air proportioning valve is combined with a clean-sweep agitator valve to draw the salt or spice from a pressurized hopper either on a continuous or an intermittent basis. The granules are then carried pneumatically to a spreader which dissipates the air and allows the material to shower down on the food

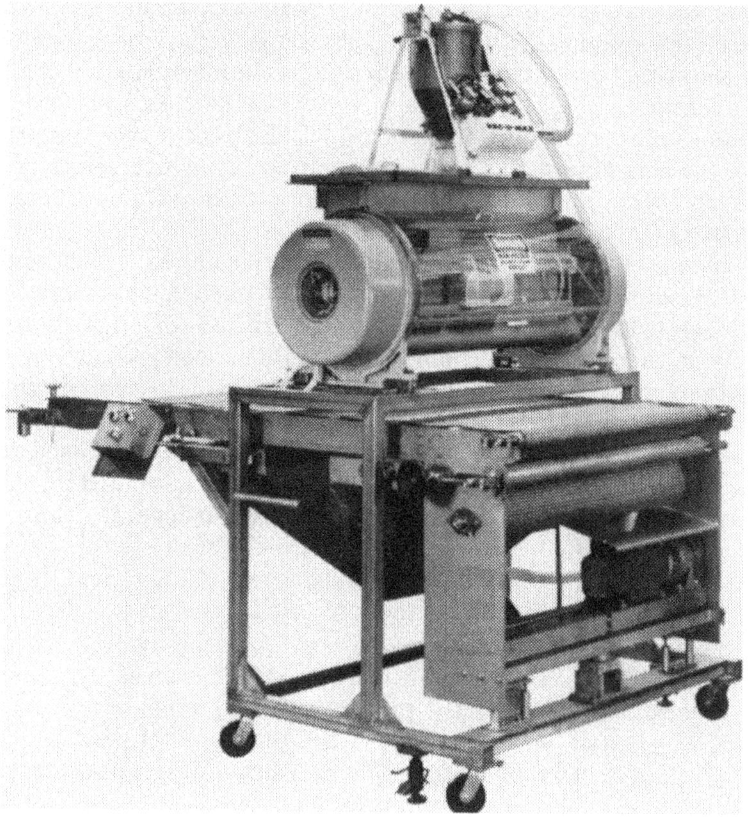

FIG. 12.1. AN ELECTROSTATIC SALTER.
SOURCE: APV (WERNER LEHARA)

product. Discharge may also be made through a variety of nozzles directly into tumblers or batch mixing vats. The dispenser can be controlled manually or automatically by photocell, timer, or other switching device. There are specialized valves which provide instant material flow at a predetermined rate, and are said to be highly accurate and easily adjustable, with complete cutoff. The equipment is claimed to be adaptable to many hygroscopic or hard to handle powdery substances (Anon. 1979). Some of the distribution control means which can be used on this machine are the wide pattern distributor, the fan pattern nozzle, and the cone pattern nozzle.

Another pneumatic salter has a low profile portable stainless steel cabinet for salt storage and the metering and operating equipment.

Inside the cabinet, salt is dispensed from a tapered hopper onto a woven wire metering belt which is driven by a variable speed motor. The metering belt strains salt evenly into a chamber where it combines with compressed air. From the metering cabinet, salt is blown to a stainless steel distribution unit mounted directly over the product area. As salt enters this unit, it is routed through a series of polyurethane tubes and onto the product from stainless steel nozzles (see Figure 12.2).

A patent (Watkins 1970) describes a more complex design of pneumatic condiment applicator. The condiment particles are mixed in an airstream which then passes through an ionized region to charge the particles, then through a chamber containing a food product having an electric charge opposite to that of the particles. The electric charges cause the particles to be attracted to the product, so that nearly all the particles are deposited on the product. The chamber is of special construction having a food product container which is rotated by the airstream. As described in the patent, this operation appears to be a batch or a semi-continuous function. The commercial status of this concept is not known to the author; the primary interest of the patent lies in its exposition of the principles of the ionization dispersion process.

Other dispensers for salt, seasonings, and toppings rely on a belt distribution method. The mesh dispensing belt, moved by a variable speed drive, is as wide as the product belt and is placed a few inches above the product. A hopper, the width of the dispensing belt, holds the seasoning and meters it onto the belt. Multiple belt shakers assure positive clearing of the mesh to prevent clogging.

Sugar sprinklers—Many of the same units recommended for salt can be used for sugar. One of the simpler units consists of two stainless steel hoppers (one feeding the other to insure uniformity of flow), with a sloping vibratory distribution plate at the base of the lower hopper. The amount of vibration can be altered to control the amount of material fed.

Sugar dispensers are found on many cookie lines. A representative example is described here. This portable sugar topper works in conjunction with a forming machine. A hinged mesh extension reaches between the die cups of the forming machine and the oven band. As products are cut from the dies, they drop onto the mesh conveyor. The belt moves the pieces under the sugar dispensing roll where they receive a deposit of topping from the continuously falling granules. Excess sugar falls through the open mesh conveyor onto a cross conveyor for collection and re-use. The sugar coated articles are then discharged

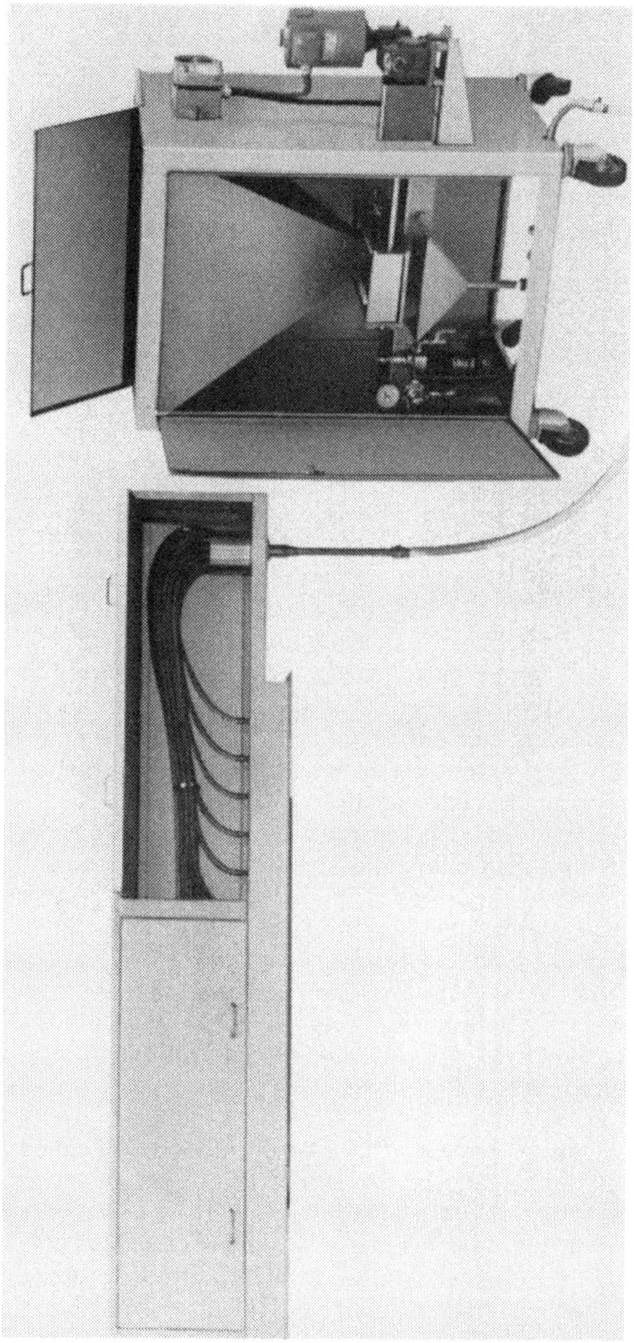

FIG. 12.2. A PNEUMATIC SALTER.

SOURCE: HEAT AND CONTROL

onto the oven band. The sugar discharge mechanism is made up of a hopper with an adjustable gate, a special fluted dispensing roller four inches in diameter, and a sugar agitator shaft controlled by a cam mounted on the dispensing roll shaft (Anon. 1986B). A typical portable sugar sprinkler with recovery conveyor for removing the excess sugar is shown in Figure 12.3.

Sugar is often applied to doughnuts in cylindrical tumblers. These come in several sizes, some suitable for small doughnut shops. The cylindrical chamber is horizontally aligned, with a slight slant downward to the exit. Generally, there is a variable speed electrical motor for the drum and another one for the sugar dispensing system. In some cases, there is an elevating conveyor for automatic recycling of the excess sugar.

Depositors for cinnamon and other spices—Cinnamon and other spices of small particle size can often be deposited satisfactorily by machines very similar to flour dusters or sugar sprinklers. Where possible, the spices are mixed with sugar or some other carrier to make it easier to deposit the minor amount of spice. The screen size of the sugar or salt dispenser may have to be modified to handle the finer grained spice. Because these spices are quite expensive relative to flour, it is important to avoid waste, and some of the depositors for them have lever arm rollers which activate the dispensing mechanism only when dough is passing below the hopper. Also, retractable tabs an inch or so wide are placed across the orifice of the hopper so the pattern of spice deposit can be closely controlled.

When spices are being applied to pieces (such as doughnuts or snack items) in a rotating tumbler, the powder is either sifted into the cylinder or blown in. If it is blown in, the equipment will consist essentially of a hopper, a vibratory feeder, a blower, and a tube extending into the tumbler cavity. Such machines are also used for applying salt to snacks such as corn curls. Sometimes, the air is heated to facilitate handling of the more difficult flavoring powders. Other models of these machines use auger feeds to delivery the powder through a long tube to the dispensing point.

Recovery units—Except in the case of crackers, the baked product will not cover the belt completely. Since the dispenser continues to drop powder on the belt in the spaces between the pieces of product, some of the sugar or salt will fail to be intercepted by product and will fall on to or through the belt. It is always advisable to apply these powders to products being carried on a wire mesh belt. An obvious solution to the

FIG. 12.3. A PORTABLE SUGAR TOPPER WITH RECOVERY CONVEYOR.
SOURCE: APV (WERNER LEHARA)

recovery problem is to let the powder fall on the floor and collect it with a broom and dustpan, but this is likely to cause raised eyebrows on the sanitation superintendent. Simply placing a box under the mesh belt is also possible, and sometimes done, but more sophisticated types of recovery equipment are available to collect the surplus powder in a sanitary and efficient manner. Some of these devices automatically recycle the topping by conveying it from the collecting vessel to the dispenser hopper.

A typical surplus recovery unit will consist of driven rollers through which surplus sugar or salt passes, and a screw recovery from the collection trough. Discharge is to a removable stainless steel box fitted on the outside of the framework. For automatic recycling, an continuously or intermittently running elevating conveyor of any appropriate type brings the material around the side of the belt and drops it into one side of the hopper.

Nuts, Raisins, Crumbs, and Other Large Pieces

Crumb and crunch can be deposited on doughnuts or other small baked products in tumbling drums. These units are always used in combination with a glaze applicator, the glaze forming the adhesive which holds the crumb on to the doughnut surface. Typical capacity of one of these units would fall in the range of 600 to 2,000 dozen per hour. Small units are available for retail shops.

Seeding units are available with patterned deposit capability. These machines can accurately place a round, square, oval, or rectangular pattern of seeds on buns or bread. The pattern plates are easily changed. The depositors are said to accurately apply grain, bran or flour as well as other toppings.

Some nutmeat feeders have a belt conveyor running beneath the hopper. A gate at the hopper exit controls the rate of nutmeat removal, in conjunction with belt speed. An agitator shaft with fingers spreads the pieces across the belt. One manufacturer advertises a unit for dispensing nutmeats directly into rotary molder cavities.

Coconut is particularly difficult to deposit uniformly, especially if it is in shredded form. The systems used to deposit the smaller size coconut particles conform in a general way to the concepts of the other chunk dispensers described here. The rate of dispensing is controlled by the stroke length of an eccentric movement. The eccentric shaft drives a screen which controls the rate at which coconut drops from the hopper onto two rotating dispersing wheels. Screens of different mesh sizes are used for different granulations of the topping material. A gate is provided at the bottom of the feed hopper to control the amount of topping material fed to and by the Syntron feeder. A blower is furnished for removal of excess coconut from the cookies at the discharge end. With proper adjustments, the same feeder can be used for chocolate sprinkles, nonpareils, etc.

Instead of dropping the nuts or other additive on to the dough piece, a layer of nuts can be placed on a belt and dough (cookies) dropped on the nuts, after which the dough is pressed down sufficiently by an idler roller to secure some adhesion of the nuts. Excess nuts are allowed to drop through a slot to a recovery bin at the transfer point. When caramel is dropped on the nuts, and the combination chocolate coated, turtles and similar confections are formed. Caramel nut cluster lines include a caramel depositor of the positive displacement piston type working out of a water jacketed stainless steel hopper. Also, a nut dispenser, conveyors, cooling tunnel, and nut recovery unit.

GLAZE AND WATER ICING APPLICATORS

Glazes are one of the main decorating and finishing adjuncts applied to doughnuts. Water icings are applied to doughnuts, cookies, and some cakes. These two materials overlap in composition and physical properties, the only real difference being that glazes are generally applied in thinner layers which are transparent or translucent, while water icngs are inclined to be present in thicker layers and are usually opaque. Water icings contain more sugar crystals, which accounts for their opaqueness and also for the greater viscosity which makes it necessary to apply them in thicker layers.

Icing of this type is applied to cookies by a thin (about 0.6 inch diameter) rotating cylinder which picks the icing (sometimes heated) off a larger roller (e.g., 5 inch diameter) rotating at the bottom of a hopper. The two rollers are normally driven at about the same surface speed in the same direction but by different motors. There is often a presser roll driven by its friction with one of the band return rollers. It helps to insure a uniform height in products passing under the icing roll. The latter is inflexibly mounted and cookies (or other pieces) which are too high can jam up the equipment. The bottom surface of the belt under the depositing roller is supported at a uniform height (adjustable) by one of the band return rollers.

Band icers give more prolonged contact of the icing applicator with the product surface to achieve maximum adhesion of the coating. The icing is transferred to a belt the width of the product conveyor. The bottom surface of this belt is at the same height above the conveyor belt as the top surface of the product. Height is adjustable, and is generally fixed so that there is a slight tilt of the surface to apply increasing pressure on the product as it moves along. The icing which is not transferred to product is automatically scraped from the belt and returned to the hopper to be mixed with fresh icing. Band icing is often whipped slightly in the mixer to make it easier to apply and to give a somewhat less tough coating.

In the retail shops, doughnuts are glazed while hot, right out of the fryer. They are transferred from the fryer screen to the glazing screen. A quick pass with the ladle of glaze over the top of the doughnuts gives a uniform covering to one side of the doughnut. The ladle is a two part scoop about the width of the glazer. It can be opened to give a slit at the bottom through which the icing flows; other forms of ladles are known. Excess glaze drops into the sloping bottom of the glazer where it collects until removed for further use.

Continuous glazers of medium to high capacity are used in the larger bakeries. These may have a temperature controlled tank which holds the stock of icing and has provisions for receiving back the excess deposit. A pump, often of the sanitary rotary type, forces the icing from the tank through a hose to a simple waterfall depositor which forms a curtain of glaze through which the products pass on a bar conveyor. Some sort of agitator should be present in the tank to mix the returned and fresh glaze. The same type of unit can be used as a half icer by removing the glazer head and pumping the icing into a shallow pan through which the doughnuts are conveyed. The iced products then move down an incline and are flipped over, so that the uncoated side is on the bottom. See Figure 12.4 for an example of a glazer-icer.

CREAM ICING DEPOSITORS

Buttercream icings and other nonflowable icings cannot be satisfactorily applied by the glazers and icers described above. They require equipment which can extrude or sheet a semisolid, plastic material. Several effective systems have been devised for handling them.

Depositors which apply wide ribbons of icing to, for example, cakes in square or rectangular pans passing beneath the depositor on a conveyor belt, can be electronically controlled to precisely position the icing so that none is left on the edges of the pan or on the belt.

Devices which deposit whipped cream type toppings on pies can be obtained from two or three manufacturers. Although small machines performing this function as a separate operation are available for medium size bakeries, the larger units are usually integrated into continuous or semi-continuous lines for forming complete pies. These large units are either linear or rotary in configuration. A typical rotary cream topping machine for semi-automatic operation would have six plates, with manual load and automatic unload. The topping is deposited through a cylindrical head having affixed to its bottom many vertically oriented tubes each of which deposits a small mound of whipped cream. The mounds can have simple patterns such as rosette patterns on the upper surface. Retention of this pattern and of individuality of the mounds depends on the viscosity of the material.

For cakes, cookies (such as brownies), and snacks which are baked in a continuous sheet on a band hearth, icings, creams, other toppings, and fillings can be extruded in continuous ribbons on to the sheet before the cake is cut into units (Freihofer 1985).

Filling depositors have been described which extrude a cylinder of cream slightly smaller in diameter than a layer cake. From the bottom of this cylinder, a slab is cut which is of the desired quantity for the filling of the cake and the slab is dropped directly on the freshly baked and partially cooled layer (Kreisky 1965).

FILLING INJECTORS

There are several kinds of fillings which are injected into bakery products. Very familiar to most readers will be the "creme" consisting mostly of sugar and fat which is injected into small cakes of the Twinkie type. Also widely used are the imitation jams and starch-based fillings injected into many kinds of doughnuts. Injectors take several

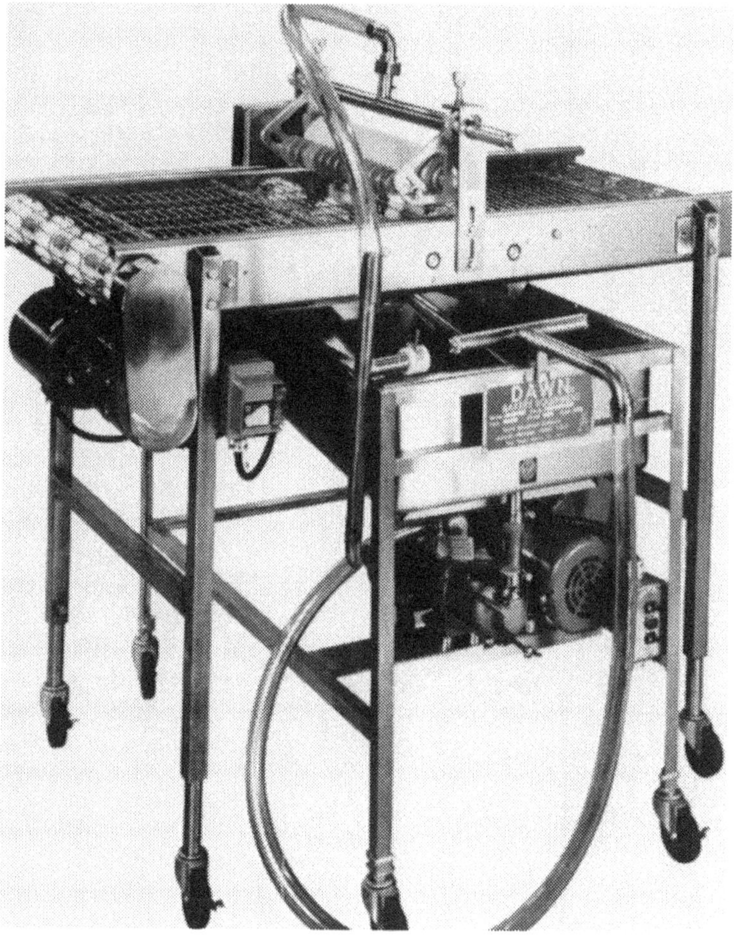

FIG. 12.4. ICING AND GLAZE CONVEYOR. SHOWN HERE WITH PRINT ROLLER AND STRING ICING TUBE FOR APPLICATION TO PRODUCTS ON PANS.
SOURCE: DAWN EQUIPMENT

forms over a very wide range of sizes, from the simple one or two tube injectors found in most doughnut shops to the continuous and automatic manifold injectors with outputs of up to 1,800 dozen pastries per hour.

The small injectors for the individual doughnut shop consist generally of a base with motor, switches, etc., a removable hopper at the bottom of which is the simple pumping mechanism, and two large hollow needles projecting from the pumping mechanism. Controls allow adjustment of the amount of filling pumped through the needles at each cycle, for example, from a fraction of an ounce up to seven ounces. In operation, the baker will push a doughnut onto each needle and activate the machine with a push button, foot pedal, or push bar.

Much larger units operate on the same general principle of inserting a spout or needle into the dough piece and forcing a measured amount of filling into the product. Registering the pieces for continuous operation is usually the most difficult part of the engineering.

The Oakes Machine Corporation describes a special feature, the suck-back capability, which is built into their creme injection manifold. Since creme is highly aerated, it tends to become compressed when put under pressure. In the injection manifold, creme flows under pressure from the distribution manifold through a slider valve and into a cavity in the lower nozzle plate assembly. Once the slider valve shuts off the source of creme, there will be a post extrusion which will continue at a decreasing rate until such time as the internal pressure in the nozzle plate cavity approaches atmospheric pressure. The concept of suck-back is to offer an alternate route within the nozzle plate cavity for the creme such that it is easier for the creme to expand into the suck-back cavity than to extrude from the needles. In order for the suck-back to work successfully, it is necessary to close the slider valve and then delay the start of the suck-back until such time as the majority of creme (and hence the pressure) has dissipated. Following is a sequence of occurrences toward the end of the horizontal motion during injection; at position #1 the slider valve has closed and vertical rise and suck-back have started simultaneously, the rate of vertical rise has been slowed to give the creme time to post extrude while the needles are still in the cakes. Also, by restricting the air bleed at the shuttle valve controlling air to and from the suck-back pistons, the start of suck-back can be further delayed until most of the internal creme pressure is dissipated between positions #2 and #3. As position #3 is reached, suck-back is now underway so that the internal creme pressure which remains now finds it easier to get into the suck-back cavities than out the needles. By delaying the rate of suck-back, it can be made to continue for most of the time during horizontal return while the needles are outside the cake. It is important that suck-back be started in time for extrusion to be stopped just prior to the needles' exit from the cakes so that the tips of the needles are wiped by the cake. Two cylinders, one at each end of the manifold, control the suck-back motion. The whole purpose of this development is to minimize the collapsing of the creme during the injection cycle by minimizing the pressure on the creme in the cake.

APPLICATORS FOR FAT

Fat applicators are available for applying this material to baked products, unbaked products, and pans. Fat can be transferred by mechanical means—brushes, rollers, etc.—or as a spray or curtain. Fat applicators are very common in the snack industry, since nearly all puffed snacks (popcorn, corn curls, etc.) are coated with oil. Most snack lines have continous drum coaters into which stainless steel pumps force metered amounts of oil through multiple nozzles. The drum has guide tracks and drive traction channels on the outside to facilitate rotation by the external motor. Inside the drum, there will be ridges, variable in size and shape according to the product being coated, which cause the pieces to tumble and expose all their surfaces to the spray. The drum is slightly tilted downward to the exit end so that the product pieces falling into the drum from a slanted chute will gradually flow to the discharge point.

Certain types of dough products are oiled at some stage in their processing. Rounded dough pieces destined for pressing into pizza crusts are nearly always oiled. Dough oilers are of three different designs: (1) Oil flowing from a hopper through a slit or gate is deposited on a roller which transfers it to the surface of a continuous dough strip, or pieces of dough; (2) Oil is sprayed in a fine mist onto a sheet or pieces of dough as they pass through an enclosure; and (3) Dough pieces pass through a curtain of oil, as in an enrober. Since oil is relatively expensive, enrober-style application is often contra-indicated, more oil being supplied by these devices than is necessary to achieve the desired result.

Some snack crackers are sprayed with oil to enhance their appearance and texture (see Figure 12.5). Large scale automatic units are available for this operation (Anon. 1985A). Sizes are available to accommodate belt widths of 800 mm, one meter, 48 inches, and 62 inches. The manufacturer reveals the following construction and operating details. There is a top header tank with thermostatically controlled heaters. Throughput of oil is controlled by a level sensor in the header tank. The oil pump is controlled through a variable speed drive unit and the oil is passed through a motorized strainer. There is a stainless steel, triple weir, filter holding tank, with heaters. Oil is dispensed by two top and one bottom manifold spray units with jets. Pipes which carry oil are traced with heaters. The belt is a stainless steel web with variable speed drive rollers geared together; it has a manual tracking unit and handwheel tensioning means. A mist collection system with air filter device prevents oil vapors from escaping.

Another self contained oil spray machine for large scale production

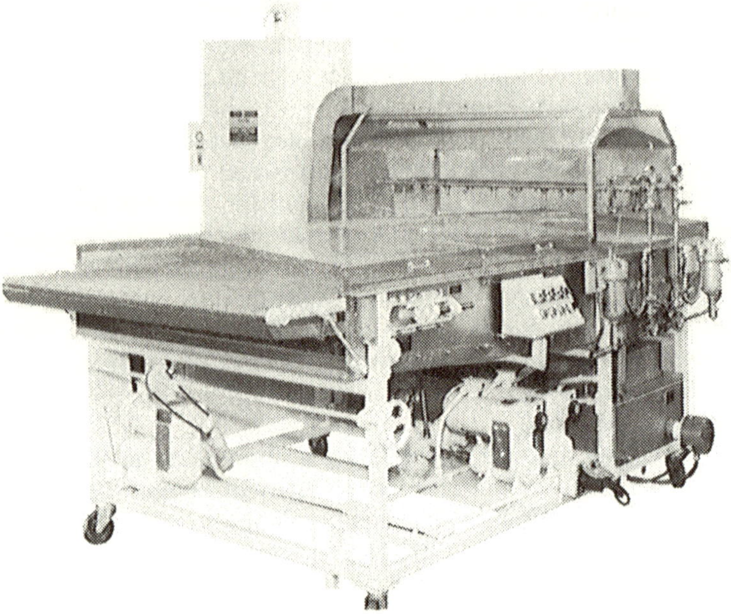

FIG. 12.5. INCLOSED OIL SPRAYING UNIT.
SOURCE: THOMAS L. GREEN

has the following characteristics. Product is carried through the spraying chamber on a balanced weave mesh belt driven by a pin roll drive pulley. A motor with variable speed drive is the prime motive force. The oil holding tank is a heated reservoir equipped with a recirculating pumping system and an immersion heater which can maintain temperatures up to 170°F. Oil is applied onto a high speed rotating disc which causes it to vaporize into a mist. The amount of oil flowing onto the disc is adjusted by control valves. A dual piping system is supplied for metering the oil. The system includes a set of flow controls for maximum oil consumption and another set of controls for minimum oil consumption. Excess oil falls into a drip collection tank and is passed through a screen before it is returned to the main holding tank. An electrostatic mist collection unit is mounted next to the spraying chamber to trap oil vapors. This unit condenses the mist into oil drops for reuse. Spray chambers can be 32, 48 or 39 inches in width. The two smaller chambers incorporate three spray systems, one above the belt and two below it, while the larger chamber has four spray systems, two above and two below the belt. Oil to product weight can be varied from 1 to 25% (weight basis).

Another type of dough oiler suitable for small or medium size bakeries consists of a number of small oil reservoirs with drip type dispensers which wet rollers that ride on the dough surface. The reservoirs resemble in their construction details the small glass drip lubricators found on many kinds of small equipment. These oil tanks are made of transparent plastic. Fluid flow is controlled by a stepless adjustment measuring screw. Magnetic valves close off the oil flow when the production line shuts down.

The fat-flour mixture used as an interleaving or laminating material in cream crackers is a rather difficult substance to handle because of its tendency to compact and cling together. Special applicators are available which use a belt to draw the mixture from a hopper which contains internal agitators.

A combination water splitter/butter applicator is offered by the Burford Corp. This unit uses a water jet to make the top cut in split loaves and apply butter to the slit immediately afterwards. The standard machine is supplied with five nozzles to split the five loaves held in the usual strap. Water coming from a 0.017 inch orifice cuts the surface about one-eighth inch deep. Pressure is adjustable, firmer doughs generally requiring more pressure than soft white loaves. The small amount of residual water has little or no effect on other crust characteristics. Jets are controlled by proximity switches which sense the presence of pan bottoms. Immediately after the pans leave the splitting area, they pass under nozzles which deposit a strip of melted butter on the opening. Oil can be deposited as well as butter, and some bakers are adding a mixture of butter and honey (Ivey 1988).

WATER AND WASH APPLICATORS

Plain water is applied to improve adhesion of dry particles (seeds, salt, crumbs, etc.) to bakery products. Normally, this water is applied by spraying it on the unbaked dough. The spray may be in the form of a mist retained in a tunnel through which the dough piece is carried. Misting can be developed by forcing water through a nozzle or using one of the centrifugal type vaporizers.

Washes can be applied by belts or brushes. A typical unit recommended for egg washes has a cloth belt carried around two small diameter rollers separated by about a foot on a frame pivoted in the middle. This belt assembly can be tilted so the bottom "nose" just touches the dough sheet passing beneath it and it is driven at a speed that approximately matches the speed of the product conveyor belt. The bottom of the wash applicator belt also contacts a metal roller which turns in a pan containing the wash liquid. The metal roller thus transfers the wash from the pan to the belt. A thumb screw adjustment changes the pressure of the belt on the roller to control the rate of wash transfer. The applicator belt is a lock-weave pre-shrunk single-ply cotton endless belt specifically woven for egg wash transfer. It can easily be removed for washing. This unit is said to be useful for water application, and presumably for other washes as well (Anon. 1971).

PIZZA TOPPING MACHINES

Rotary pizza topping machines have recently been offered by one of the leading manufacturers of pie filling and topping machines (Pluta 1988). On a six station rotary machine, sauce is deposited in a uniform pattern on the surface of a pizza crust, then fresh sausage or other topping is applied before a pre-measured amount of grated or shredded cheese is manually added. The semi-automatic unit consists of a positive piston filler, automatic sausage depositor, and cheese spreader mounted on a stainless steel rotary indexing table. Crusts of 9 to 16 inch diameter can be processed at rates of up to 20 pizzas per minute, with appropriate change parts. The pizza sausage depositor is operated in the following manner: (1) Put ground sausage meat in hopper; (2) Set the control bar for the proper size; (3) Put pizza under end of depositor using the self centering peel; (4) Push operating lever at side of depositor to move the large meat die plate under the hopper and force ground sausage into the die holes; (5) Pull the operating lever to return the plate to the depositing area completing the filling cycle; (6) Press down on the spring loaded handle to drop the sausage patties on to the pizza; and (7) Remove the finished pizza.

JELLY AND MARSHMALLOW TOPPERS

This section deals with equipment which deposits small amounts of jelled toppings and the like on cookies and cookie doughs. Most of these devices rely on some sort of piston arrangement for precise measuring and placement of the deposit. A unit for small scale production is shown in Figure 12.6.

In one commercial unit, a positive placement piston and rotary valve system deposites jellies and jams onto the surface of preformed products. The operating mechanism is driven at a 1:1 ratio by the product forming equipment. Proper placement of the deposited mass is accomplished by linear movement of the entire dispensing unit in relationship to band conveyor direction. This topper can achieve up to 120 row deposits per minute. The minimum center-to-center distance of

FIG. 12.6. A DEPOSITOR FOR FRUIT FILLING.
SOURCE: MOLINE DIV. PILLSBURY

items across the width of the band is 54 mm. In operation, the material to be deposited is placed into the hopper of the unit and gravity fed to the piston-valve mechanism. Valves rotate to open and shut in synchronization with the piston stroke. A manual adjustment of the piston stroke determines the volume of the deposited mass. Horizontal head movement in relation to band speed and in coordination with deposits is controlled by an eccentric and linkage.

The Oven Pacer Depositor (Oakes 1983) is particularly adapted to the placing of marshmallow on base cakes (cookies) for later enrobing. The basic principle is that the depositing manifold travels with the base cake but does not interfere with the movement of the cookies. To accomplish this, the continuously fed pressure-depositing manifolds are mounted on a yoke assembly over the cookie conveyor. The yoke can move simultaneously in both the horizontal and vertical plane so that the depositing nozzle can be positioned over the center of the base cake at the time the slider valve is opened. The deposit is formed while the manifold is tracking at the same speed as the conveyor. A synchronous vertical profile is used to form not only the diameter but also the height and shape of the deposit. At a predetermined height above the base cake, the slider valve is closed and the horizontal motion is reversed, causing a rapid return to the start position. This unit is shown in Figure 12.7.

ENROBERS FOR CHOCOLATE AND OTHER FAT BASED COATINGS

The application of fat-based coatings, of which a typical example would be an eating type chocolate such as sweet chocolate or milk chocolate, is an excellent way of increasing the consumer acceptability of bakery products. These coatings are frequently applied to cupcakes, swiss rolls, doughnuts, and cookies, and can also be found on many other products. They have even been applied to pretzels, crackers, croissants, fried pies, and other unlikely candidates for this treatment. Analogous coatings with no chocolate content are also available, and are cheaper but meet with considerably less demand. The chief problems encountered by the user of these coatings are cost, difficult processing requirements, and susceptibility of the coatings to damage by normal ambient temperatures. In the following section, the processing requirements and how they are satisfied by existing commercial equipment will be discussed.

Need for Special Treatment of Chocolate

Enrobing lines for applying fat-based moisture-free coatings to small baked goods are common in the baking industry. Chocolate-flavored coatings are the most common, but other types such as white vanilla or pastel colored coatings are used occasionally. Coatings which contain vegetable fats other than cocoa butter are usually called compound coatings. Processing requirements for coatings containing cocoa butter are substantially different from those based on most other vegetable fats. Manufacturers of edible fats have spent much research time and money in trying to duplicate the organoleptic properties of chocolate with coatings which are cheaper and which do not have the temperature sensitivity of cocoa butter, but complete success has not been achieved.

In the enrobing process, the cooled baked items proceed from a textile or plastic belt onto a wire mesh belt which is an integral part of the coating machine. The mesh belt carries the product through a curtain of liquid coating which covers the top and sides of the articles. A roller under the belt coats the undersides. A tank holding a supply of tempered coating is located somewhere in the vicinity, often under the belt. A pump circulates the coating to a flow pan which forms the curtain by

FIG. 12.7. DEPOSITOR FOR MARSHMALLOW.
SOURCE: OAKES MACHINE

allowing the liquid to flow over one side. An air blower removes the excess coating and helps to intensify the gloss, but it may leave slight ripples on the product. A vibrator shakes a section of the mesh conveyor belt and smooths out some of the ripples on the coating. The detailer rod (anti-tailing device) helps to control the amount of coating on the bottom and remove strings and tails.

Enrobing bakery goods with chocolate coatings presents a number of problems not encountered in any other finishing operation. Obtaining and retaining optimum texture (mouthfeel) and appearance of chocolate coatings depends upon carefully following a prescribed course of temperature treatments. This "tempering", as it is called, is very intolerant to departures from the acceptable temperature ranges. Furthermore, products covered by even the best tempered coating will be irreparably damaged by exposure to temperatures above about 95°F—the exact temperature depending somewhat on the composition of the coating; there is also a time factor involved.

All of these problems arise from the characteristics of the fatty substances which make up about 50% of pure chocolate. Melted chocolate liquor (bakers chocolate, bitter chocolate, pure chocolate) is made up of a continuous phase of fat (cocoa butter) and a dispersed phase consisting of very fine particles of the nonfatty portions of the cocoa bean. When mixed with sugar, milk fat, or other components of coating chocolate (sweet chocolate, semi-sweet chocolate, milk chocolate, etc.), the basic system of continuous phase fat (cocoa butter perhaps with milk fat) and discontinuous phase (microscopic particles of sugar, cocoa, milk solids, etc.) persists. On cooling, the liquid chocolate behaves largely according to the crystalline status of the fat in this continuous phase. The fat can crystallize in many forms, only one of which (the beta form) is stable. The unstable crystalline forms, if present, will revert in time to the stable form, but products manufactured from chocolate containing these unstable fat crystals will be of poor appearance, suffer from fat bloom, and (in the case of molded products) will be difficult to extract from the mold due to reduced shrinkage. The purpose of tempering chocolate coatings is first to remove unstable forms by melting the chocolate and then to form a maximum percentage of stable crystal forms by cooling the liquid chocolate slowly according to an appropriate schedule. Ideally, only the beta crystalline form will be present in the finished article.

If the tempered chocolate is placed on a hot product, all of the preceding effort will go for naught. Baked items should reach the enrober at a temperature of 75° to 85°F, and the room in which the enrober is located should be at the same temperature. The coating, after having been through the tempering process, should be maintained at the appropriate holding temperature which for pure chocolate is about 90°F. Compound coatings, which may include cocoa but normally not chocolate, may be either require or not require tempering, depending on their composition.

Equipment for Tempering Chocolate Coatings

According to one authority (Anon. 1987), five construction features are of great importance to the tempering of chocolate so that there is a perfect microcrystallizaton of the mass. These are (1) many tempering sections, (2) large cooling surfaces, (3) perfect scraping and efficient mixing, (4) cooling time, and (5) accurate temperature control. When all these requirements are satisfied, it is possible to get a finished product with high deep gloss, fine-grained and crispy break, good keeping qualities, short solidification time, and maximum contraction. For coatings, contraction is not much of a factor, since the product will not be removed from a mold, but there is some opinion that poor or inadequately controlled tempering can lead to spontaneous cracking of coatings, especially in products such as chocolate coated marshmallow deposits.

The initial handling of chocolate coating will depend upon whether it is received from the supplier in melted condition or in slabs. All small users and most medium sized users will obtain the material in the solid form. The slabs must first be melted. In one type of conditioner, a melting drum maintained at a predetermined temperature (about 100°F) by a thermostatic control continuously rotates against a block of coating (or chunks) and melts a thin film of chocolate from it (see Figure 12.8). This film is stripped from the roll by a blade and drops through a strainer into a water-jacketed storage tank. A specially-designed agitator in the storage tank thoroughly mixes the melted chocolate. Continuous delivery eliminates the need for large holding tanks for liquid coating.

The melted chocolate must be cooled to the proper temperature to establish stable seed crystals. Various types of heat exchange equipment are used for this purpose. In one type, chocolate is pumped through a spiral-shaped, close tolerance gap between inner and outer water-cooled jackets of a heat exchanger. Use of thin layers allows complete equilibration of the chocolate because of the minimum temperature differentials necessary to reach the desired temperature.

A specific tempering unit available commercially is described as follows. The unit is a free standing, self-contained machine consisting of the tempering tube with its drive, a control panel, and a water system, all mounted onto a fabricated steel framework with stainless steel covers. The unit requires a supply of cold water (about 55°F) and of hot water. Hot water is usually supplied from the chocolate feed pipe jacket. Four standard sizes cover a throughput range of 1,000 to 3,250 lb per hr. Capacity depends on the type of chocolate; as a general rule, to induce correct crystal growth chocolates containing milk fats must be cooled to a lower temperature and held longer than chocolates containing only cocoa butter.

The tempering tube of the unit mentioned above is vertically mounted and divided into three sections—cooling, retention, and final temperature adjusting. The cooling section consists of a number of water cooled discs, each working in conjunction with a rotating scraper

FIG. 12.8. CHOCOLATE MELTING AND CONDITIONING UNIT.
SOURCE: J. W. GREER

and arranged so that a thin film of chocolate passes across a disc, cools, and is scraped off. Cooling is followed by the retention section which is a water jacketed holding tank with a central low energy stirrer and stators.

The final, temperature adjusting section is made up of a pair of water jacketed discs with a rotating scraper. There are two temperature controlled water systems. In one, the flow of water for the discs in the cooling stage is controlled by an electronic thermometer and magnetic valve used in conjunction with pre-set throttle valves. The thermometer senses the temperature of the chocolate leaving the cooling section. Throttle valves control the delivery of cooling water to each disc to achieve the necessary cooling. In the second system, which is a circulation system with heating and cooling abilities, water is circulated at a controlled temperature through the jacket of the retention section and final stages. An electronic thermometer measures the temperature of the chocolate leaving the unit. Heating of the water is by electric immersion elements and cooling is by tap water. A second temperature controller overrides the first to prevent water temperature from going outside the preset extremes. During plant shutdown, all circulating water is switched to a higher temperature. The cooling sections automatically connect into hot water circulating through the chocolate feed pipe jackets, and the retention and re-heat stages switch into the water heating unit.

Using the tempering unit just described, untempered chocolate is continuously tempered during production. Liquid untempered chocolate is metered into the tempering tube where it is cooled in a swept film heat exchanger to a temperature where stable beta crystals can grow. The chocolate then passes to the retention section, where these crystals grow and multiply. Without this retention time, it would be necessary to cool the chocolate in the first stage to a much lower temperature, and this could result in the growth of unstable crystals. The long retention time also means that crystal growth can be maximized without shortcuts such as mixing tempered chocolate with untempered coating. As crystals multiply in the retention section, there is an increase in mass temperature. In the last stage, another swept film heat exchanger is used to control the final chocolate temperature to within 1.8°F of the required coating temperature for the enrober. Precise temperature control and a low temperature variation throughout the mass enables the machine to raise chocolate temperature without destroying any crystals or disrupting the state of temper. High temperature results in lower viscosity which allows thinner coatings and easier enrobing. Chocolates of lower fat content can also be handled better by higher temperature retention (Anon. 1984).

Another commercial tempering unit of somewhat different type has been described in company literature (Anon. 1987A). By means of a pump, chocolate mass is transported through the unit. The built-in motor drives through V-belts a strong worm gear and the main shaft. Under continuous scraping of the cooling surfaces, the chocolate gives off heat without being exposed to any shock treatment. The unit is equipped with up to seven cooling zones controlled by digital electronic equipment set at the required temperatures. Scraping elements wipe all heat exchange surfaces. To prevent the tempering unit from freezing when stopped, a switch is activated, causing hot water to circulate around all tempering spaces.

Tanks, Pumps, and Pipes

Chocolate holding tanks can be of many different sizes and designs. A typical holding tank for use between the tempering equipment and the depositing machine might be comprised of a water jacketed tank fitted with a stirrer and a vibratory sieve driven by independent motors. It also includes a level control switch for activating a valve or pump in the chocolate supply line so as to maintain the prescribed quantity of coating in the tank. The tank itself is a welded steel trough supported on four legs and provided with a water jacket heated by three immersion elements. The bottom of the tank is hemispherical in shape and is swept by a reciprocating stirrer driven by a motor acting on a fixed speed reduction drive. On top of the tank there is a vibratory sieve connected to a fixed speed drive by a V-belt. There is a bottom outlet with a length of water jacketed pipe terminating in a flange suitable for coupling the tank to a chocolate pump.

The piping in these systems, which may be of black iron, must be wound with industrial heating cable and insulated. An even better system of temperature control is to water jacket the pipes. For chocolate coating, pipes of two inch diameter are normally recommended, and these would be encased in three inch diameter pipes for hot water circulation. One authority recommends that temperature of the circulating water be kept at 1° to 2°F above the depositing temperature. In one version of the water temperature control and recirculating system, a unit consists of a circulating pump fitted on top of a water tank; heating takes place by means of electric elements which are controlled by a thermostat inserted in the outlet pipe (Anon. 1985B).

Pumps recommended for chocolate systems are slow moving rotary displacement pumps. Description of a typical pump follows (Anon. 1986A). The pump has two moving parts, a rotor on the main shaft and a pawl. The rotor, pawl, and pump body are made from high grade cast iron. The rotor is housed behind a front cover plate which can be quickly removed to give access to the pump cavity. Rotor shaft seals are neoprene high pressure seals in cartridge form so that seal changing is simplified. Rotors have one or two lobes depending on the output and application. A single lobe should be used only when delivering against minimum head pressure. Two-lobed rotors are limited to a maximum speed of 50 rpm. The pump is water jacketed.

There are available three different sizes of pump body; within these bodies different rotors can be fitted. Output can also be varied by fitting different drives. Standard pumps have fixed speed drives, but variable speed drives can be supplied. The three sizes of pumps have maximum outputs of 2,045 lb per hr, 4,090 lb per hr, and 5,455 lb per hr.

When the pump is operating, chocolate is drawn through a side entry port by the rotor movement. Chocolate is carried around the inside of the pump by the rotors until the pawl is reached. The pawl follows the contour of the rotor lobe and directs the chocolate upward to the discharge port. Efficiency of the pump is high, usually around 95%. Maintenance is infrequently needed, but is quick and easy. It can be done with the pump in position. Seals are in cartridge form for quick removal and replacement. It is not necessary to disturb the drive shaft or bearings to change the seals; the seals are reached by removing the front cover plate and rotor.

Enrobers

General features—The great majority of continuous enrobers used for bakery foods are of the curtain or waterfall type, in which the product is moved through a vertical sheetlike stream of liquid chocolate with coverage of the bottoms by other methods (see Figure 12.9). The usual sequence of operations in these machines is: (1) Properly tempered coating supplied by the conditioning unit is delivered to the enrober as required; (2) Product is carried on a wire mesh or rod conveyor belt through a curtain of coating in an enrober consisting of, as a minimum, a jacketed coating tank with an agitator and a flow pan, an agitator device for the take-away section of the conveyor belt, a water-jacketed heat exchanger, a bottomer, a detailer rod, and a blower; and (3) All of this will be serviced by the necessary pumps, piping, motors, and controls. A considerable amount of variation is possible in designs meeting these minimum requirements and some of the machines which are available will be discussed in the following section so the reader will be familiarized with the possibilities.

Example 1—A specific design of enrober is described in the following section. Most of the details are applicable to enrobers in general. The coating tank and a drip pan extension, which together form an integral machine, are made from continuously formed and welded stainless steel sheet. Surrounding this frame is a stainless steel air jacket. Thermostatically regulated, electrically heated air circulates continuously within this jacket to maintain the temperature of the tank. During shutdown periods, the hot air system increases the temperature within the jacket to maintain the coating in a liquid state.

A stainless steel wire mesh belt with a 0.25 inch pitch conveys product through the coater. A shaker system adjustable in frequency and amplitude is provided to agitate the enrobed product. This vibratory action distributes the coating more or less evenly over the product and also contributes to the removal of excess coating. The detailer rod affixed at the end of the wire belt can be adjusted up or down to eliminate trailing bits of coating.

Stainless steel flow pans allow a double curtain of coating to flow

FIG. 12.9. CHOCOLATE ENROBER COATING MARSHMALLOW COOKIES.
SOURCE: J. W. GREER

onto the belt. The melted coating either flows through a slit at the bottom of the pan or over one edge of the pan. The latter design is less likely to clog up at some point along the stream. The height of the pan can be adjusted in some models, while in others it is fixed at about a five inch product clearance. Coating is pressure fed from the coating tank to a distributor which dispenses the melted material across the width of the pan. Pans can be designed to pour many evenly separated narrow streams of coating over the product to give stripes.

Beneath the belt, a bottoming pan receives the excess coating. A supply of coating is piped into this pan when bottomed goods are being made. The belt carrying the cookies or other product across the bottomer is generally composed of thin metal rods (or wires) transversely disposed, and it travels just below the surface of the chocolate so that the coating will be able to contact the bottoms of the pieces. The pan can be moved toward or away from the belt to affect the extent of bottoming. A bottoming roller can be placed in the pan to pick up the coating and apply it to the bottom of the cookie.

Excess coating can be removed by a high-pressure blower system. A damper on the blower intake controls the flow of air through the nozzles. This nozzle, which is a tubular plenum chamber, gives uniform air distribution across the width of the belt. Nozzle height and angle, and width of air discharge are adjustable. The detailer, a simple powered roller of small diameter located so as to contact the bottom of the piece, removes "tails" of coating which drop down from the product as it leaves the enrober belt.

It is necessary to have precise control of chocolate temper if glossy coatings are to be obtained and bloom avoided. Temperature control is assisted by use of a water-jacketed heat exchanger of the swept-wall type in the coating line subsequent to the sump. Automatic tempering cycles can be programmed. In tempering, the coating is first heated high enough to destroy all fat crystal structures, then cooled to the temperature at which stable crystals develop, and finally reheated to production temperature and maintained there during the run. Overnight holding is preceded by a heating step to melt all fat crystals which have developed, after which the pump is reversed to clear all lines of the seed forms.

Example 2—Another type of enrober, the "advanced coating system" (ACS) coater, is designed to be used with an outboard tempering unit. It is said to make it possible to handle chocolate at a higher temperature and lower viscosity than usual. The unit is constructed in two pieces, a top section and a lower section. The upper unit contains a wire mesh conveyor, bottoming bath, licking rolls, anti-tail rod ("detailer"), hood heater, and top and side covers. The wire mesh belt is driven by a variable speed motor and housed in stainless steel side frames and two stainless steel water jacketed trays. Mesh tension is adjusted by a handwheel control.

The bottoming bath of the ACS Coater has a drive stirrer to prevent chocolate buildup and can be drained for removal. Included are bottoming roll, adjustable scraper, and surge plate. The topflow pan has two adjustable outlets and gives a double curtain. Overflows at each end feed chocolate to the bottoming roller. The shaker frame is driven by a pneumatic oscillator and its frequency and amplitude can be adjusted. Vibration to other components is reduced by special mountings. There are two licking rolls with scrapers. The anti-tailer has a reversible drive and is mounted in swivel blocks around the terminal rod so its position can be adjusted.

Blower air temperature in the ACS Coater can be adjusted by use of a quadrant arm, whether the air is drawn from outside the machine or inside the hood. The blower is adjustable for both height and angle. Air passes from the blower fan through flexible ductwork to the blower outlet. This duct has a built-in pressure equalization system to assure even distribution of air in the curtain. A heater on the hood maintains the proper air temperature within the machine.

In the lower section of the ACS Coater, the main coating tank is mounted on a steel framework fitted with casters so it can be withdrawn from beneath the top section. Flexible connections are used between the top and bottom halves so that the semicircular tank can be removed. This tank is equipped with a stirring mechanism having a fixed speed drive. Mounted on the same framework is a jacketed chocolate circulation pump and a delivery pipe. This is a positive lobe pump with a fixed speed drive. The water jacket of the pump and the delivery pipe are connected into the coating tank water circulation system. On the side of the tank are two overflow points and a drain. The overflow hopper is water jacketed and has two handwheel controls to open or close the coating tank intermediate level drain. When the latter is open, the tank runs half full and when it is closed the tank runs completely filled and overflows from the top. An overflow pump is optional to pump excess chocolate back to a holding tank.

The enrober described above can be supplied as a half coating machine in which the top flow pan is replaced by an optional diversion chute which directs the circulating chocolate into the bottoming bath. A variable speed roller curtain is available for elimination of air bubbles in difficult coatings. The enrober can also be modified for handling compound coatings. Certain simplifications are possible when these materials are being handled. The constant overflow feature is no longer required; a drip feed principle is used. Also, the main chocolate tank stirrer is removed and the tank level controls regulate the coating drip feed.

Operational details—The operation of another enrober described by its manufacturer as a "Series D Coater" varies somewhat from the preceding examples and will be discussed further. There is a constant feed of cooled chocolate from the tempering tube into the main tank of the coater in excess of the product coating requirements. This ensures that there is a constant overflow into the detempering tank, which supplies the infeed of chocolate to the tempering tube. The detempering tank has a built-in low level control which provides a signal to the fresh chocolate supply source to activate either a valve or pump to keep the level correct.

During production, overflow chocolate is at a minimum and fresh chocolate is brought into the detempering tank by the level control.

During a production stoppage, however, no chocolate is taken away by the product so overflow becomes equal to throughput. Therefore, the amount of chocolate and its state of temper remain constant regardless of production conditions.

The overflow control on the main tank is in the form of an adjustable weir plate which keeps the main tank between half full and full. This enables the retention time to be altered to suit the type of chocolate being used.

During plant shutdown, all the chocolate in the machine is untempered. When preparing chocolate for production, valves are set so that chocolate passing through the tempering tube is taken from the main tank—the detempering tank is bypassed. This condition remains in effect until the chocolate passing through the circulating pump is down to its required temperature. Then, the valves are changed over so that the tempering tube receives untempered chocolate directly from the detempering tank during production.

The following adjustments can be made to achieve optimum production conditions: mesh speed, curtain height and thickness, curtain position relative to the bottoming roll, thickness of chocolate on the bottoming roll and bottoming bath drum, bath depth, blower height and angle, shaker amplitude and frequency, licking roll adjustment, anti-tailer height and direction, and tank retention time (Anon. 1987B).

Spray Enrobers

A completely different method for applying confectionery coatings is based on using large orifice multidirectional spray nozzles through which the coatings are sprayed from the bottom and top onto the products to be coated. As contrasted with the standard enrobing curtain, this system has a large coating area which provides a very uniform coating on all surfaces with controlled rates. It is said that light, medium, or heavy coatings can be "dialed in" and uniform covering achieved. Enrobing speeds can be increased up to 50%, it is claimed. Energy savings are also possible, since up to ten times more coating is applied to the product relative to the volume of coating recirculated, as compared to the standard waterfall enrober (Anon. 1983).

Cooling Tunnels, Slabs, and Conveyors

Cooling of enrobed products is necessary to solidify the coating so the pieces can be packed and to put the fat crystals in a more stable condition. Proper cooling of a well tempered coating will give a product having a good gloss and a firm texture, and the gloss and texture will persist through a long storage period, assuming storage conditions are suitable. If the coating is not firm, it will smear or retain fingerprint impressions from the packing operation.

Sufficient cooling capacity must be available to solidify all of the fat if bloom is to be avoided. A considerable amount of heat must be removed. This consists of two forms, sensible heat and latent heat. The former comes from cooling of the coating and the baked product while latent heat is released when the fat crystallizes. Lack of adequate refrigeration capacity can limit the output of the entire enrobed goods line.

The cooling equipment for enrobed baked goods generally takes the form of a tunnel with refrigeration units to chill the air or surfaces and a conveyor belt to carry the product pieces through the tunnel. Often, there are fans to move the chilled air. Suppliers may offer these tunnels in module form so that plants can be assembled with any required capacity. It is important to cool the enrobed product promptly to fix the crystal composition and to make it possible to pack the items without marking the coating or smearing the packaging material. Air temperature at the entrance to the tunnel should be about 65°F. The temperature should gradually drop to approximately 55°F at the discharge end. These recommendations are for the older type tunnel where air movement is opposite to the movement of the product, i.e., counter current cooling. In the more recent types, where zone cooling is used, it is possible to cool to much lower temperatures in the middle of the tunnel because each section is a separate cooling environment which can be individually adjusted. In the zone type, air temperature may be brought down from about 65°F at the entrance to 40°F toward the middle or the last two-thirds, and then warmed up again in the succeeding zones to about 55°F so that the products, as they emerge, are not cold enough to cause moisture to condense on them. Duration of time in the tunnel varies according to size of product, thickness of the coating, efficiency of the cooling, etc., but will normally be only a few minutes.

Multizone cooling tunnels are preferred. Some makes are in module form with each section being self-contained and automatically regulated. The product is carried through the assembled modules on a belt with a nonstick surface. Single-or double-row evaporator coils and the air-circulation means are placed between the belt supports in some designs. Inspection doors should be installed at frequent intervals to permit observation of the product and withdrawal of samples.

The Werner Lehara Turboflow Cooling Tunnel Type 1 is one option. Each cooling tunnel can consist of independently controlled cooling zones. These zones can be constructed to give top cooling by either radiation or jet convection, and bottom cooling by either air or water. Zones are constructed from steel frames covered by insulated reinforced plastic. The top covers are fixed but the side covers are removable for access to the product zone. The duct supports and air flow ducts are mounted onto the steel framework. The duct and belt support bases are made from steel sheets and have an etched, black mat finish.

Generally, the first third of the cooling tunnel is set up for radiation cooling. In the remainder, jet convection cooling is used. Below the belt, flotation jets provide air cooling (see Figure 12.10). For uncoated products, jet convection cooling is used above the belt the total length of the conveyor. Radiant cooling is accomplished with an overhead air-cooled absorption plate. At the end of the radiant zone, side vents direct the cooling air from the top radiant duct to the return air duct beneath the product zone. No air drafts come in contact with the product. Product is cooled from the inside to the outside to make certain all latent heat is absorbed. The radiated heat is absorbed without cooling the surrounding air, allowing the chocolate to crystallize fully without the risk of unstable crystal growth. There will be no temperature differential within the coating, so there will not be non-uniform shrinking and stressing of the chocolate, thus minimizing spontaneous cracking of the coating.

For jet (convection) cooling, air is distributed along the zone by an overhead duct. The lower part of this duct is perforated so that jets of air are directed vertically downward onto the product and then into the return air duct beneath the product zone. Jet cooling is more efficient than regular air cooling because it completely removes the surface layer of air in contact with the product and breaks up the pockets of air between the pieces. With the low air flow rate, there is no disturbance of the coating before it is set. The same temperature air is used throughout the cooling zone, unlike contraflow or crossflow methods of cooling, where there is an increase in air temperature from one end of the zone to the other.

Contact cooling occurs below the conveyor by means of an air cooled base or a water cooled system. With the air cooled base, the belt floats on a cushion of air which escapes through holes in a perforated bed plate. Since the belt is supported on the air, friction between the band and the table is slight, and a light highly conductive belt can be pulled over a great distance. Belt life is increased.

The belt is driven by a variable speed motor. Automatic belt tracking units can be used at the discharge and feed ends of the conveyor. At the discharge end, it is housed in the drive frame while at the feed end, it is

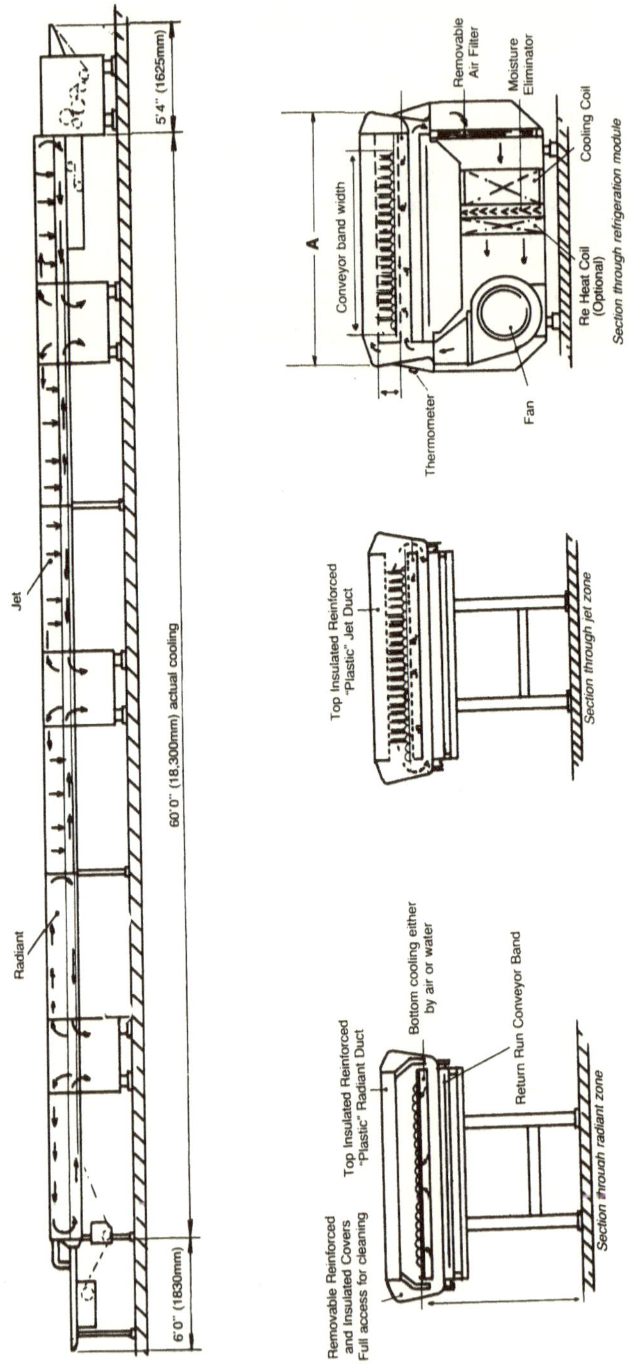

FIG. 12.10. A TURBOFLOW COOLING TUNNEL AND ITS MAJOR COMPONENTS.

SOURCE: APV (WERNER LEHARA)

attached to the decorating table. In the belt tracking system, sideways movement of the band is sensed by a counter-balanced lever which actuates an air pilot valve. This in turn operates an air cylinder coupled to a tracking roller. Sideways movement of the band leads to movement of the tracking roller, forcing the band to be returned to the center line.

Services required for the cooling tunnel described above include a drive motor (typically 1.5 hp), one 2 hp motor per module, one refrigeration compressor per module (size depends on the application), water at the rate of 2.5 gal per min per ton of refrigeration and 30 psi, and compressed air at 80 psi (for the automatic band tracking).

Contraflow cooling tunnels can consist of independently controlled cooling zones. They provide top cooling by radiation or linear convection and bottom cooling by conduction from a water cooled slab. Zones are constructed from steel frames and covered by insulated and reinforced plastic hoods which are removable for sanitation and complete accessibility to the conveyor and cooling module. Cooling module components require 8 ft of length within the tunnel, so the length of the tunnel zone will be 8 ft as a minimum, but it can be longer.

Each section of the contraflow cooling tunnel incorporates a water cooled table for bottom cooling. For top cooling, each section is split horizontally by a steel plate to form a supply return duct and product cooling area. The belt conveyor passes through the product cooling area while resting directly on the water-cooled slab. Cooling air above the conveyor belt is horizontally distributed along the product zone against the product travel from the mouth of the tunnel to its middle, and with the product flow from the middle to the tunnel end. Chilled water is circulated through the bottom bed and the cooling coils of each module. Individual valves are fitted for the regulation of cooling water to the coils or bottom bed from a central source. A completely self-contained water chilling and circulating unit is available. It is located beneath the band in the center of the cooler and includes a water-cooled condensor and compressor coupled to evaporator coils housed in a reinforced plastic chilled water tank and water circulating pump. A radiation cooling system is available for cooling chocolate coated articles. Generally, the first third of the cooling tunnel is set up for radiation cooling. Black-finished, water-cooled radiant panels are installed, replacing the steel sheeting above the product, and chilled water is circulated through them. Radiant cooling allows chocolate to cool at a natural rate. As chocolate crystallizes and sets, latent heat is radiated from it. This heat is absorbed by the chilled panels before the product reaches the convection cooling zones. Convection cooling can then provide quicker cooling in the final stages.

Cooling slab conveyors are used to cool and set bottoms of chocolate coated bakery products. They are simpler and cheaper devices than cooling tunnels. The cooling slab conveyor normally receives products from the prebottomer unit and transfers them into the main coating machine. The conveyor bed is an assembly of stainless steel or carbon steel plates. Channels having a serpentine pattern are formed between the plates so that water or freon can be circulated for cooling. There will be adjustable nose pieces at each end of the conveyor to assist in the smooth transfer of product. The delivery end is hinged to give access to the chocolate coater, and the feed end is hinged to facilitate access to a prebottomer unit or other preceding equipment. A single variable speed drive motor drives the conveyor belt. Tension of the belt is manually adjustable by moving an idler roll. A manual belt tracking system fitted at both ends of the conveyor is standard, and an automatic system can be obtained as an option.

PRINTING DESIGNS ON PRODUCT SURFACES

Designs can be formed on surfaces of baked products by using shaping containers which leave lines on the top. This is most successful with rotary molded cookies, with which good results can be obtained if the cookie normally bakes out to a light color and enough radiant heat can be applied to brown the raised lines. Some rather complex designs—animals, clowns, etc.—can be formed in this manner. Alternatively, the raised portions can be emphasized by contact applications of water icing, although this is seldom entirely satisfactory because of the smearing, dripping, and running characteristics of the icing. Of course, as everyone knows, very elaborate designs can be applied to cakes and large cookies by hand application of colored icings. There appears to be a demand, however, for the printing of designs on (particularly) cookies. This demand has been at least marginally met by systems which transfer edible inks either to the unbaked dough pieces or to the baked units. High speed forming operations can be serviced in this way. Sometimes, as in the patent of Krubert (1986), the cookies are first iced and the design then transferred to the hard surface from a printing pad. Prior patents have discussed other methods.

BIBLIOGRAPHY

ANON. 1971. Egg-wash Applicator. Anetsberger Brothers, Inc., Northbrook, IL

ANON. 1972. Christy Industrial Dispensing Machines. Christy Machine Co., Fremont, OH

ANON. 1979. AFM Salt/spice Dispenser. American Foods Machinery Corp., Memphis, TN

ANON. 1983. New Spray Enrober. Spray-Dynamics, Newport Beach, CA
ANON. 1984. ACS Tempering Unit. APV (Werner Lehara), Grand Rapids, MI
ANON. 1985A. Oil Spray. Vicars Group, Ltd., Merseyside, England
ANON. 1985B. Water Recirculating System. Aasted International, Farum, Denmark
ANON. 1986A. Chocolate Pump, Model 117 CP. APV (Werner Lehara), Grand Rapids, MI
ANON. 1986B. Portable sugar topper. APV (Werner Lehara), Grand Rapids, MI
ANON. 1987A. DMW Temperer. Aasted International, Farnum, Denmark
ANON. 1987B. Series D Coater. APV (Werner Lehara), Grand Rapids, MI
BOLLENBECK, G. N. 1965. Latest formulas and techniques for fondant and icing production. Proc. Am. Soc. Bakery Engineers *1965*, 266-270.
CLIFFORD, F. J. 1972. Production of icings and creme fillings with continuous equipment. Proc. Am. Soc. Bakery Engineers *1972*, 159-162.
FREIHOFER, W. D. 1985. New trends in small cake production. Proc. Am. Soc. Bakery Engineers *1985*, 134-141.
GUCKENBERGER, J. D. 1977. Icings and fillings for cakes. Proc. Am. Soc. Bakery Engineers *1977*, 81-87.
ILLFELDER, B. 1964. Preparation and application of icings for sweet yeast raised products. Proc. Am. Soc. Bakery Engineers *1964*, 254-259.
IVEY, D. 1988. Personal communication. Burford Corp., Maysville, OK
KREISKY, K. 1965. European baking equipment today. Proc. Am. Soc. Bakery Engineers *1965*, 118-129.
KRUBERT, G. J. 1986. Printing of foods. U. S. Patent 4,578,273.
MINIFIE, B. W. 1980. Chocolate, Cocoa, and Confectionery. Second Edition. AVI Publishing Co., Westport, CT
OAKES, W. P. 1983. Oakes Pacer Depositer. Frito-Lay Cookie Seminar, AIB, Manhattan, KS.
PLUTA, R. 1988. Personal communication. Colborne Corp., Glenview, IL
PORCELLO, S. J., MANNS, J. M., PLAYER, K. W., and WILSON, L. L. 1987. Cookie filler composition. U. S. Patent 4,711,788.
STRIETELMEIER, D. M. 1969. On electrostatic salting. Snack Food *58*, No. 9, 60-61.
WATKINS, H. E. 1970. Apparatus for flavoring of snack foods and the like. U. S. Patent 3,536,035.
WELCH, R. C. 1968. Chocolate and hard butter coatings. Proc. Am. Soc. Bakery Engineers *1968*, 242-263.
WING, D. H. 1975. Enrobing of bakery products. Proc. Am. Soc. Bakery Engineers *1975*, 136-142.

INDEX

www.ingramcontent.com/pod-product-compliance
Lightning Source LLC
Chambersburg PA
CBHW030103010526
44116CB00005B/76
9780942849271